PERGAMON INTERNATIONAL LIBRARY
of Science, Techology, Engineering and Social Studies
The 1000-volume original paperback library in aid of education,
industrial training and the enjoyment of leisure
Publisher: Robert Maxwell, M.C.

INTRODUCTION TO CROP HUSBANDRY
(including GRASSLAND)

FIFTH EDITION

Other Titles of Interest

BUCKETT	Introduction to Livestock Husbandry, 2nd Edition
	An Introduction to Farm Organisation and Management
DILLON	The Analysis of Response in Crop and Livestock Production, 2nd Edition
DODSWORTH	Beef Production
GARRETT	Soil Fungi and Soil Fertility, 2nd Edition
GORDON	Controlled Breeding in Farm Animals
HILL	Introduction to Economics for Students of Agriculture
KENT	Technology of Cereals, 3rd Edition
LAWRIE	Meat Science, 3rd Edition
LOWE	Milking Machines
NASH	Crop Conservation and Storage
NELSON	An Introduction to Feeding Farm Livestock, 2nd Edition
PRESTON & WILLIS	Intensive Beef Production, 2nd Edition
SHIPPEN & TURNER	Basic Farm Machinery, 3rd Edition
WAREING & PHILLIPS	Growth and Differentiation in Plants, 3rd Edition

INTRODUCTION
TO CROP HUSBANDRY

(including GRASSLAND)

by

J. A. R. LOCKHART

and

A. J. L. WISEMAN

Royal Agricultural College, Cirencester

FIFTH EDITION

PERGAMON PRESS

OXFORD · NEW YORK · TORONTO · SYDNEY · PARIS · FRANKFURT

U.K.	Pergamon Press Ltd., Headington Hill Hall, Oxford OX3 0BW, England
U.S.A.	Pergamon Press Inc., Maxwell House, Fairview Park, Elmsford, New York 10523, U.S.A.
CANADA	Pergamon Press Canada Ltd., Suite 104, 150 Consumers Road, Willowdale, Ontario M2J 1P9, Canada
AUSTRALIA	Pergamon Press (Aust.) Pty. Ltd., P.O. Box 544, Potts Point, N.S.W. 2011, Australia
FRANCE	Pergamon Press SARL, 24 rue des Ecoles, 75240 Paris, Cedex 05, France
FEDERAL REPUBLIC OF GERMANY	Pergamon Press GmbH, Hammerweg 6, D-6242 Kronberg-Taunus, Federal Republic of Germany

First edition 1966
Second edition 1970
Third edition 1975, Reprinted 1976
Fourth edition 1978, Reprinted (with additions) 1980
Fifth edition 1983

Library of Congress Cataloging in Publication Data

Lockhart, J. A. R.
  Introduction to crop husbandry.
  (Pergamon international library of science, technology, engineering, and social studies)
  Includes bibliographies and index.
  1. Agriculture. 2. Field crops. 3. Agriculture—
Great Britain. 4. Field-crops—Great Britain.
I. Wiseman, A. J. L. II. Title. III. Series.
SB98.L64   1983   631'.0941   82-16568

British Library Cataloguing in Publication Data

Lockhart, J. A. R.
  Introduction to crop husbandry.—5th ed.—
  (Pergamon international library)
  1. Field crops—Great Britain
  I. Title   II. Wiseman, A. J. L.
  663'00941   SB187G/
  ISBN 0-08-029793-5 Hard cover
  ISBN 0-08-029792-7 Flexi cover

Printed in Great Britain by A. Wheaton & Co. Ltd., Exeter

FOREWORD

IN commending this book I would like to stress the importance of making the rudiments of agriculture available at Farm Institute and Day Release level. It is our technicians, our foremen and stockmen, as much as our farmers and farm managers, who will require intellectual assurance as well as intuitive skill if agriculture is to match in technological advance the manufacturing industries of the future. Agriculture has peculiar problems of its own which lie in the fields of engineering, animal nutrition or agronomy. Mr. Lockhart and Mr. Wiseman have produced a book which deals comprehensively with the last category in a manner that should not be too advanced for the arable foreman of tomorrow. They give the subject a modern slant by incorporating such matters as the selective control of weeds, the principles of crop storage, and field meteorology within the traditional framework of geology, botany and chemistry. In short they take much of the scientific mystery out of the subject by describing in basic terms those forces which promote and those factors which inhibit the growth of economic plants. This then is a grammar without which modern farming will not again become an art.

Whitehall Place, S.W.1.

February 1966

JOHN GREEN
Chairman of the
Agricultural Advisory Council
for England and Wales

PREFACE TO THE
FIRST EDITION

THIS book is an introduction to the science and practice of crop husbandry. It is written in simple language without losing its technical value. Young people doing their practical training will find it helpful for explaining modern farming practices of growing and harvesting crops. All aspects of the subject are dealt with, such as the growth and development of plants, types and management of soils, drainage and irrigation, modern practices of growing and harvesting crops, management of grassland—including conservation, typical life-cycles of common pests and diseases, and the latest developments in the use of chemicals as fertilizers and in the control of weeds, pests and diseases. Suggestions for classwork are also included at the end of sections.

Students taking the City and Guilds General Agriculture (Part 1) and similar examinations can use it as a textbook, and those taking higher examinations will find the book a valuable source of basic information which will be enlarged on in their courses.

The authors acknowledge with very grateful thanks the valuable help given by their wives in preparing and typing the manuscript.

Cirencester
February 1966

J. A. R. LOCKHART
A. J. L. WISEMAN

PREFACE TO THE
FIFTH EDITION

WHEN the fourth, and considerably revised, edition of *Introduction to Crop Husbandry* was published, followed just two years later by a reprint with additions, the authors believed that any fifth edition would not need to be a major revision. This has not been the case. Quite large sections have had to be revised, thus instancing again the rapid technological progress which these days so characterizes the agricultural industry.

It is hoped that this book will continue to form the basis of crop husbandry syllabi at all levels of agricultural teaching. Depending on their course, students will generally find that further reading of the subject will be necessary. References are given where appropriate, and it should be noted that the MAFF Booklets are now replacing some of the Short Term Leaflets (STL), and the L Leaflet is replacing the Advisory Leaflet (AL).

As in previous editions, we are glad to acknowledge the help given by Mr. R. Churchill and Mr. W. Heatherington in the revision of the Pest and Diseases chapter.

Cirencester
June 1982

J. A. R. LOCKHART
A. J. L. WISEMAN

CONTENTS

Introduction   xi

Chapter 1. Plants

What they are; What they do; and How they live—photosynthesis, transpiration, osmosis   1
Plant groups—annuals, biennials, perennials   4
Seed structure—dicotyledons and monocotyledons, germination   4
Plant structure, seed formation   8
What plants need; plant nutrients; the nitrogen cycle   12
Suggestions for Classwork   16
Further Reading   16

Chapter 2. Soils

Topsoil, subsoil, soil profile   17
Soil formation—rocks, weathering   18
Composition of the soil—mineral and organic matter, water and air in the soil, soil micro-organisms,
    earthworms   20
Soil texture and structure   24
Soil fertility and productivity   25
Types of soil—clay, sand, loam, calcareous, silt and organic soils   26
Soil improvement—liming, drainage, irrigation, warping, claying   31
Tillage and cultivations—seed-beds, implements, pans, soil loosening, soil capping, weed control 40
Suggestions for Classwork   46
Further Reading   46

Chapter 3. Fertilizers and Manures

Nutrients removed by crops—the need for and effects of nitrogen, phosphorus, and potassium; trace
    elements   47
Units of plant food—unit values   49
"Straight" and compound fertilizers; plant food ratios   51
Storage and application of fertilizers   55
Organic manures—farmyard manure, slurry, cereal straws, leys, green manuring, seaweed, poultry
    manure, waste products, e.g. shoddy   55
Residual values of manures and fertilizers; organic farming   60
Suggestions for Classwork   61
Further Reading   61

Chapter 4. Cropping

Climate and weather and their effects on cropping   62
Rotations, continuous cereals   63
Cereals—wheat, barley, oats, rye, triticale and maize   66
Pulses—beans, peas, vetches, lupins, soya beans   82
Oil-seed rape, linseed and flax; sunflowers   90
Potatoes   94
Sugar-beet, mangels and fodder beet   100
Turnips and swedes   107
Kale, cabbage and Brussels sprouts   108
Carrots and bulb onions   114
Vegetable production and seed production   116
Suggestions for Classwork   120
Further Reading   120

ix

Chapter 5. Grassland

Classification of grassland	121
Plants used in leys—grasses, legumes and herbs	122
Grass and clover identification	122
Varieties and strains of grasses and legumes; herbs	126
Herbage digestibility	133
Seeds mixtures	135
Establishment of leys	136
Management—manuring, grazing systems, stocking rates	139
Pasture renovation and renewal	147
Conservation—haymaking, ensilage, green-crop drying	149
Suggestions for Classwork	164
Further Reading	165

Chapter 6. Weeds

Harmful effects	166
Spread of weeds	166
Assessing weed problems in the field	167
Control methods	167
Chemical control—contact, residual, total, translocated and hormone herbicides	168
Weed control in cereals	169
Weed control in potatoes, roots and kale	175
Weed control in grassland	176
Spraying with herbicides	178
Suggestings for Classwork	179
Further Reading	179

Chapter 7. Pests and Diseases of Farm Crops

Structure of an insect	180
Life-cycles	181
Methods of pest control, including use of insecticides	182
Control of the more important pests	185
Other pests—birds and mammals	194
Main agencies of disease—fungi, viruses, bacteria, deficiency	194
Control of the more important diseases	196
Control of plant diseases, including use of fungicides	207
Suggestions for Classwork	212
Further Reading	212

APPENDIX

1. World Crop Production	213
2. Metrication	214
3. Agricultural Land Classification in England and Wales	216
4. Factors Affecting the Application (and Mixing) of Crop Protection Chemicals	220
5. Latin Names for Crops	223
6. Latin Names for Weeds	225
7. Latin names for Diseases	229
8. Latin names for Insect Pests	231
9. Crop Seeds	233
10. The Decimal Code for Growth Stages of all Small Grain Cereals	235
11. The Gross Margin System of Analysis	237
12. Estimating Crop Yields	239
13. Farming and Wildlife Conservation	240
14. The European Economic Community (EEC) or Common Market	242

INDEX

	247

INTRODUCTION

Crops are plants which have been carefully selected and developed to produce food for man and animals.

Crop husbandry is the practice of growing and harvesting crops. The main objective is to produce good crops as economically as possible without impoverishing the land.

The methods used have been developed over the past centuries from practical experience and experiments. In recent years there have been many sweeping changes as the result of:

(a) introduction of many new and improved varieties,
(b) better use of fertilizers,
(c) better control of pests and diseases,
(d) chemical weed control,
(e) rapid improvements in the mechanization of such operations as seed-bed preparation, planting, harvesting and storage.

An understanding of how plants grow, and what they need, is a useful guide when providing for their requirements.

Good crop husbandry is really good management of crop plants so that they are provided with the best possible conditions for growth.

PLANTS

WHAT THEY ARE;
WHAT THEY DO;
AND HOW THEY LIVE

PLANTS are living organisms consisting of innumerable tiny cells. They differ from animals in many ways but the most important difference is that plants can build up valuable organic substances from simple materials. The most important part of this building process, which is called *photosynthesis*, is the production of *carbohydrates* such as *sugars, starches* and *cellulose*.

Photosynthesis

In photosynthesis a special green substance called *chlorophyll* uses *light* energy (normally sunlight) to change *carbon dioxide* and *water* into *sugars* (carbohydrates) in the *green* parts of the plant. The daily amount of photosynthesis is limited by the duration and intensity of sunlight. The amount of carbon dioxide available is also a limiting factor. Shortage of water and low temperatures can also reduce photosynthesis.

The cells which contain chlorophyll also have yellow pigments such as *carotene*. Crop plants can only build up chlorophyll in the light and so any leaves which develop in the dark are yellow and cannot produce carbohydrates.

Oxygen is released during photosynthesis and the process may be set out as follows:

This process not only provides the basis for all our food but it also supplies the oxygen which animals and plants need for respiration.

The simple carbohydrates, such as *glucose*, may build up to form *starch* for storage purposes, or to *cellulose* for building cell walls. *Fats* and *oils* are formed from carbohydrates. *Protein* material, which is an essential part of all living cells, is made from carbohydrates and nitrogen compounds.

Most plants consist of *roots, stems, leaves* and *reproductive parts* and need *soil* in which to grow.

The *roots* spread through the spaces between the particles in the soil and anchor the plant. In a plant such as wheat the root system may total many miles.

The *leaves*, with their broad surfaces, are the main parts of the plant where photosynthesis occurs (see Fig. 1).

A very important feature of the leaf structure is the presence of large numbers of tiny pores (*stomata*) on the surface of the leaf (see Fig. 2). There are usually thousands of stomata per square cm of leaf surface. Each pore (stoma) is oval-shaped and surrounded by two guard cells. When the guard cells are turgid (full of water) the stoma is open and when they lose water the stoma closes.

The carbon dioxide used in photosynthesis *diffuses* into the leaf through the stomata and most of the water vapour leaving the plant, and the oxygen from photosynthesis diffuses out through the stomata.

$$\text{Carbon dioxide} + \text{water} + \text{energy} \xrightarrow{\text{chlorophyll}} \text{carbohydrates} + \text{oxygen}$$

$$nCO_2 \qquad nH_2O \qquad \text{(light)} \qquad = \qquad (CH_2O)n \qquad nO_2$$

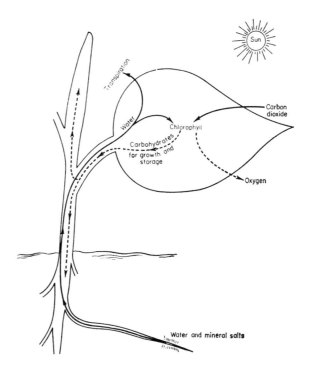

FIG. 1. Photosynthesis illustrated diagrammatically.

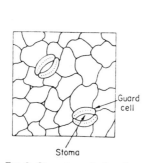

FIG. 2. Stomata on leaf surface.

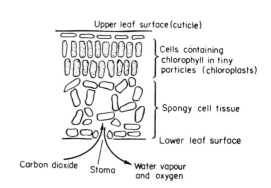

FIG. 3. Cross-section of green leaf showing gaseous movements during daylight.

Transpiration

The evaporation of water from plants is called *transpiration*. It mainly occurs through the stomata and has a cooling effect on the leaf cells. Water in the cells of the leaf can pass into the pore spaces in the leaf and then out through the stomata as water vapour (see Fig. 3).

The rate of transpiration varies considerably. It is greatest when the plant is well supplied with water and the air outside the leaf is warm and dry. In very hot or windy weather water evaporates from the guard cells and so the stomata close and reduce the rate of transpiration. The stomata also close in very cold weather, e.g. 0°C.

The rate of loss is reduced if the plant is short

of water because the guard cells then lose water and close the stomata; it is also retarded if the humidity of the atmosphere is high.

The stomata guard cells close (and so transpiration ceases) during darkness. They close because photosynthesis ceases and water is lost from the guard cells (osmosis) when some of the sugars present change to starch.

Respiration

Plants, like animals, breathe, i.e. they take in oxygen which combines with organic foodstuffs and this releases energy, carbon dioxide and water. Farm crops are likely to be checked in growth if the roots are deprived of oxygen for respiration as might occur in a waterlogged soil.

Translocation

The movement of materials through the plant is known as *translocation*.

The *xylem* or *wood vessels* which carry the water and mineral salts (*sap*) from the roots to the leaves are tubes made from dead cells. The cross walls of the cells have disappeared and the longitudinal walls are thickened with *lignum* to form wood. These tubes help to strengthen the stem.

The *phloem tubes* (*bast*) carry organic material through the plant, for example, sugars and amino acids from the leaves to storage parts or growing points. These vessels are chains of living cells, not lignified, and with cross walls which are perforated—hence the alternative name—*sieve tubes*.

In the stem the xylem and phloem tubes are usually found in a ring near the outside of the stem.

In the root, the xylem and phloem tubes form separate bundles and are found near the centre of the root.

Uptake of water by plants

Water is taken into the plant from the soil. This occurs mainly through the root hairs near the root tip. There are thousands (perhaps millions) of root tips (and root hair regions) on a single healthy crop plant (see Fig. 4).

The absorption of water into the plant in this way is due to suction pull which starts in the leaves. As water transpires (evaporates) from the cells in the leaf more water is drawn from the xylem tubes which extend from the leaves to the root tips. In these tubes the water is stretched like a taut wire or thread. This is possible because the tiny particles (molecules) of water

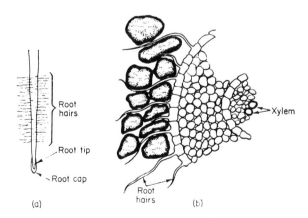

FIG. 4. (a) Section of root tip and root hair region, (b) cross-section of root showing the root hairs as tube-like elongations of the surface cells in contact with soil particles.

hold together very firmly when in narrow tubes. The pull of this water in the xylem tubes of the root is transferred through the root cells to the root hairs and so water is absorbed into the roots and up to the leaves. In general, the greater the rate of transpiration, the greater is the amount of water taken into the plant. The rate of absorption is slowed down by:

(a) shortage of water in the soil,
(b) lack of oxygen for root respiration (e.g. in waterlogged soils),
(c) a high concentration of salts in the soil water near the roots.

Normally, the concentration of the soil solution does not interfere with water absorption. High soil water concentration can occur in salty soils and near bands of fertilizer. Too much fertilizer near developing seedlings may damage germination by restricting the uptake of water.

Osmosis

Much of the water movements into and from cell to cell in plants is due to *osmosis*. This is a process in which a solvent, such as water, will flow through a *semi-permeable* membrane (e.g. a cell wall) from a weak solution to a more concentrated one. The cell wall may allow only the water to pass through. The force exerted by such a flow is called the *osmotic pressure*. In plants, the normal movement of the water is into the cell. However, if the concentration of a solution outside the cell is greater than that inside, there is a loss of water from the cell, and its contents contract (shrivel); this is called *plasmolysis*.

Uptake of nutrients

The absorption of chemical substances (nutrients) into the root cells is partly due to a *diffusion* process but it is mainly due to ability of the cells near root tips to *accumulate* such nutrients. The process is complicated and not fully understood. It is slowed up if root respiration is checked by a shortage of oxygen.

PLANT GROUPS

Plants can be divided into annuals, biennials and perennials according to their total length of life.

Annuals

Typical examples are wheat, barley and oats which complete their life history in one growing season, i.e. starting from the seed, in 1 year they develop roots, stem and leaves and then produce flowers and seed before dying.

Biennials

These plants grow for 2 years. They spend the first year in producing roots, stem and leaves, and the following year in producing the flowering stem and seeds, after which they die.

Sugar-beet, swedes and turnips are typical biennials, although the grower treats these crops as annuals, harvesting them at the end of the first year when all the foodstuff is stored up in the root.

Perennials

They live for more than 2 years and, once fully developed, they usually produce seeds each year. Many of the grasses and legumes are perennials.

STRUCTURE OF THE SEED

Plants are also classified as *dicotyledons* and *monocotyledons* according to the structure of the seed.

Dicotyledons

A good example of a dicotyledon seed is the broad bean because it is large and easy to study.

If a pod of the broad bean plant is opened when it is nearly ripe it will be seen that each seed is attached to the inside of the pod by a short stalk called the *funicle*. All the nourishment which the developing seed requires passes through the funicle from the bean plant.

When the seed is ripe and has separated from the pod a black scar, known as the *hilum*, can be seen where the funicle was attached. Near one end of this hilum is a minute hole called the *micropyle* (see Fig. 5).

If a bean is soaked in water the seed coat can be removed easily and all that is left is largely made up of the *embryo* (*germ*). This consists of two seed leaves, or *cotyledons*, which contain the food for the young seedling.

Lying between the two cotyledons is the *radicle*, which eventually forms the *primary root*, and a continuation of the radicle the other end, the *plumule* (see Fig. 6). This develops into the young *shoot*, and is the first *bud* of the plant.

Monocotyledons

This class includes all the cereals and grasses and it is, therefore, very important.

The wheat grain is a typical example. It is not a true seed (it should be called a single-seeded fruit). The seed completely fills the whole grain, being practically united with the inside wall of the grain or fruit.

This fruit wall is made up of many different layers which are separated on milling into varying degrees of fineness, e.g. bran and pollards, and these are valuable livestock feed.

Most of the interior of the grain is taken up by the floury *endosperm*. The embryo occupies the small raised area at the base. The *scutellum*, a shield-like structure, separates the embryo from the endosperm. Attached to the base of the scutellum are the five roots of the embryo, one primary and two pairs of *secondary* rootlets. The roots are enclosed by a sheath called the *coleorhiza*. The position of the radicle and the plumule can be seen in the diagram (Fig. 7).

The scutellum can be regarded as the cotyledon of the seed. There is only *one* cotyledon present and so wheat is a monocotyledon.

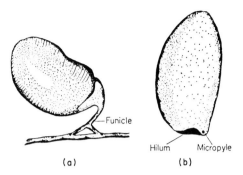

FIG. 5. (a) Bean seed attached to the inside of the pod by the funicle, (b) bean seed showing the hilum and micropyle.

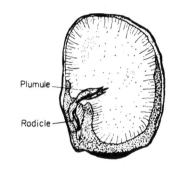

FIG. 6. Bean seed with one cotyledon removed.

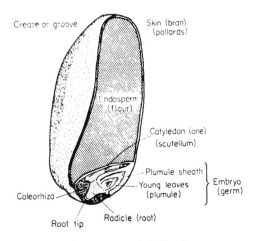

FIG. 7. Wheat grain cut in half at the crease.

Germination of the bean—the dicotyledon

Given suitable conditions for germination, i.e. water, heat and air, the seed coat of the dormant but living seed splits near the micropyle, and the radicle begins to grow downwards through this split to form the main, or primary root, from which lateral branches will soon develop (see Fig. 8).

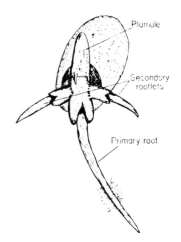

FIG. 9. Germination of the wheat grain.

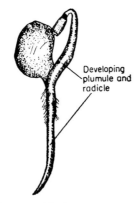

FIG. 8. Germination of the bean, one cotyledon removed.

same time the roots break through the coleorhiza (see Fig. 9).

The primary root is soon formed, supported by the two pairs of secondary rootlets, but this root system (the seminal roots) is only temporary and is soon replaced by *adventitious roots* (see Figs. 10 and 14). As the first root system is being

When the root is firmly held in the soil, the plumule starts to grow by pushing its way out of the same opening in the seed coat. As it grows upwards its tip is bent to protect it from injury in passing through the soil, but it straightens out on reaching the surface, and *leaves* very quickly develop from the plumular shoot.

With the broad bean the cotyledons remain underground—gradually giving up their stored food materials to the developing plant, but with the French bean, and many other dicotyledon seeds, the cotyledons are brought above ground with the plumule.

Germination of wheat—the monocotyledon

When the grain germinates the coleorhiza expands and splits open the seed coat, and at the

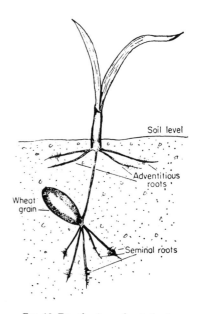

FIG. 10. Developing wheat plant.

formed at the base of the stem so the plumule starts to grow upwards, and its first leaf, the *coleoptile*, appears above the ground as a single pale tube-like structure.

From a slit in the top of the leaf there appears the first *true leaf* which is quickly followed by others, the younger leaves growing from the older leaves (see Fig. 11).

As the wheat embryo grows so the floury endosperm is used up by the developing roots

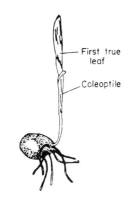

First true leaf

Coleoptile

FIG. 11. Seedling wheat plant.

and plumule, and the scutellum has the important function of changing the endosperm into digestible food for the growing parts.

With the broad bean, the cotyledons provide the food for the early nutrition of the plant, whilst the wheat grain is dependent upon the endosperm and scutellum, and in both cases it is not until the plumule has reached the light and turned green that the plants can begin to be independent.

This point is important in relation to the depth at which seeds should be sown. Small seeds, such as the clovers and many of the grasses, must, as far as possible, be sown very shallow. Their food reserves will be exhausted before the shoot reaches the surface if sown too deeply. Larger seeds, such as the beans and peas, can and should be sown deeper.

When the leaves of the plant begin to manufacture food by photosynthesis (see page 1) and when the primary root has established itself sufficiently well to absorb nutrients from the soil (see page 4) then the plant can develop independently, provided there is sufficient moisture and air present.

The main differences between the two groups of plants can be summarized as follows:

Dicotyledons	*Monocotyledons*
The embryo has two seed leaves.	The embryo has one seed leaf.
A primary root system is developed and persists	A primary root system is developed, but is replaced by an adventitious root system.
Usually broad-leafed plants, e.g. clovers, cabbage and potato.	Usually narrow-leaved plants, e.g. the cereals and grasses, and most bulbous plants.

These two great groups of flowering plants can be further divided in the following way:

Families or orders, e.g.	The legume family, potato, the grasses and cereals.
Genus	Clovers of the legume family, and wheat of the cereal family.
Species	Red clover.
Cultivar or Variety	Late-flowering red clover.
Strain	S123 late-flowering red clover.

PLANT STRUCTURE

The plant can be divided into two parts:

1. The root system

The root system is concerned with the parts of the plant growing in the soil and there are two main types:

(a) *The tap root or primary system.* This is made up of the primary root called the tap root with *lateral secondary* roots branching out from it, and from these *tertiary* roots may develop obliquely to form, in some cases, a very extensive system of roots (see Fig. 12).

The root of the bean plant is a good example of a tap root system, and if this is split it will be seen that there is a slightly darker central woody core; this is the *skeleton* of the root. It helps to anchor the plant, and also transports foodstuffs. The lateral secondary roots arise from this central core (see Fig. 13).

Carrots, and other true root crops, such as sugar-beet and mangolds, have very well-developed tap roots. These biennials store food in their roots during the first year of growth to be used in the following year for the production of the flowering shoot and seeds. However, they are normally harvested after one season and the roots are used as food for man and stock.

(b) *The adventitious root system.* This is found on all grasses and cereals, and it is, in fact, the main root system of most monocotyledons. The primary root is quickly replaced by adventitious roots, which arise from the base of the stem (see Fig. 14).

Actually, these roots can develop from any part of the stem, and they are found on some dicotyledons as well, but not as the main root system, e.g. underground stems of the potato.

Root hairs (see Fig. 4). These are very small white hair-like structures which are found near the tips of all roots. As the root grows, the hairs on the older parts die off, and others develop on the younger parts of the root.

Root hairs play a very important part in the life of a plant (see page 3).

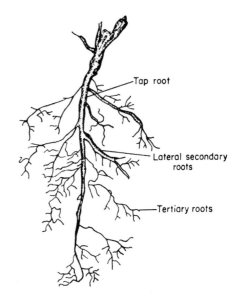

FIG. 12. Tap root or primary root system.

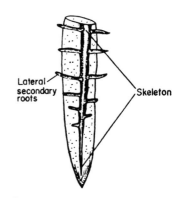

FIG. 13. Tap root of the bean plant.

FIG. 14. Adventitious root system.

2. The stem

The second part of the flowering plant is the shoot which normally grows upright above the ground. It is made up of a main stem, branches, leaves and flowers.

Stems are either soft (*herbaceous*) or hard (*woody*) and in British agriculture it is only the soft and green herbaceous stems which are of any importance. These usually die back every year.

How stems grow. All stems start life as *buds* and the increase in length takes place at the tip of the shoot called the *terminal* bud.

If a Brussels sprout is cut lengthwise and examined, it will be seen that the young leaves arise from the bud *axis*. This axis is made up of different types of cell tissue, which is continually making new cells and thus growing (see Fig. 15).

Stems are usually jointed, each joint forming a *node*, and the part between two nodes is the *internode*. At the nodes the stem is usually solid and thicker, and this swelling is caused by the storing up of material at the base of the leaf (see Fig. 16).

The bud consists of closely packed leaves arising from a number of nodes. It is, in fact, a condensed portion of the stem which develops by a lengthening of the internodes.

Axillary buds are formed in the angle between the stem and leaf stalk. These buds, which are similar to the terminal bud, develop to form lateral *branches, leaves* and *flowers*.

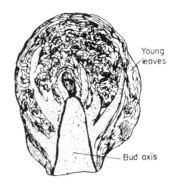

FIG. 15. Longitudinal section of a Brussels sprout.

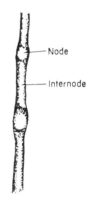

FIG. 16. Jointed stem.

Some modification of stems

(1) A *stolon* is a stem which grows along the ground surface. Adventitious roots are produced at the nodes, and buds on the runner can develop into upright shoots, and separate plants can be formed, e.g. strawberry plants (see Fig. 17).

(2) A *rhizome* is similar to a stolon but grows under the surface of the ground, e.g. couch grass (see Fig. 17).

(3) A *tuber* is really a modified rhizome. The end of the rhizomes swell to form tubers. The tuber is therefore a swollen stem. The potato is a

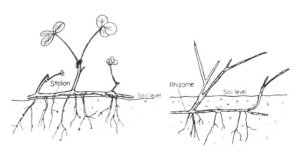

FIG. 17. Modified stems.

well-known example, and it has "eyes" which are really buds and these develop shoots when the potato tuber is planted.

(4) A *tendril* is found on certain legumes, such as the pea. The terminal leaflet is modified as in the diagram. This is useful for climbing purposes to support the plant (see Fig. 18).

Corms and *suckers* are other examples of modified stems.

FIG. 18. Modified stems.

The leaf

Leaves in all cases arise from buds. They are extremely important organs, being not only responsible for the manufacture of sugar and starch from the atmosphere for the growing parts of the plant, but they also are the organs through which transpiration of water takes place.

A typical leaf of a dicotyledon consists of three main parts:

(1) The *blade*.
(2) The *stalk* or *petiole*.
(3) The *basal sheath* connecting the leaf to the stem. This may be modified as with legumes into a pair of wing-like *stipules* (see Fig. 19(a)).

The blade is the most obvious part of the leaf and it is made up of a network of veins.

There are two main types of dicotyledonous leaves:

(1) A prominent central *midrib*, from which lateral veins branch off on either side. These side veins branch into smaller and smaller ones, as in the diagram (see Fig. 19(a)).
(2) No single midrib, but several main ribs spread out from the top of the leaf stalk; between these the finer veins spread out as before, e.g. horse-chestnut leaf (see Fig. 19(b)).

The veins are the essential supply lines for the process of photosynthesis. They consist of two main parts, one for bringing the required raw material up to the leaf (*xylem*), and the other part being concerned with carrying the finished product away from the leaf (*phloem*).

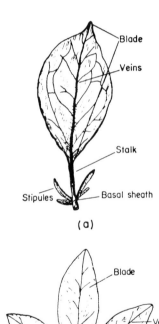

(a)

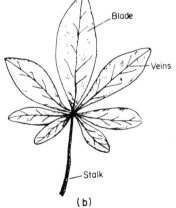

(b)

FIG. 19. (a) Simple leaf, (b) Compound leaf.

Leaves can show great variation in shape and type of margin, as in Fig. 19. They can also be divided into two broad classes as follows:

(1) *Simple* leaves. The blade consists of one continuous piece (see Fig. 19(a)).
(2) *Compound* leaves. Simple leaves may become deeply lobed and when the division between the lobes reaches the midrib it is a compound leaf, and the separate parts of the blade are called the *leaflets* (see Fig. 19(b)).

The blade surface may be *smooth* (glabrous) or *hairy*, according to variety, and this is important in legumes because it can affect its palatability to stock.

Monocotyledonous leaves are dealt with in the chapter on "Grassland".

Modified leaves

(a) *Cotyledons* or seed leaves are usually of a very simple form.

(b) *Scales* are normally rather thin, yellowish to brown membranous leaf structures, very variable in size and form. On woody stems they are present as *bud scales* which protect the bud, and they are also found on rhizomes such as couch.

(c) *Leaf tendrils*. The terminal leaflet like the stem can be modified into thin threadlike structures, e.g. pea and vetch.

Other examples of modified leaves are *leaf-spines* and *bracts*.

The flower

In the centre of the flower is the *axis* which is simply the continuation of the flower stalk. It is known as the *receptacle*, and on it are arranged four kinds of organs:

(1) The lowermost is a ring of green leaves called the *calyx*, made up of individual *sepals*.
(2) Immediately above the calyx is a ring of *petals* known as the *corolla*.
(3) Above the corolla are the *stamens*, again arranged in a ring. They are similar in

appearance to an ordinary match, the swollen tip being called the *anther* which, when ripe, contains the *pollen grains*.
(4) The highest position on the receptacle is occupied by the *pistil* which is made up of one or a series of small green bottle-shaped bodies—the *carpel*, which is itself made up of three parts: the *stigma*, *style*, and the *ovary* (containing *ovules*). It is within the ovary that the future *seeds* are produced (see Figs. 20 and 21).

Most flowers are more complicated in appearance than the above, but basically they consist of these four main parts.

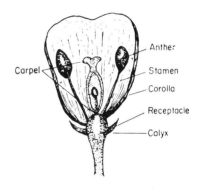

FIG. 20. Longitudinal section of a simple flower.

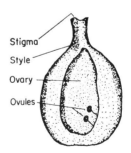

FIG. 21. Carpel.

The formation of seeds

Pollination precedes *fertilization*, which is the union of the male and female reproductive cells. When pollination takes place the pollen grain is transferred from the anther to the stigma. This

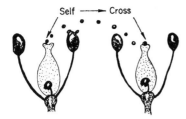

FIG. 22. Self- and cross-pollination.

(a)

FIG. 23(a). Indefinite influorescence.

may be *self-pollination* where the pollen is transferred from the anther to the stigma of the same flower, or *cross-pollination* when it is carried to a different flower (see Fig. 22).

With fertilization the pollen grain grows down the style of the carpel to fuse with the ovule. After fertilization, changes take place whereby the ovule develops into the embryo, and endosperm may be formed according to the species. This makes up the *seed*. The ovary also changes after fertilization to form the *fruit*, as distinct from the seed.

With the grasses and cereals there is only one seed formed in the fruit and, being so closely united with the inside wall of the ovary, it cannot easily be separated from it.

The one-seeded fruit is called a *grain*.

The inflorescence

Special branches of the plant are modified to bear the flowers, and they form the inflorescence. There are two main types of inflorescence:

(1) Where the branches bearing the flowers continue to grow, so that the youngest flowers are nearest the apex and the oldest farthest away—*indefinite* inflorescence (see Fig. 23(a)).

A well known example of this inflorescence is the *spike* found in many species of grasses.

(2) Where the main stem is terminated by a single flower and ceases to grow in length; any further growth takes place by lateral branches, and they eventually terminate in a single flower and growth is stopped—*definite* inflorescence, e.g. stitchworts (see Fig. 23(b)).

There are many variations of these two main types of inflorescence.

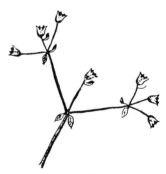

FIG. 23(b). Definite influorescence.

WHAT PLANTS NEED

To grow satisfactorily a plant needs *warmth, light, water, carbon dioxide* and about a dozen other *chemical elements* which it can obtain from the soil.

Warmth

Most crop plants in this country start growing when the average daily temperature is above 6°C. Growth is best between 16°C and 27°C. These temperatures apply to thermometer readings taken in the shade about 1.5 m above ground. Crops grown in hotter countries usually have higher temperature requirements.

Cold frosty conditions may seriously damage

plant growth. Crop plants differ in their ability to withstand very cold conditions. For example, winter rye and wheat can stand colder conditions than winter oats. Potato plants and stored tubers are easily damaged by frost. Sugar-beet may bolt (go to seed) if there are frosts after germination; frost in December and January may destroy crops left in the ground.

Light

Without light, plants cannot produce carbohydrates and will soon die. The amount of photosynthesis which takes place daily in a plant is partly due to the length of daylight and partly to the intensity of the sunlight. Bright sunlight is of most importance where there is dense plant growth.

The lengths of daylight and darkness periods vary according to the distance from the equator and also from season to season. This can affect the flowering and seeding of crop plants and is one of the limiting factors in introducing new crops into a country. Grasses are now being tested in this country which will remain leafy and not produce flowering shoots under the daylight conditions here.

Water

Water is an essential part of all plant cells and it is also required in extravagant amounts for the process of transpiration. Water carries nutrients from the soil into and through the plant and also carries the products of photosynthesis from the leaves to wherever they are needed. Plants take up about 200 tonnes of water for every tonne of dry matter produced.

Carbon dioxide (CO_2)

Plants need carbon dioxide for photosynthesis. This is taken into the leaves through the stomata and so the amount which can go in is affected by the rate of transpiration. Another limiting factor is the small amount (0.03%) of carbon dioxide in the atmosphere. The percentage can increase just above the surface of soils rich in organic matter where soil bacteria are active and releasing carbon dioxide. This is possibly one of the reasons why crops grow better on such soils. (See STL 45.)

Chemical elements required by plants

In order that a plant may build up its cell structure and function as a food factory many simple chemical substances are needed. These are taken into the roots from the soil solution and the clay particles. Those required in fairly large amounts—a few kilogrammes to one or more hundred kilogrammes per hectare—are called the *major* nutrients; those required in small amounts—a few grams to several kilogrammes per hectare—are the *minor* nutrients or *trace elements*.

Deficiencies of boron and manganese are often caused by using too much lime.

Other trace elements are chlorine, iron, molybdenum and zinc, but these rarely cause trouble on most farm soils (see page 49).

Cobalt is not considered necessary for plant growth, but animals feeding on plants deficient in cobalt (e.g. on some all-grass areas) waste away ("pine"). The remedy would be a few kilogrammes of a cobalt salt per hectare or in a salt lick.

Sodium does not appear to be essential, but some crops such as sugar-beet and mangolds grow better if it is applied (e.g. as common salt). It may partly replace potassium.

The effects of nitrogen, phosphorus and potassium are discussed more fully in the chapter on "Fertilizers".

The trace elements are shown in Table 2.

LEGUMES AND THE NITROGEN CYCLE

Legumes are plants which have several interesting characteristics such as:

(1) A special type of fruit called a legume, which splits along both sides to release its seeds, e.g. pea pod.

ITCH - B

TABLE 1

The major nutrients	Use	Source
Carbon (C) Hydrogen (H) Oxygen (O)	Used in making carbohydrates	The air and water
Nitrogen (N)	Very important for building proteins	Organic matter (including FYM) Nodules on legumes (page 15) Nitrogen fertilizers such as ammonium and nitrate compounds and urea Nitrogen-fixing soil micro-organisms
Phosphorus (P) (phosphate)	Essential for cell division and many chemical reactions	Small amounts from the mineral and organic matter in the soil Mainly from phosphate fertilizers, e.g. superphosphate, ground rock phosphate, basic slag and compounds, and residues of previous fertilizer applications
Potassium (K) (potash)	Helps with formation of carbohydrates and proteins Regulates water in and through the plant	Small amounts from mineral and organic matter in the soil. Potash fertilizers, e.g. muriate and sulphate of potash
Calcium (Ca)	Essential for development of growth tissue, e.g. root tips	Usually enough in the soil. Applied as chalk or limestone to neutralize acidity
Magnesium (Mg)	A necessary part of chlorophyll	If soil is deficient, may be added as magnesium limestone or magnesium sulphate, also FYM
Sulphur (S)	Part of many proteins and some oils	Usually enough in the soil. Added in some fertilizers (e.g. sulphate of ammonia and superphosphate)

TABLE 2

	Deficiency symptoms	Remedy
Boron (B)	*Heart-rot* in sugar-beet and mangolds *Brown-heart* (*raan*) in turnips and swedes	20 kg/ha borax evenly spread, e.g. with fertilizer
Copper (Cu) Deficiency often associated with organic soils	Commonest in cereals—leaf tips and edges white-yellowish-grey and in twisted spiral. Head may be distorted or fail to emerge. Yields are very seriously reduced Found on deep fen peats and on the black soils of the chalk downs	25–50 kg/ha of copper sulphate applied to soil to last 3–5 yrs or 1–2 kg/ha copper oxychloride applied to growing crop or copper chelate
Manganese (Mn) Deficiency associated with high pH and soils rich in organic matter	Grey patches on leaves of cereals *"Marsh-spot"* in peas *Speckled yellowing* of leaves of sugar-beet	Manganese sulphate applied to soil 60–120 kg/ha or 10 kg sprayed on young crop, or 14 kg combine drilled (cereals) or manganese chelate Some wheat varieties, e.g. *Maris Dove* (S), *Maris Butler* (S), *Mega* (W) and *Champlein* (W) are resistant to manganese deficiency

(2) The flowers closely resemble pea flowers.

(3) Nodules (lumps) on their roots contain special types of bacteria which can "fix" (convert) nitrogen from the air into nitrogen compounds. These bacteria enter the plant through the root hairs from the surrounding soil.

This "fixation" of nitrogen is of considerable agricultural importance. Many of our farm crops are legumes, for example, *peas, beans, vetches, lupins, clovers, lucerne (alfalfa), sainfoin* and *trefoil*. The bacteria obtain carbohydrates (energy) from the plant and in return they supply nitrogen compounds. The nodules can release nitrogen compounds into the soil. These compounds are changed to nitrates and taken up by neighbouring plants (e.g. by grasses in a grass and clover sward) or by the following crop, e.g. wheat after clover or beans. The amount of nitrogen which can be "fixed" by legume bacteria varies widely; estimates of 56–450 kg/ha of nitrogen have been made. Some of the reasons for variations are:

(a) *The type of plant.* Some crop plants "fix" more nitrogen than others, e.g. lucerne and clovers (especially if grazed) are usually better than peas and beans.

(b) *The conditions in the soil.* The bacteria usually work best in soils which favour the growth of the plant on which they live.

A good supply of calcium and phosphate in the soil is usually beneficial, although lupins grow well on acid soils.

(c) *The strains of bacteria present.* Most soils in this country contain the strains of bacteria required for most of the leguminous crops which are grown. Lucerne (alfalfa) is an exception and it is common practice to coat the lucerne seed with the proper bacterial culture before sowing; these bacteria will later enter the roots of the young plant. French beans do not "fix" their own nitrogen in this country at present.

THE NITROGEN CYCLE

The circulation of nitrogen (in various compounds such as nitrates and proteins) as found

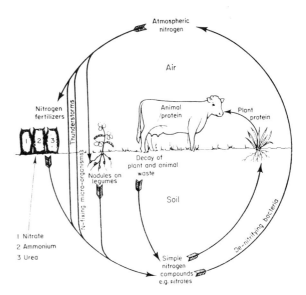

FIG. 24. The nitrogen cycle.

on the farm is illustrated diagrammatically in Fig. 24.

Atmospheric nitrogen is "fixed" (combined) in compounds by *legume nodule bacteria*, by various *nitrogen-fixing micro-organisms*, by *thunderstorms* and in the manufacture of *nitrogen fertilizers*.

Simple nitrogen compounds (mainly nitrates) are taken up by plants to form plant proteins which may then be eaten by animals to form animal proteins. Dead plants and animals, and the faeces and urine of animals are broken down by decay micro-organisms to leave simple nitrogen compounds in the soil.

The de-nitrifying bacteria change nitrogen compounds back to free nitrogen. This is most likely to happen where nitrates are abundant and oxygen is in short-supply, e.g. in waterlogged soils.

Nitrate pollution of rivers

Nitrate–nitrogen compounds, some of which may have been applied as fertilizer, leach into drainage water and so into rivers. This has caused some alarm in recent years when the nitrate level in drinking water has reached such high levels as to cause possible health hazards, especially to young children. Whilst nitrogen fertilizer usage may be responsible to a small extent, the main cause of the trouble is excessive amounts of sewage (treated or untreated) which enters rivers which are used as a source of drinking water. The problem is really associated with densely populated areas where insufficient attention is paid to proper waste disposal, but the blame is directed at farmers. The nitrate in water encourages the growth of micro-organisms, such as algae, which can de-oxygenate the water and so fish may die. The water becomes tainted and is difficult to filter and purify.

SUGGESTIONS FOR CLASSWORK

1. Compare and contrast the dicotyledon seed and the monocotyledon seed.

2. Germinate the bean seed, and study its development.

3. Dig up and carefully examine the root system of different plants.

4. Examine different types of modified stems.

5. Examine different leaves and modified leaves.

6. With a lens, examine carefully the parts of a flower.

FURTHER READING

Gill and Vear, *Agricultural Botany*, Duckworth.
Lowson, *Textbook of Botany*, Oxford Univ. Tutorial Press.

SOILS

SOILS are very complex natural formations which make up the surface of the earth. They provide a suitable environment in which plants may obtain *water*, *nutrients* and *oxygen* for root respiration, and firm *anchorage*. Soils are formed by the weathering of rocks, followed by the growth and decay of plants, animals, and soil micro-organisms. If a farmer is to provide the best possible conditions for crop growth, it is desirable that he should understand what soils are, how they were formed and how they should be managed.

The *topsoil* or *surface soil* is a layer about 8–45 cm deep which may be taken as the greatest depth which a farmer would plough or cultivate and in which most of the plant roots are found.

Loose, cultivated, topsoil is sometimes called *mould*.

The *subsoil*, which lies underneath, is an intermediate stage in the formation of soil from the rock below.

A *soil profile* is a section taken through the soil down to the parent rock. In some cases this may consist of only a shallow surface soil 10–15 cm on top of a rock such as chalk or limestone. In other well-developed soils (about a metre deep) there are usually three or more definite layers (or horizons) which vary in colour, texture and compaction (see Fig. 25).

The soil profile can be examined by digging a trench or by taking out cores of soil from various depths with a *soil auger*.

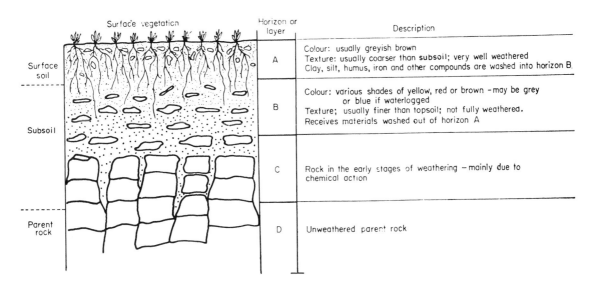

FIG. 25. Soil profile diagram showing the breakdown of rock to form various soil layers (horizons).

A careful examination of the layers (horizons) can be useful in forming an opinion as to how the soil was formed, its natural drainage and how it might be farmed. Some detailed soil classifications are based on soil profile (see Appendix 3).

SOIL FORMATION

There are very many different types of soils and subsoils. The differences are mainly due to the variety of rocks from which they are formed. However, other factors such as *climate, topography, plant and animal life,* the *age of the developing soil material* and *farming operations* also affect the type of soil which develops.

The more important rock formations

Igneous or primary rocks, e.g. *granite* (coarse crystals) and *basalt* (fine crystals). These rocks were formed from the very hot molten material which made up the earth, millions of years ago. The minerals (chemical compounds) in these rocks are mostly in the form of crystals and are the primary source of the minerals found in all our soils. Igneous rocks are very hard and weather very slowly. Clay and sand are breakdown products.

Sedimentary or transported rocks. These have been formed from weathered material (e.g. clay, silt and sand) carried and deposited by water and wind. The sediments later became compressed by more material on top and cemented to form new rocks such as *sandstones, clays* and *shales.*

The *chalks* and *limestones* were formed from the shells and skeletons of sea animals of various sizes. These rocks are mainly calcium carbonate but in some cases are magnesium carbonate. The calcareous soils are formed from them (see page 29).

Metamorphic rocks, e.g. *marble* (from limestone) and *slate* (from shale). These are rocks which have been changed in various ways.

Organic matter

Deep deposits of *organic matter* (humus) are found in places where waterlogged soil conditions did not allow the breakdown of dead plant material by micro-organisms and oxidation.

Peats have been formed in water-logged acidic areas where the vegetation is mainly mosses, rushes, heather and some trees.

Black fen (muck soil) has been formed in marshy river estuary conditions where the water was hard (lime rich) and often silty; the plants were mainly reeds, sedges, rushes and some trees.

Good drainage is necessary before preparing these areas for cropping.

Glacial drift

Many soils in the British Isles are not derived from the rocks underneath but are deposits carried from other rock formations by glaciers, e.g. boulder clays. This makes the study of our soils very complicated.

Alluvium is material which has been deposited recently, for example, by river flooding. This material is very variable in composition.

Weathering of rocks

The breakdown of rocks is mainly caused by the *physical* and *chemical* effects of the weather.

Physical weathering. Changes of temperature cause the various mineral crystals in rocks to expand and contract by different amounts, and so cracking and shattering often occurs.

Water can cause pieces of rock surfaces to split off when it freezes and expands in cracks and crevices. Also, the molecules (small particles) of water in the pores and fissures of the rock exert expansion and contraction forces similar to those of freezing and thawing. The pieces of rock broken off are usually sharp-edged, but if they are carried and knocked about by glaciers, rivers or wind, they become more rounded in shape, e.g. sand and stones in a river bed.

Wetting and drying of some rocks such as clays and shales causes expansion and contraction and results in cracking and flaking.

Chemical weathering. Chemical breakdown of the mineral matter in a developing soil is brought about by the action of water, oxygen, carbon dioxide and nitric acid from the atmosphere; and by carbonic and organic acids from the biological activity in the soil. The soil water, which is a weak acid, dissolves some minerals and allows chemical reactions to take place.

Water can also unite with substances in the soil (hydration) to form new substances which are more bulky and so can cause shattering of rock fragments.

Clay is produced by chemical weathering. In the case of rocks such as granite, when the clay-producing parts are weathered away the more resistant quartz crystals are left as sand or silt.

In the later stages of chemical weathering the soil minerals are broken down to release plant nutrients—this is a continuing process in most soils.

In badly drained soils, which become waterlogged from time to time, various complex reactions (including a reduction process) occur and are referred to as *gleying*. This process, which is very important in the formation of some soils—especially upland soils in the U.K.—results in ferrous iron, manganese and some trace elements moving around more freely and producing colour changes in the soil. *Gley* soils are generally greyish in colour (may also be green or blue) in the waterlogged regions, but rusty-coloured deposits of ferric iron also occur in root and other channels, and along the boundaries between the waterlogged and aerated soil, so producing a mottled appearance (hence the Russian name "gley"). Glazing or coating of the soil structure units with fine clay is also associated with gleying.

OTHER FACTORS IN SOIL FORMATION

Climate

The rate of weathering partly depends on the climate. For example, the wide variations in temperature and the high rainfall of the tropics makes for much faster soil development than would be possible in the colder and drier climatic regions.

Topography

The depth of soil can be considerably affected by the slope of the ground. Weathered soil tends to erode from steep slopes and build up on the flatter land at the bottom. Level land is more likely to produce uniform weathering.

Biological activity

Plants, animals and micro-organisms, during their life-cycles, leave many organic substances in the soil. Some of the substances may dissolve some of the mineral material; dead material may partially decompose to give *humus.*

The roots of plants may open up cracks in the soil.

Vegetation such as mosses and lichens can attack and break down the surface of rocks.

Holes made in the soil by burrowing animals such as earthworms, moles, rabbits, etc., help to breakdown soft and partly weathered rocks.

Farming operations

Deep ploughing and cultivation, artificial drainage, liming, etc., can speed up the soil formation processes very considerably.

THE PHYSICAL MAKE-UP OF SOIL AND ITS EFFECT ON PLANT GROWTH

The farmer must consider the soil from the point of view of its ability to grow crops. To produce good crops the soil must provide suitable conditions in which plant roots can grow. It must also supply nutrients, water and air; and the temperature must be suitable for the growth of the crop.

The soil is composed of:

Solids	Mainly *mineral matter* (stones, sand, silt, clay, etc.) and *organic matter*—remains of plants and animals.
Liquids	Mainly *soil water* (a weak acid).
Gases	*Soil air* (competes with water to occupy the spaces between the particles).
Living organisms	Micro-organisms such as bacteria, fungi, small soil animals, earthworms, etc.

Mineral matter

This weathered material, and especially the clay part, is mainly responsible for making a soil difficult or easy to work. It may provide many plant nutrients—but *not* nitrogen. The farmer normally cannot alter the mineral matter in a soil (but see "*Claying*").

The amounts of clay, silt and sand which a soil contains can be measured by a *mechanical analysis* of a sample in the laboratory (see Table 4).

Organic matter (o.m.)

The organic matter content of a soil can vary from time to time. It usually increases with clay content, e.g. ordinary heavy soils have about 3–4% o.m. compared with 1–1.5% in very light soils: most fertile soils contain about 3–5% (on a dry weight basis). "Peaty" soils contain 20–35% o.m. and "peats" (including black fen) have over 35% (some are entirely o.m.). Organic matter may remain for a short time in the undecayed state and as such can help to "open-up" the soil—this could be harmful on sandy soils. However, the organic matter is soon attacked by all sorts of soil organisms—bacteria, fungi, earthworms, insects, etc. When they have finished eating and digesting it and each other a complex, dark-coloured, structureless material called *humus* remains: materials produced during the breakdown process are very beneficial in restoring and stabilizing a good soil structure.

The amount of humus formed is greatest from plants which have a lot of strengthening (lignified) tissue (e.g. straw). Humus is finally broken down by an oxidation process which is not fully understood.

The amount of humus which can remain in a soil is fairly constant for any particular type of soil. The addition of more organic matter often does not alter the humus content appreciably because the rate of breakdown increases. Organic matter is broken down most rapidly in warm, moist soils which are well limed and well aerated. Breakdown is slowest in waterlogged, acid conditions.

The chemical make-up of humus is not fully understood but its effects on the soil are well known.

Like clay, it is a *colloid* (i.e. it is a gluey substance which behaves like a sponge—it absorbs water and swells up when wetted and shrinks on drying). The humus colloids are not so gummy and plastic as the clay colloids but they can improve light (sandy) soils by binding groups of particles together. This reduces the size of the pores (spaces between the particles) and increases the water-holding capacity. Humus can also improve clay soils by making them less plastic and by assisting in the formation of a crumb structure—lime must also be present. Earthworms help in this soil improvement by digesting the clay and humus material with lime.

Plant nutrients—particularly nitrogen and phosphorus—are released for uptake by other plants when organic matter breaks down. The humus colloids can hold bases such as potassium and ammonia in an available form. In these ways it has a very beneficial effect in promoting steady crop growth.

Organic matter in the soil may be maintained or increased by growing *leys*, working-in *straw* and similar *crop residues, farmyard manure* (FYM), *composts*, etc. The roots and stubble are usually sufficient to maintain an adequate humus content in a soil growing cereals continuously.

In areas where erosion by wind and water is common, mineral soils are less likely to suffer damage if they are well supplied with humus.

Where it is possible to grow good leys and utilize them fully, this is one of the best ways of maintaining a high level of organic matter and a good soil structure.

Increasing the organic matter (humus) content of a soil is the best way of increasing its water-holding capacity: 50–60 tonnes/ha of well-rotted FYM may increase the amount of water which can be held by 25% or more.

WATER IN THE SOIL

Soils vary in their capacity to hold water, and to understand why, it is necessary to understand some of the differences between soils.

The soil is a mass of irregular-shaped particles forming a network of spaces or channels called the *pore space*, which may be filled with air or water or both. If the pore space is completely filled with water the soil is *waterlogged* and unsuitable for plant growth because the roots need oxygen for respiration. Ideally, there should be about equal volumes of air and water.

When the soil particles are small (e.g. clay) then the spaces between the particles are also small, and when they are large (e.g. sand) the spaces are large. However, although the spaces are small in a clay soil there are very many more spaces than in the same volume of a sandy soil. In a clay soil about half of the total volume is pore space whereas in a sandy soil only about one third is pore space. These volumes refer to dry soils. The pore space may be altered by a change in:

(a) grouping of the soil particles (i.e. structure),
(b) amount of organic matter (humus) present,
(c) compaction of the soil.

The fact that clay soils have a greater pore space than sandy soils partly explains why the clay soil can hold more water.

Another important factor is the *surface area* of the particles.

Water is held as a thin layer or film around the soil particles. The smaller the particles the stronger are the attractive forces holding the water. Also, the smaller the particles the greater is the surface area per unit volume. (Compare boxes filled with billiard balls, marbles and small ball-bearings.) A comparison for pure materials is set out in Table 3.

TABLE 3

Material	Particle size (mm)	Surface area
Coarse sand	2.0	×
Finest sand	0.02	100 ×
Finest silt	0.002	1000 ×
Finest clay		100,000 ×

The surface area of the particles in a cubic metre of clay may be over 1000 hectares.

The organic matter (humus) in the soil also holds water.

The water in the soil comes from rainfall, or, in dry areas, from irrigation.

When water falls on a dry soil it does *not* become evenly distributed through the soil. The topmost layer becomes saturated first and as more water is added the depth of the saturated layer increases. In this layer most of the pore space is filled with water. However, a well-drained soil cannot hold all of this water for very long. After a day or so some of the water will soak into the lower layers or run away in drains. The amount of water which is then retained by the soil is called the *moisture-holding capacity* or *field capacity*. The amount of water which can be held in this way varies according to the texture, and structure of the soil (see pages 24 and 37). The weight of water held by a clay soil may be equal to the weight of the soil particles, whereas a sandy soil may hold less than one-tenth of the weight of the particles. The water-holding capacity of a soil is usually expressed in mm, e.g. a clay soil may have a field capacity of 4 mm/cm in depth.

The ways in which water is retained in the soil can be summarized as follows:

(a) as a film around the soil particles,
(b) in the organic matter,

(c) filling some of the smaller spaces,

(d) chemically combined with the soil minerals.

Most of this water can be easily taken up by plant roots but as the soil dries out the remaining water is more firmly held and eventually a stage is reached when no more water can be extracted by the plant. This is called the *wilting point* because plants wilt permanently and soon die. This *permanent wilting* should not be confused with the *temporary wilting* which sometimes occurs on very hot days because the rate of transpiration is greater than the rate of water absorption through the roots; in these cases the plants recover during the night or earlier. The water which can be taken up by the plant roots is called the *available water*. It is the difference between the amounts at field capacity and wilting point. In clay soils only about 50–60% of "field capacity" water is available; in sandy soils up to 90% or more may be available. Although plants may not die until the wilting point is reached, they will suffer from shortage of water as it becomes more difficult to extract (see Fig. 26).

Water in the soil tends to hold the particles together and lumps of soil may stick together.

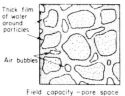

Field capacity – pore space filled with water and air which is ideal for plant growth

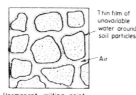

Permanent wilting point – plants wilt and will soon die due to lack of water

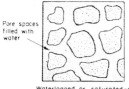

Waterlogged or saturated – no air present and so crop plants die or grow very slowly

Dry – no water present and so plants die – this is unlikely to happen in a field

FIG. 26. Highly-magnified particles and pores showing how water and air may be found in the soil.

When a loam or heavy soil is at or above half field capacity it is possible to form it into a ball which will not fall apart when handled and tossed about. At wilting point, the soil is crumply and will not hold together. So, if irrigation is economically possible, it should be used before the soil dries out to a state in which it will not hold together.

Some of the water in soils with very small pores and channels can move through the soil by *capillary forces*, i.e. surface tension between the water and the walls of the fine tubes or capillaries. This is a very slow movement and may not be fast enough to supply plant roots in a soil which is drying out. Heavy rolling of a soil may reduce the size of the pores and so set up some capillary action.

Water is lost from the soil by *evaporation* from the surface and by *transpiration* through plants. It moves very slowly from the body of the soil to the surface, so after the top 20 to 50 mm have dried out the loss of water by evaporation is very small. Cultivations increase evaporation losses. Most of the available water in a soil is taken up by plants—during the growing season—and air moves in to take its place. This movement of air is easy where the soil has large pore spaces but the movement into the very small pore channels in clay soils is slow until the soil shrinks and cracks—vertically and horizontally—as the water is removed by plants.

The water which enters the soil soon becomes a dilute solution of the soluble soil chemicals. It dissolves some of the carbon dioxide in the soil and so becomes a weak acid.

Soil aeration

Plant roots and many of the soil animals and micro-organisms require oxygen for respiration and give out carbon dioxide. The air found in the soil is really atmospheric air which has been changed by these activities (and also by various chemical reactions), and so contains less oxygen and much more carbon dioxide. After a time this reduction in oxygen and increase in carbon dioxide becomes harmful to the plant and other organisms.

Aeration is the replacement of this stagnant soil air with fresh air. The process is mainly brought about by the movement of water into and out of the soil, e.g. rain water soaks into the soil filling many of the pore spaces and driving out the air. Then, as the surplus water soaks down to the drains or is taken up by plants, fresh air is drawn into the soil to refill the pore spaces.

Also, oxygen moves into the soil and carbon dioxide moves out of the soil by a diffusion process similar to what happens through the stomata in plant leaves.

The aeration process is also assisted by:

(1) changes in temperature,
(2) changes in barometric pressure,
(3) good drainage,
(4) cultivations—especially on clay soils and where a soil cap has formed,
(5) open soil structure.

Sandy soils are usually well aerated because of their open structure. Clay soils are usually poorly aerated—especially when the very small pores in such soils become filled with water. Good aeration is especially important for germinating seeds and seedling plants.

Soil micro-organisms

There are thousands of millions of very small organisms in every gramme of fertile soil. Many different types are found but the main groups are:

(1) *Bacteria*—the most numerous group. Bacteria are the smallest type of single-celled organisms and can only be seen with a microscope. There are many kinds in the soil. Most of them feed on and break down organic matter. They obtain energy from the carbohydrates (e.g. sugar, starches, cellulose, etc.) and release carbon dioxide in the process. They also need nitrogen to build cell proteins. If they cannot get this protein from the organic matter they may use other sources such as the nitrogen applied as fertilizers. When this happens (e.g. where straw is ploughed in) the following crop may suffer from shortage of nitrogen unless extra fertilizer

is applied. Some types of bacteria can convert (fix) the nitrogen from the air into nitrogen compounds which can be used by plants (see Legumes and the Nitrogen Cycle, page 15). Soil bacteria are most active in warm, damp, well-aerated soils which are not acid.

(2) *Fungi*. Fungi are simple types of plants which feed on and break down organic matter. They are mainly responsible for breaking down lignified (woody) tissue. They have *no* chlorophyll or proper flowers. The fungi usually found in arable soils are very small, but larger types are found in other soils, e.g. peats. Fungi can live in acid conditions and in drier conditions than bacteria. (Mushrooms are fungi, and "fairy rings" are produced by fungi.) Sometimes disease-producing fungi develop in some fields, e.g. those causing "take-all" and "eyespot" in cereals.

(3) *Actinomycetes*. These are organisms which are intermediate between bacteria and fungi and have a similar effect on the soil. They need oxygen for growth and are more common in the drier, warmer soils. They are not so numerous as bacteria and fungi. Some types can cause plant disease, e.g. common scab in potatoes (worst in light, dry soils).

(4) *Algae*. Soil algae are very small simple organisms which contain chlorophyll and so can build up their bodies by using carbon dioxide from the air and nitrogen from the soil. Algae grow well in fertile damp soils exposed to the sun. Algae growing in swampy (waterlogged) soils can use dissolved carbon dioxide from the water and release oxygen. This process is an important source of oxygen for crop plants such as rice. Algae are important in colonizing bare soils in the early stages of weathering.

(5) *Protozoa*. These are very small, single-celled animals. Most of them feed on bacteria and similar small organisms. A few types contain chlorophyll and so can produce carbohydrates like plants.

The activities of the micro-organisms in the soil are rather complex and as yet not fully understood, but we do know that they improve the productivity of the soil. In general, the more fertile the soil the more organisms there are present.

Earthworms

It is generally believed that earthworms have a beneficial effect on the fertility of soils, particularly those under grass, but there is very little definite proof that they do any good on arable land. There are several different kinds found in our soils but most of their activities are very similar. They live in holes in the soil and feed on organic matter—either living plants or, more often, dead and decaying matter. They carry down into the soil fallen leaves and twigs, straw and similar materials. Earthworms do not thrive in acid soils because they want plenty of calcium (lime) to digest with the organic matter they eat. Their casts, which are usually left on the surface, consist of a useful mixture of organic matter, mineral matter and lime. This material may weigh 25 tonnes/ha. The greatest numbers are found in loam soils (under grass) where there is usually a good supply of air, moisture, organic matter and lime. Various methods have been tried to estimate the numbers present in a soil but with limited success.

The many holes they make allow water to enter and drain from the soil very easily and this in turn draws fresh air in as it soaks downwards. This may not always be a good thing, because the holes often have a smooth and, in places, impervious lining which may allow the water to go through to the drains too easily instead of soaking into the soil.

Earthworms are the main food of the mole which does so much damage by burrowing and throwing up heaps of soil.

Other soil animals

In addition to earthworms there are many species of small animals present in most soils. They feed on living and decaying plant material and micro-organisms. Some of the common ones are: slugs, snails, millipedes, centipedes, ants, spiders, eelworms, beetles, larvae of various insects such as cutworms, leather jackets and wireworms. The farmer is only directly concerned with those which damage his crops or livestock. The more troublesome crop pests are dealt with in Chapter 7.

SOIL TEXTURE AND STRUCTURE

Soil texture is that characteristic which is determined by the amounts of clay, silt, sand and organic matter which the soil contains. This property normally cannot be altered by the farmer (but see "Claying"). Soil texture can be measured by a *mechanical analysis* of a sample in the laboratory and classified accordingly (see Table 4).

Soil structure is the arrangement of the soil particles individually (e.g. grains of sand), in groups (e.g. crumbs or clods) or as a mixture of the two. It can be altered by: *weather conditions* (e.g. lumps changed to crumbs by frost action or alternate wetting and drying), *penetration of plant roots, cultivations*, etc. It is not possible to measure soil structure satisfactorily, but an experienced person can easily assess its quality at any one time by its appearance and "feel".

It is possible to classify the texture of a soil by rubbing a moist sample of it between the thumb and fingers:

Clay is sticky, will take a polish, and can be moulded.
Silt feels silky, smooth and slightly sticky.
Sand feels gritty.
Organic matter usually feels soft and slightly sticky.

The dominant "feel" indicates the texture group of the soil; if there is no dominant feeling then the soil is a loam. This method of field classification is set out more fully in Appendix 3 (page 218).

Soils are often classified by farmers and others as heavy, medium and light (not a weight measure). The terms "heavy" and "light" refer to the amount of power required to draw a plough or cultivator through the soil. A heavy (clayey) soil consists mainly of very small particles which pack tightly together whereas a light (sandy) soil consists mainly of large particles which are loosely held together because of the relatively large pore spaces.

Crumb structure is formed by the grouping together (aggregation) of the particles of clay, sand and silt. This aggregation is possible because there are positive and negative electric charges (forces) acting through the surface of the particles. These forces are strongest in clay and very weak in sand. This strong adhesive property of clay particles makes clay soils more difficult to work than sandy soils but it also enables them to form crumbs easily.

Water has special electric properties and its presence is necessary for the grouping (crumbing) of soil particles. The electric forces in the water and in the soil particles make the water stick as a thin film around the particles of soil. As this film becomes thinner (e.g. when soil is drying out) the particles are drawn closer together to form groups (crumbs). The particles in the crumbs may come apart again if the soil becomes very wet.

There must be lime present in the water if clay particles are to stick together to form porous crumbs. This partly explains why liming benefits clay soils.

If organic matter or an iron compound (ferric hydroxide) is present then the particles in the crumbs may remain cemented together and have a more lasting effect on soil structure. Too much ferric hydroxide can have a harmful effect because tightly cemented crumbs are very difficult to wet again after they have dried out. This condition occurs in the so-called "drummy" soils found in the fen district.

Where there is very little organic matter or ferric hydroxide the stability of the crumbs depends mainly on the amount of clay present. The more clay there is, the stronger will be the forces holding the particles together.

Some soil structures are more stable than others, e.g. clays usually have a more stable structure than silts. Soils containing fine sand and silt easily lose their structure and are difficult to work if they are low in organic matter. This is because under wet conditions the sand and silty materials flow very easily and block the aeration and drainage channels in the soil.

Tilth is a term used to describe the condition of the soil in a seed-bed. For example, the soil may be in a finely-divided state or it may be rough and lumpy; also, the soil may be damp or it may be very dry. Whether a tilth is suitable or not partly depends on the crop to be grown. In general, small seeds require a finer tilth than large seeds.

SOIL FERTILITY AND PRODUCTIVITY

Soil fertility is a rather loose term used to indicate the potential capacity of a soil to grow a crop (or sequence of crops). The productivity of a soil is the combined result of fertility and management.

The fertility of a soil at any one time is partly due to its natural make-up (inherent or *natural fertility*) and partly due to its *condition* (variable fertility) at that time.

Natural fertility has an important influence on the *rental* and *sale value* of land. It is the result of factors which are normally beyond the control of the farmer, such as:

(1) the texture and chemical composition of the mineral matter,
(2) the topography (natural slope of the land)—this can affect drainage, temperature and workability of the soil,
(3) climate and local weather—particularly the effects on temperature, and rainfall (quantity and distribution).

Soil condition is largely dependent on the management of the soil in recent times. It can be built up by good husbandry but if this high standard is not maintained the soil will soon return to its natural fertility level. The application of fertilizers can raise soil fertility by increasing the quantities of plant food in the growth and decay cycle.

Management can control the following production factors:

(1) the amount of organic matter in the soil (see "Soils"),
(2) artificial drainage and irrigation (see "Soil Improvement"),
(3) erosion (removal of soil by wind and water) (see "Claying"),

(4) pH of the soil (see "Liming"), and the plant nutrients applied (see "Fertilizers" and "Crops"),
(5) cultivations and time of planting (see "Cultivations"),
(6) variety and plant spacing (see individual crops),
(7) sequence of cropping (see "Rotations"),
(8) weeds, pests and diseases (see separate chapters).

Good management of the above factors should maintain or increase soil fertility and at the same time be commercially profitable. These subjects are dealt with in more detail in other chapters.

TYPES OF SOIL

There are wide variations in the types of soil found on farms. They may be classified in various ways but here they are grouped according to texture. The amount of clay, silt and sand which they contain can be found by a mechanical analysis. This is an elaborate separation of the particles by settling from a water suspension and sieving in the laboratory which can give accurate measurements of the amount of sand, silt and clay particles present. Gravel and stones are not included in a sample for mechanical analysis. The generally accepted size of particles for each material is given in Table 4.

TABLE 4

Material	Diameter of particles
Clay	less than 0.002 mm
Silt	0.02–0.002 mm
Fine sand	0.2–0.02 mm
Coarse sand	2.0–0.2 mm
Gravel	more than 2.0 mm

The "farm-soil" groups to be considered in more detail are: *Clay, sand, loam, silt, calcareous, peat* and *black fen*.

The approximate mechanical analyses of some soil types are shown in Table 5.

TABLE 5 TEXTURAL GROUPING OF SOILS (ON DRY WEIGHT)

Soil type	Texture	Clay (%)	Silt (%)	Sand (%)
Clay	Fine (heavy)	over 50	15–25	up to 35
Clay loam	Fine (heavy)	30–50	15–25	35–45
Silt loam	Medium	20–30	30–50	30–35
Loam	Medium	20–30	20–30	about 50
Sandy loam	Coarse (light)	10–20	15–25	55–75
Sand	Coarse (light)	0–10	0–10	80–100

A given amount of clay has a very much greater effect on the characteristics of a soil than the same amount of sand or silt.

Clay soils

These soils have a high proportion of clay and silty material—usually over 60%; of this, at least half is *pure clay*, which is mainly responsible for their characteristic qualities. The particles of pure clay are so small that they cannot be seen under an ordinary microscope but they have several very important *colloidal* and base-exchange properties, e.g.

They are gluey and plastic (can be moulded).
They will *swell* when wetted and *shrink* when dried.
They can group together into small clusters (flocculate) or become scattered (deflocculated).
They can combine with various chemical substances (base-exchange) such as calcium, sodium, potassium and ammonia and in this way may hold plant nutrients in the soil.

Grouping or flocculation of the particles is very important in making clay soils easy to work. Clay particles combined with calcium (lime) will flocculate easily whereas those combined with sodium will not and so salt (sodium chloride) must be used very carefully on clay soils. Deflocculation can occur when clays are worked in a wet condition. The adhesive properties of clay are very beneficial to the soil structure when the groups of particles are small (like crumbs) but can be very harmful when large lumps (clods) are formed. Frost action, and alternating periods of wetting and drying will help to restore them to the flocculated crumb condition.

Characteristics.

(1) Clay soils feel very *sticky* when wet and can be moulded into various shapes.
(2) They can hold more total water than most other soil types and although only about half of this is available to plants, crops seldom suffer from drought.
(3) They *lie wet in winter* so stock should be taken off the land to avoid *poaching*.
(4) They are very *late* in warming up in springtime because water heats up much slower than mineral matter.
(5) They are normally fairly *rich in potash*, but are deficient in phosphates.
(6) Lime requirements are very variable—a clay soil which is well limed usually has a better structure and so is easier to work; over-liming will not cause any troubles such as trace-element deficiency.

Management. They should not be worked in spring when wet because they become puddled and later dry into hard lumps, which can only be broken down by well timed cultivations following repeated wetting (swelling) and drying (shrinking). Some air is drawn into cracks caused by shrinkage, and remains when the clod is wetted again and so lines of weakness are formed which eventually allow the clod to be broken. In dry weather irrigation may be used to wet the clods.

In prolonged dry weather, wide and deep cracks are formed which may break animals' legs but which are very beneficial for drainage later.

Clays are often called *heavy* soils because, compared with light (sandy) soils, for ploughing and cultivating two to four times the amount of tractor power is required. All cultivations must be very carefully timed (often restricted to a short period) so that the soil structure is not damaged. This means that more tractors and implements are required than on similar sized loam or sandy soil farms. Autumn ploughing, to allow for a frost tilth, is essential if good seed-beds are to be produced in the spring.

Good drainage is essential. Many clay fields are still in "ridge-and-furrow". This was set up by ploughing—making the "openings" and "finishes" in the same respective places until a distinct ridge and furrow pattern was formed. The direction of the furrows is the same as the fall on the field so that water can easily run off into ditches. This practice also increases the grazing area of a field and for this reason is sometimes found on other types of soil! If these ridges and furrows are levelled out then a mole-drainage system using tiled main drains should be substituted (see "Drainage"). This change is well worthwhile where arable crops are grown.

In many clay-land areas—especially where rainfall is high—the fields are often small and irregular in shape because the boundaries were originally ditches which followed the fall of the land. The hedges and deciduous trees, which were planted later, grow very well on these fertile, wet soils.

The close texture and an adequate water supply often restrict root development on clay soils.

Organic matter, such as strawy farmyard manure, ploughed-in straw or grassland residues make these soils easier to work.

Cropping. Because of the many difficulties to be overcome in growing arable crops on these soils they are often left in *permanent grass* and only grazed during the growing season. Where arable crops are grown, a 3- or 4-year ley is often included in the rotation. *Winter wheat* is the most popular arable crop; *winter beans* are also grown in some areas. Both these crops are planted in the autumn (preferably October) when more liberties can be taken with seed-bed preparation than would be permissible in the springtime. *Mangolds* and *cabbage* grow well on clay soils but are declining in popularity in many areas. *Sugar-beet* and *potatoes* are grown in some districts but are very troublesome because of the difficulties in seed-bed preparation, weed control and harvesting—especially in a wet autumn. The best place to take either of these crops is after a period under grass when the soil structure is more stable and the soil easier to work.

Sandy soils

Characteristics.

(1) In many ways these are the opposite of clays and are often called *light* soils because comparatively little power is required to draw cultivation implements.
(2) They can be worked at any time—even in wet weather—without harmful effects.
(3) They are normally free-draining but a few drains may be required where there is clay or other impervious layer underneath.
(4) They have a high proportion of sand and other coarse particles but very little clay—usually less than 5%—(they feel gritty).
(5) They warm up early in spring but crops are very liable to "burn-up" in a dry period because the water-holding capacity is low.

Management. Sandy soils are naturally very low in plant nutrients and fertilizers are easily washed out, so adequate amounts of fertilizer must be applied to every crop. *Liming* is necessary but must be used carefully—a little and often is the rule here.

Organic matter—especially as *humus*—is very beneficial because it helps to hold water and plant nutrients in the soil. On properly limed fields it breaks down very rapidly because the soil micro-organisms are very active in these open-textured soils which have a good air supply.

Irrigation can be very important if the rainfall is low or not well distributed over the growing season.

In some sandy areas the surface soil is liable to "blow" in dry, windy weather and so could destroy a young crop. Where possible, the remedy is to add about 400 tonnes/ha of clay (see "Claying"). Shelter belts are helpful where clay is not readily available.

Cropping. A wide range of crops can be grown but yields are very dependent on a good supply of water and adequate fertilizers. *Market gardening* is often carried on where a good sandy area is situated near a large population; e.g. Sandy, Potton, Biggleswade area (Bedfordshire). Here growers are prepared to use irrigation and apply plenty of manures and fertilizers on these *very early, easily worked soils.*

On the lighter sands in low rainfall areas and where irrigation is not possible, the main crops grown are *rye, carrots, sugar-beet,* and *lucerne; lupins* are grown in a few areas where the soil is very poor and acid.

On the better sandy soils, and particularly where the water supply (from rain or irrigation) is reasonably good, the main arable farm-crops grown are *barley, peas, rye, sugar-beet, potatoes* and *carrots.*

Because of the poor quality of this type of land, the farms and fields are usually larger than on better-land farms. Hedges are not very common because there is not enough water for good growth. The trees are usually drought-resistant coniferous types.

Stock can be out-wintered on sandy soils with very little risk of damage by poaching even in wet weather.

Loams

Characteristics.

(1) These are intermediate in texture between the clays and sandy soils and, in general, have most of the advantages and few of the disadvantages of these two extreme types. They may feel gritty but also somewhat sticky.
(2) The amount of clay present varies considerably and so this group is sometimes divided into *heavy* or *clay loams* (resembling clays in many respects), *medium loams,* and *sandy* or *light loams* (resembling the better sandy soils).
(3) These soils warm up reasonably early in spring and are fairly resistant to drought.

Management. Loams are easily worked but should not be worked when wet—especially clay loams. They usually require to be drained but this is not difficult using tile or plastic drains.

Cropping. They are regarded by most farmers as the *best all-round soils* because they are naturally fertile and can be used for growing any crop provided the depth of soil is sufficient. Crop yields do not vary much from year to year.

Farms with loamy soils can be used for most types of arable or grassland farming but in general, mixed farming is carried on. Cereals, potatoes and sugar-beet are the main cash crops and leys provide grazing and winter bulk foods for dairy cows, beef cattle or sheep.

Calcareous soils

Characteristics. These are soils derived from *chalk* and *limestone* rocks and contain various amounts of calcium carbonate—usually 5–50%. The depth of soil and subsoil may vary from 8 cm to over a metre; in general, the deep soils are more fertile than the shallow ones. The ease of working and stickiness of these soils depends on the amount of clay and chalk or limestone present; they usually have a loamy texture. Sharp-edged *flints* of various sizes, found in soils over-lying some of the chalk formations, are very wearing on cultivation implements and rubber tyres, and are rather destructive when picked up by harvesting machinery. In some places the flints are found mixed with clay, e.g. *clay-with-flints* soils.

The soils are free-draining except in a few small areas where there is a deep clayish subsoil. Dry valleys are characteristic of these downlands and wolds. The few rivers rise from underground streams.

There are very few hedges and most of the trees have been planted for various reasons—they are mainly beech and conifers.

Walls of local stone form the field boundaries in some limestone areas, e.g. the Cotswolds.

Management. The soils are usually deficient in phosphates and potash but only the deeper ones are likely to need liming (see "Liming"). Organic matter can be beneficial but it breaks down fairly rapidly and may be expensive to replace.

The farms and fields on this type of land are usually large—especially on the thinner soils.

Some areas are still unfenced and have no water laid on for stock but this state of affairs is changing as mixed farming systems with grazing animals replace the folded-sheep flocks.

The flooding of water-meadows used to be a common practice but is not done now because labour costs are too high.

Cropping

Barley and wheat (on the deeper soils) are the best crops for these soils. The combine-drill for sowing cereals has been very useful in producing good crops—particularly on the poorer, thinner soils. Continuous barley growing is now common on many farms and is likely to become a widely accepted practice. Roots, such as *sugar-beet* and *mangolds*, and *potatoes*, are grown on some of the deeper soils. *Leys* for grazing and seed production provide a rotational break with cereal growing. *Kale* is grown on some farms for stock, and pheasant cover! Apart from some parkland, only the poorest, thinnest soils remain as permanent grassland.

Silts

Characteristics. These are soils which contain a high proportion of silt (up to 80% or more). The particles (between clay and sand in size) pack together very closely and retard the movement of water.

Bad drainage is one of the main problems with these soils. They do not have a stable subsoil structure such as is found in clay soils. The particles do not group together readily and firmly and so quickly block up drainage cracks and tile drains. Unlike clay, the silt particles cannot take part in chemical reactions, so adding lime is not helpful; it is very difficult to create an easy-working soil structure. Frost has very little useful effect.

Management and cropping. Arable cropping is very difficult and these areas are best left down to permanent grass. Deep-rooted plants, such as *lucerne*, left growing for several years, are likely to be helpful in opening up the subsoil with their roots and so facilitating drainage.

There is not much land of this type in the British Isles—the best example is part of the Lower Weald in Sussex.

Note. The Fenland "silts" are alluvial material consisting mainly of clay, silt and fine sand. They vary in texture from sandy to medium loam. These soils are very fertile and fairly easy to work. They are cropped intensively with all kinds of arable crops—the main ones are *wheat, potatoes, sugar-beet, peas, seed production from root crops and grasses, bulbs* and *market gardening crops.*

Peats and peaty soils

Characteristics. Peaty soils contain about 20–25% of organic matter whereas there is about 50–90% in true peats.

The *acid* or *peat-bog* peats and peaty soils have been formed in waterlogged areas where plants such as mosses, cotton grass, heather, molinia and rushes grew. The dead material from these plants was only partly broken down by the types of bacteria which can survive under these acidic waterlogged conditions. This "humus" material built up slowly—probably 30 cm every century.

When reclaimed, these soils break down easily to release nutrients, particularly nitrogen, but they are very low in phosphates and potash. Old tree trunks have to be dug out from time to time as the level of the soil falls due to organic matter breakdown.

Management. Before reclaiming this land for cropping, much of the peat is often cut away for fuel or sold as peat moss for horticultural purposes or bedding. Good drainage must then be carried out by cutting deep ditches through the area. Deep ploughing also helps to drain the soil. Heavy applications—up to 25 tonnes/ha—of ground limestone may be required to neutralize the acidity.

In the first year, about 15 tonnes/ha of farmyard manure improves the yields of pioneer crops (usually potatoes, sometimes oats or rye); the reason for this may be that the FYM introduces beneficial types of bacteria.

Cropping. In exposed areas they are often sown down to good grasses and clovers. Good swards can be established but these must not be over grazed or "poached" in wet weather otherwise the field will quickly go back to rushes and weed grasses. Under cultivation, most arable crops can be grown but potatoes and oats are the most suitable.

Black Fen soils

Characteristics. Black Fen soils are found in part of the Fen district of East Anglia and are amongst the most fertile soils in the British Isles (The "muck soils" of North America are somewhat similar.)

These soils were formed in marshy river estuary conditions where the water came from limestone and chalk areas and so carried calcium carbonate and, in flood times, considerable amounts of silty material. The remains of the vegetation (mainly reeds, sedges and other estuary plants) did not break down completely because of the waterlogged conditions and so built up as humus. The soils vary a lot from district to district but most of them consist almost entirely of organic matter.

Management. After building strong sea walls, the area has been reclaimed by draining with deep ditches and underground drains. Most of the land is below sea level and so the water in the ditches has to be pumped over the sea walls or into the main drainage channels.

The soil breaks down readily and the level is falling about 2 cm per year and eventually will reach the clay or gravel subsoil. Tree trunks have to be dug out occasionally.

"Blowing" in spring is a serious problem on these dry sooty-black, friable soils. Several plantings of crop seedlings together with the top 5–8 cm of soil and fertilizers may be blown into the ditches. This can be prevented by applying 400–750 tonnes of clay per hectare (see "Claying"), or by deep ploughing or cultivation 1–1.5 metres to mix the underlying clay and organic topsoil.

These soils are rich in nitrogen, released by the breakdown of the organic matter, but are very poor in phosphates and potash and also trace elements such as manganese and copper.

Cropping. This is an intensive arable area where the main crops are *wheat, potatoes* and *sugar-beet*; also smaller areas of *celery, peas, carrots* and market garden crops.

In some parts leys have been introduced—with limited success—in an attempt to check the rapid rate of breakdown of the soils.

SOIL IMPROVEMENT

Liming

Most farm crops will not grow satisfactorily if the soil is very acid (sour). This can be remedied by applying one of the commonly used liming materials.

Soil reaction. All substances in the presence of water are either acid, alkaline or neutral. The term *reaction* describes the degree or condition of acidity, alkalinity or neutrality. Acidity and alkalinity are expressed by a pH scale on which pH 7 is neutral, numbers below 7 indicate acidity and those above 7 alkalinity. Most cultivated soils have a pH range between 4.5 and 8.0 and may be grouped as in Table 6.

TABLE 6

pH	Reaction
Over 7	Alkaline
7	Neutral
6.0–6.9	Slightly acid
5.2–5.9	Moderately acid
Below 5.2	Very acid

Lime requirement. This is the amount of lime required to raise the pH to approximately 6.5 in the top 15 cm layer of soil. This amount varies considerably with the degree of acidity or "sourness", and the type of soil. Heavy (clay) soils and soils rich in organic matter require more lime to raise the pH than other types of soil. For example, to raise the pH from 5.5 to 6.5 on a sandy loam may require about 5 tonnes/ha of ground lime-

stone, but on a clay soil 10–12.5 tonnes/ha of ground limestone may be required. The actual lime requirement can be calculated from chemical tests in the laboratory. It is unnecessary to lime soils which have a pH of more than 6.5.

Indications of soil acidity (i.e. a need for liming)

(a) Crops failing in patches—particularly the acid-sensitive ones such as *barley* and *sugar-beet*. The plants usually die off or are very unthrifty in the seedling stage.
(b) On grassland, there are poor types of grasses present such as *bents*. Often a *mat* of undecayed vegetation builds up because the acidity reduces the activities of earthworms and bacteria which break down such material.
(c) On arable land, weeds such as *sheep's sorrel, corn marigold* and *spurrey* are common.
(d) *Soil analysis.* Chemical and electrical methods may be used to determine the pH and lime requirements of a soil. Portable testing equipment, using colour charts, are sometimes used to test for pH.

The main benefits of applying lime are:

(1) It neutralizes the acidity or sourness.
(2) It supplies calcium (and sometimes magnesium) for plant nutrition.
(3) It improves soil structure. In well limed soils, plants usually produce more roots and grow better; bacteria are more active in breaking down organic matter. This usually results in a better soil structure and the soil can be cultivated more easily (see also "Soil Structure").
(4) It affects the availability of plant nutrients. The main plant nutrients such as nitrogen, phosphates and potash are freely available on properly limed soils. Too much lime in the soil is likely to make some minor nutrients unavailable to plants, e.g. *manganese, boron, copper* and *zinc*—this is least likely to happen in clay soils.

pH and crop growth. To give crops the best opportunity to grow well the soil pH should be near or above the following.

	pH
Barley, sugar-beet and lucerne	6.5
Red clover, maize, oil-seed rape	6.0
Wheat, beans, peas, turnips and swedes	5.5
Oats, potatoes	5.0
Rye and lupins	4.5

Lime is removed from the soil by:

(1) *Drainage*. Lime is fairly easily removed in drainage water. 125–2000 kg/ha of calcium carbonate may be lost annually. The rate of loss is greatest in industrial, smoke-polluted areas, areas of high rainfall, well-drained soils and soils rich in lime.

(2) *Fertilizers and manures*. Every 1 kg of sulphate of ammonia removes about 1 kg of calcium carbonate from the soil. Poultry manure may also remove some lime.

(3) *Crops*. The approximate amounts of calcium carbonate removed by crops are:

Cereals	1–3 kg per tonne of grain.
	5–7 kg/tonne of straw.
Potatoes	28 kg/40 tonnes of tubers.
Sugar-beet	58 kg/40 tonnes of roots.
	230 kg/35 tonnes of tops.
Swedes	100 kg/60 tonnes of crop.
Kale (carted off)	450 kg/50 tonnes of crop.
Lucerne hay	580 kg/tonne of hay.

(4) Stock also remove lime, for example, a 500-kg bullock sold off the farm removes about 16 kg of calcium carbonate in its bones. A 40-kg lamb about 1.3 kg of calcium carbonate and 5000 litres of milk about 18 kg of calcium carbonate.

Materials commonly used for liming soils

Ground limestone or chalk (also called carbonate of lime and calcium carbonate, $CaCO_3$).

This is obtained by quarrying the limestone or chalk rock and grinding it to a fine powder. It is the commonest liming material used at present.

Burnt lime (also called quicklime, lump lime, shell lime and calcium oxide, CaO). This is produced by burning lumps of limestone or chalk rock with coke or other fuel in a kiln.

Carbon dioxide is given off and the lumps of burnt lime which are left are sold as lumps, or are ground up ready for mechanical spreading. This "concentrated" form of lime is especially useful for application to remote areas where transport costs are high. Burnt lime may scorch growing crops because it readily takes water from the leaves. When lumps of burnt lime are wetted they break down to a fine powder called *hydrated* or *slaked lime* [$Ca(OH)_2$].

Hydrated lime is a good liming material but is usually too expensive for liming the soil.

Waste limes. These are liming materials which can sometimes be obtained from industrial processes where lime is used as a purifying material. These limes are cheap but usually contain a lot of water. Some of the sources are: sugar-beet factories, waste from manufacture of sulphate of ammonia, soap works, bleaching, tanneries, etc. Care is needed when using these materials because some may contain harmful substances. Sugar-beet waste lime is also a valuable source of plant nutrients.

A comparison of the various liming materials is as follows:

1 tonne of burnt lime (CaO) is equivalent to

1.37 tonne hydrated lime $Ca(OH)_2$
1.83 tonne ground limestone $CaCO_3$

or at least 2.5 tonnes waste lime (usually $CaCO_3$)

The supplier of lime must give a statement of the *neutralizing* value (N.V.) of the liming material—this is really the same as the calcium oxide equivalent.

Magnesian or dolomite limestone. This limestone consists of magnesium carbonate ($MgCO_3$) and $CaCO_3$ and is commonly used as a liming material in areas where it is found. Magnesium carbonate has a better neutralizing value (about one-fifth better) than calcium carbonate. In addition, the magnesium may prevent magnesium deficiency diseases in crops (e.g. interveinal yellowing of leaves in potatoes, sugar-beet, oats and stock (e.g. "grass-staggers" in grazing animals).

Cost. The cost of liming is largely dependent on the transport costs from the lime works to the farm. By dividing the cost per ton of the liming material by the figure for the neutralizing value, the *unit cost* is obtained. In this way it is possible to compare the costs of various liming materials.

Most farmers now use ground limestone or chalk and arrange for it to be spread mechanically by the suppliers. Where large amounts are required (over 7 tonnes/ha) it is sometimes best to apply it in two dressings, e.g. half before ploughing and half after ploughing.

Rates of application. 2.5–25 tonnes/ha of calcium carbonate ($CaCO_3$) or its equivalent may be needed to satisfy the lime requirements of a soil. Afterwards, about 2.5–4 tonnes/ha $CaCO_3$ every 4 years should be enough to replace average losses (see MAFF Booklet 2191—*Lime and Fertilizer recommendations* and MAFF Ref. book 35—*Lime and Liming*).

2. Drainage (see also "Water in the soil")

Normally, the soil can only hold some of the rainwater which falls on it. The remainder either runs off or is evaporated from the surface or soaks through the soil to the subsoil. If surplus water is prevented from moving through the soil and subsoil it soon fills up all the pore-spaces and this will kill or stunt the crops growing there.

The *water-table* is the level in the soil or subsoil below which all the pore space is filled with water. This is not easy to see or measure in clay soils but can be seen in open textured soils (see Fig. 27). The water-table level fluctuates through the year and in the British Isles is usually highest in February and lowest in September; the greater amount of evaporation, much more transpiration and lower rainfall (usually) in summer allow the level to fall in late summer and early autumn. This is shown by the water level falling in ponds and shallow wells, some springs drying up and wet parts of fields drying out.

In chalk and limestone areas, and in most sandy and gravelly soils, water can drain away easily into the porous subsoil. These are *free-draining* soils. Steep slopes also drain freely.

On most other types of farmland some sort of artificial *field drainage* is necessary to carry away the surplus water and so keep the water-table at a reasonable level. For most arable farm crops the water-table should be about 0.6-metre or more below the surface; for grass land 0.3–0.45 m is sufficient.

Some of the signs of bad drainage are:

(a) Machinery is easily "bogged down" in wet weather.
(b) Stock grazing pastures in wet weather easily damage and trample holes in the sward (poaching).
(c) Water lies about in pools on the surface for many days after heavy rain.
(d) Weeds such as rushes, sedges, horsetail, tussock grass and meadowsweet are common in grassland. Peat forms in places which have been very wet for a long time.
(e) Young plants are pale green or yellow in colour and unthrifty, when compared with the greener and more vigorous plants on drier land nearby.
(f) Subsoil is often various shades of blue, or grey compared with shades of reddish brown, yellow and orange in well-drained soil.

Some of the practical advantages of good drainage are:

(a) Well-drained land is better aerated and the crops grow better and are less likely to be damaged by root-decaying fungi.
(b) The soil dries out better in spring and so warms up quicker and can be worked early.

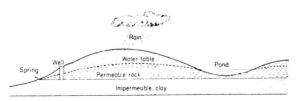

Fig. 27. Diagram showing position of a water-table and its effect on the water levels in the well and pond.

(c) Plants are encouraged to form a deeper and more extensive root system. In this way they can often obtain more plant food.

(d) Grassland is firmer—especially after wet periods. Good drainage is essential for high density stocking and where cattle are out-wintered if serious poaching of the pastures is to be avoided. Rushes and other moisture-loving weeds usually disappear after draining wet land.

(e) Disease risk from parasites is reduced. A good example is the liver fluke—this must pass part of its life cycle in a water snail found on badly drained land.

(f) Inter-row cultivations and harvesting of root crops and potatoes can be carried out more efficiently.

(g) Fertilizers and manures will give better results because the crop can grow more vigorously.

(h) Crops grown under contract, e.g. vining peas, can usually be planted and harvested at the proper time on well-drained land.

The main methods used to remove surplus water and control the water-table are:

(a) Open channels or ditches.
(b) Underground pipe drains—tiles and plastic.
(c) Mole drains.

Ditches and open drains. Ditches may be adequate to drain an area by themselves but they usually serve as outlets for underground drains. They are capable of dealing with large volumes of water in very wet periods. The size of a ditch varies according to the area it serves (see Fig. 28).

Ditches should be kept cleaned out to their original depth as often as necessary—usually annually or biennially. The spoil removed should be spread well clear of the edge of the ditch. Many different types of machines are now available for making new ditches and cleaning neglected ones.

Small open channels 10–60 metres apart are used for draining hill grazing areas. These are either dug by hand using a special spade or made with a special type of plough, drawn by a crawler tractor. Similar open channels are used on low lying meadow land where underground drainage is not possible.

Underground drains—general. The distance between drains which is necessary for good drainage depends on the soil texture. In clay soils the small pore spaces restrict the movement of water and so the drains must be spaced much closer together than on the lighter types of soil where water can flow freely through the large pore spaces (see Figs. 29 and 30).

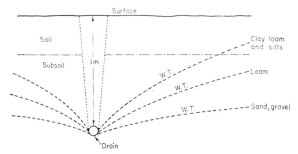

FIG. 29. Diagram to show how the steepness of the water-table (W.T.) varies with different types of soil.

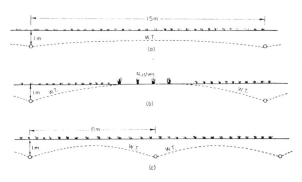

FIG. 30. Diagram showing the effect of spacing of drains on the water table (W.T.) (a) sandy loam, (b) and (c) silt or clay loam.

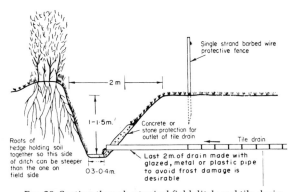

FIG. 28. Section though a typical field ditch and tile drain.

TABLE 7

Soil type	Depth of drains metres	Distance between pipe drains metres	chains
Sand, gravel	0.90–1.20	40–100	2–5
Fine sand	0.75–0.90	20–50	1–2.5
Loams	0.75–0.90	10–35	0.5–1.7
Peats	1.00–1.50	7–50	0.7–2.5
Silts and clays	0.60–0.75	7–20	0.7–1.0
Clays (with mole drains)	0.60–0.75	50–150	2.5–7.5

Underground drains—pipe drains (tiles or plastic). The correct spacing of underground drains depends on the permeability of the soil and subsoil. The small pore spaces in fine-textured soils restrict the movement of water and so the drains must be placed closer together than in coarse-textured and creviced soils where water can flow more freely. The approximate spacings in various soils is set out in Table 7.

Tile drainage is the commonest type of underground drainage. It can be used on all types of soil but on clay soils it is usually restricted to main drains only because of the high cost. Tiles are made of burnt clay, but sometimes pipes made of concrete are used. They are usually 300 mm long and of various diameters. 75 mm diameter tiles are used for the ordinary side or lateral drains: 100, 150 and 220 mm diameter tiles are used for the main drains. The size of tile required will depend on the rainfall, area to be drained, fall, and soil structure.

When tile drainage is used on clays and heavy loam soils, a porous material such as gravel or clinker should be used as backfill to allow the water to move down to the tiles easily. The cost of this material is reduced when narrow trenches are dug for the tiles.

There are many types of trenching and pipe laying machines available. Most pipe drainage work is done by specialist contractors.

Various types of *plastic* and *polythene* pipes are now being used instead of tiles for underground drains. They are supplied in 10-chain lengths and are laid in the soil by special machines—some of which are modified mole-ploughs; 50-mm diameter pipes are used for the side drains and 100-mm diameter pipes are used for the main drains. The drainage water enters through holes in the pipes which may have smooth or corrugated walls.

Some of the machines which lay the plastic pipes can also lay porous filling above the pipes. Most plastic pipe drains are laid by contractors. Although plastic pipes cost more than tiles the overall cost of pipes plus laying is less for the plastic system.

After laying a tile or plastic pipe system it may be necessary to use a subsoiler at 1-m intervals to crack pans and allow water free movement to the drains.

Underground drains—mole drainage. This is a cheap drainage method which can be used in some fields. Although the method is sometimes used on peat soils it is normally used in fields which have:

(a) *clay subsoil* (no stones, sand or gravel patches),
(b) *suitable fall* 5–10 cm/chain,
(c) *reasonably smooth surface.*

A mole plough, which has a torpedo or bullet-shaped "mole" attached to a steel coulter or blade, forms a cylindrical channel in the subsoil. The three main types are:

(1) mounted on three-point linkage,
(2) on wheeled carrying frame and adjustable for depth by winch,
(3) simple skid type (see Fig. 31).

The best conditions for mole draining are when the subsoil is damp enough to be plastic and forms a good surface on the mole channel, and also sufficiently dry to form cracks as the mole plough passes (see Fig. 32). If the surface is dry the tractor hauling the plough can get a better grip. The plough should be drawn slowly

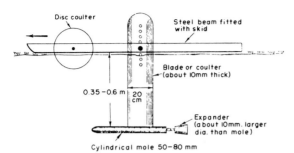

FIG. 31. Diagram of a simple (skid) type of mole plough.

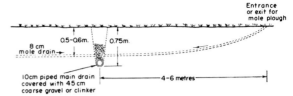

FIG. 34. Section through tile drain shown in Fig. 33 to show how water from the mole drain can enter the pipe drain through the porous backfilling.

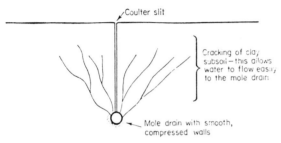

FIG. 32. Section through mole drain and surrounding soil.

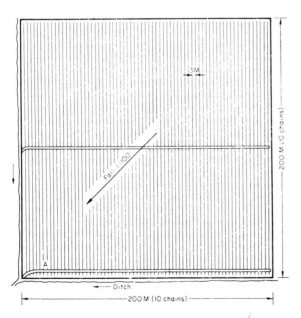

FIG. 33. Layout of mole drainage (with tiled main) in a 4-hectare area. Section through drains A shown in Fig. 34.

about 3 km.p.h. otherwise the vacuum created is likely to spoil the mole. Reasonably dry weather after moling will allow the surface of the mole to harden and so it should last longer.

Mole drains are drawn 3–4 metres apart except on "ridge and furrow" land where one or more drains are drawn along the furrows. For best results the moles should be drawn through the porous backfilling of a tiled or plastic main drain (see Figs. 33 and 34). 4–5 chains of tiled main are usually required per hectare. The mains are semi-permanent and a new set of mole drains can be drawn every 3–10 years as required. A large-wheeled tractor can pull a 5-cm mole 34–45 cm deep. For larger and deeper drains a crawler tractor is required.

In fields where the clay subsoil is not very stable, filling the mole channels with gravel (10–20 mm size) allows them to work longer and better.

(See also MAFF *Field Drainage* leaflets Nos. 1–29.)

3. Irrigation (see also "Water in the soil")

Irrigation is the term used when water is applied to crops which are suffering from drought. If water is not applied at such times the crops will be checked and may die and this means that yields will be low and quality may be poor.

It is usual to measure irrigation water, like rainfall, in millimetres (mm):

1 mm on 1 hectare = 10 m^3 = 10 tonnes;

25 mm on 1 ha = 250 m³ (a common application at one time);

100 mm on 1 ha = 1000 m³ = 1 km³ (about a season requirement);

100 mm on 50 ha = 50 km³ (a guide to amount of water required).

Green plants take up water from the soil and transpire it through their leaves (a small proportion is retained to build up the plant structure). Most green crops which are covering the ground have the same water requirements and this varies from day to day and place to place according to climate and weather conditions (it is called potential transpiration).

The average green crop requirements for water in the southern half of England are set out in Table 8.

TABLE 8

Period	per day		per month	
	mm	m³/ha	mm	m³/ha
April and September	1.5	15	50	500
May and August	2.5	25	80	800
June and July	3.5	35	100	1000

When the rainfall is well distributed crops are less likely to suffer than when it comes as very wet periods followed by long dry periods; during such dry periods the crops have to survive on the available water held in the soil (see below).

Soils differ considerably in their capacity to hold water. Table 9a gives the available water capacity (AWC) of various soil types (textures) expressed as mm per cm depth and also as % depth.

TABLE 9a

Soil texture	Available water capacity	
	mm/cm	% depth
Coarse sandy loam	1.3	13
Sandy clay	1.4	14
Sand	1.5	15
Clay and loam	1.7	17
Clay loam	1.8	18
Silt loam	1.9	19
Very fine sandy loam	2.2	22
Very fine sand	2.3	23
Peat	3.6	36

The amount of organic matter in the soil can alter the water holding capacity very considerably, e.g. it is increased by the organic matter produced by leys and by the application of FYM. Soil which is consolidated holds more water per unit depth than loosely packed soil.

The water capacities of the soil and subsoil are usually different.

The amount of water available to a crop at any one time depends on the capacity of the soil per unit depth, and the depth of soil from which the roots take up water. The figures in Table 9b are average rooting depths—these would be greater on deep, well-drained soils and less on shallow, badly-drained or panned soils; this can be checked in the field by digging a soil profile. Usually most of the plant roots are concentrated in the top layers and only a small proportion penetrate deeply into the soil.

TABLE 9b

Crop	Rooting depth
Cereals, pasture	35–45 cm
Potatoes	30–60 cm
Beans, peas, conserved grass	45–60 cm
Sugar-beet	60–75 cm
Lucerne (alfalfa)	over 120 cm

From the above tables (9a and 9b) it can be calculated that a crop of peas, for example, growing on a silty loam soil would have about 100–150 mm of water available to it when the soil is at field capacity and this should be sufficient to keep it growing satisfactorily for a few weeks if no rain falls. If irrigation is possible, it is advisable not to let the amount of available water, within root range, fall below 50% of the maximum at field capacity. This could be calculated from average transpiration rates (about 2–4 mm per day) or would be indicated by finding that it is difficult to form the soil into a ball which will hold together (this test does not apply to sandy soils).

The greatest need for irrigation in the British Isles is in the south-east, where the lower rainfall and higher potential evaporation and transpiration means that irrigation would be beneficial about 9 years out of 10. In the wetter western and northern areas the need is much less. Potatoes and intensively managed grassland usually produce the best profits from irrigation.

In many parts of this country very limited amounts of water are available for irrigation although much could be done to improve this situation by the construction of reservoirs to conserve the winter surplus. Grants are available for permanent works of this kind (see *MAFF Bulletin* No. 202—*Water for Irrigation*).

More experimental work is required to determine the stages of growth at which crops are most responsive to irrigation, e.g.:

Grass and other leaf-crops respond at all times when there is a shortage of water.

It is known that peas are most responsive at "start of flowering" and "pod swelling" stages.

Also, if a maincrop potato variety which normally produces a lot of tubers (e.g. King Edward) is irrigated before the tubers are marble size too many small tubers and a low ware yield may result. Excessive amounts of water are likely to produce poor-quality tubers. (See also page 97.)

Irrigation of sugar-beet before the leaves meet across the rows encourages surface root development instead of deep rooting which would be less dependent on irrigation.

TABLE 10

Soil type	Time (hr) required to absorb 25 mm of water	Depth of soil (at wilting point) which is wetted by 25 mm of water
Sandy	1–2	300–600 mm
Loams	3	100–150 mm
Clay	4–5	120–150 mm

Table 10 is a guide to the rate at which various soil types can absorb water and so the rate at which irrigation water should be applied when using ordinary sprinklers and rain-guns. These rates could be considerably increased if very fine droplets were used but this is considered to be uneconomical at present. When irrigation water, or rain, falls on a dry soil it saturates the top layers of soil to full field capacity before moving

further down—unless the soil is deeply cracked (e.g. clay), when some of the water will run down the cracks. Very dry soils can absorb water at a faster rate than shown in Table 10, but only for a short period.

Irrigation water is applied by:

(a) *Rotary sprinklers*. This is the commonest method used. Each rotating sprinkler covers an area about 21–36 metres in diameter; the sprinklers are usually spaced at 10.5 metres apart along the supply line and the line is moved about 18 metres for each setting. Fifty such sprinklers would cover 1 hectare and each sprinkler applies about 18–45 litres per minute. Special nozzles can be used to apply a fine spray 2.5 mm per hour for frost protection of fruit crops and potatoes; icicles form on the plants but the latent heat of freezing protects the plant tissue from damage.

(b) *Rain-guns*. These are used on grassland and crops such as sugar-beet which are covering the ground; the droplets are large; diameter of area covered may be 60–120 metres. Some types are used for organic irrigation (slurry).

(c) *Spraylines*. These apply the water gently and are used mainly for horticultural crops.

(d) *Surface channels*. This method requires almost level or contoured land. It is wasteful of land, water and labour, but capital costs are very low.

(e) *Underground pipes*. On level land, water can be dammed in the ditches and allowed to flow up the drainage pipes into the subsoil and lower soil layers. This is drainage in reverse!

Systems for using sprinklers and rain-guns

1. *Conventional hand-moved laterals and sprinklers*. Sprinklers permanently fixed to lateral pipes. Each lateral dismantled and moved forward to new setting three or four times per day. Instead of sprinklers, more widely spaced rain-guns can be used for grassland or sugar-beet irrigation.

pros and cons

   (i) high labour cost, with men committed to
        continuous irrigation work all day;
  (ii) type of work is disliked by workers;
 (iii) lowest capital cost.

2. *Alternate-outlet hand-moved laterals and
sprinklers.* Small-diameter laterals moved once
only per day. Sprinklers on detachable
standpipes with self-sealing couplings moved
along the laterals to give several settings per day.

pros and cons

   (i) less labour intensive than system 1;
  (ii) moving sprinklers alone less arduous than
        moving integral laterals plus sprinklers;
 (iii) slightly higher capital cost than 1.

3. *Semi-permanent grid sprinkler systems.*
Small-diameter laterals laid out over the whole
field at the start of the season. Then sprinklers on
standpipes with self-sealing couplings moved
several times a day, as in system 2.

pros and cons

   (i) considerably reduced labour demand;
  (ii) poor versatility, unable to move equipment
        quickly from crop to crop;
 (iii) very high capital cost.

4. *Mobile machines, rotating-boom type.* Each
machine is driven to a new setting three or four
times a day. It waters a circular area of 0.5–1.2
hectares at each setting (depending upon model)
and settings are overlapped.

pros and cons

   (i) very great labour economy;
  (ii) wind can distort the spread pattern;
 (iii) each machine can be expensive, but area
        covered spreads this cost very well.

5. *Mobile machines, automatic continuously-
moving fixed-boom or rain-gun type, e.g. hose-reel
irrigators.* Each machine is moved to a new run
once a day and then set to move forward continuously during the ensuing 23 hours. The rate of
water applied is determined by the forward
speed setting.

pros and cons

   (i) very great labour economy with only one
        "setting" per day;
  (ii) ability to apply a very low rate of water
        quickly over a large area, if required;
 (iii) wind can distort the spread pattern of the
        rain-gun models.

6. *Power-moved pipelines with sprinklers.* A
continuous pipe with sprinklers is laid across
the field. The pipe forms the axle for large-
diameter wheels bolted round it at intervals. To
move the whole assembly forward to a new
setting (several per day) an engine is used to
rotate the whole pipeline.

pros and cons

   (i) fair labour economy;
  (ii) reasonable evenness of water distribution.

7. *Baars Irrigator.* Flexible hose with sixteen
folding sprinklers. Can be wound onto reel for
storage and transport.

8. *Centre-pivot.* This is a very large irrigator
with a boom of 560 m or more in ten or more
sections and rotating round a centre pivot which
is also the source of the water supply. It covers
100 ha or more and can operate on undulating
land, being best suited to large arable areas;
modified versions operate at right angles to a
canal and move along automatically. Centre-
pivots are expensive to buy and set up but
labour requirements are very low.

   With all the above systems, it is very important
that all nozzles operate at the correct pressure to
ensure uniform water distribution. (See also
MAFF Bulletin No. 138—*Irrigation*, and Booklet
2067—*Irrigation Guide*.)

4. Warping

   This is a process of soil formation where land,
lying between high and low water levels,
alongside a tidal river, is deliberately flooded
with muddy water. The area to be treated is
surrounded by earth banks fitted with sluice

gates. At high tide, water is allowed to flood quickly onto the enclosed area and is run off slowly through sluice gates at low tide. The fineness of the material deposited will depend on the length of time allowed for settling—the coarse particles will settle very quickly but the finer particles may take one or more days to settle. The depth of the deposit may be 0.4 m in one winter. When enough alluvium is deposited it is then drained and prepared for cropping. Sections dug through the subsoil clearly show separate deposition layers.

These soils are very fertile and are usually intensively cropped with arable crops such as *potatoes, sugar-beet, peas, wheat* and *barley*.

Part of the land around the Humber estuary is warpland, and the best is probably that in the lower Trent valley. Most of this work was done last century.

5. Claying

The texture of "blow-away" sandy soils and Black Fen soils can be improved by applying 250–750 tonnes/hectare of clay or marl (a lime-rich clay). If the subsoil of the area is clay this can be dug out of trenches and roughly scattered by a dragline excavator. In other cases the clay is dug in pits and transported in special lorry spreaders. Rotary cultivators help to spread the clay. If the work is done in late summer or autumn, the winter frosts help to break down the lumps of clay. (See MAFF leaflet, No. 18.)

TILLAGE AND CULTIVATIONS

Cultivations are field operations which attempt to alter the soil structure. The main object is to provide a suitable seed-bed in which a crop can be planted and will grow satisfactorily; sometimes cultivations are used to kill weeds, or bury the remains of previous crops. The timing of cultivations—particularly with regard to the weather and on the heavier soils—is more an art than a science and is largely based on experience. The cost of the work can be considerably reduced by good timing and use of the right implements. Ideally, a good seed-bed should be prepared with the minimum amount of working and the least loss of moisture. On heavy soil and in a wet season, some loss of moisture is sometimes desirable. On the medium and heavy soils full advantage should be taken of weathering effects; for example, ploughing in the autumn will allow frost to break the soil into a crumb structure; wetting and drying alternately will have a similar effect. The workability of a soil is very dependent on its consistency, i.e. its firmness or plasticity due to the amount of water present.

Seed-bed requirements for various crops

(1) *Cereals.* (a) *Autumn planted.* The object here is to provide a tilth (seed-bed condition) which consists of fine material and lumps about fist-size. It should allow for the seed to be drilled and easily covered and the surface should remain rough after planting. The lumps on the surface will prevent the heavier soils from "capping" easily in a mild wet winter and they also protect the base of the cereals from the harmful effects of very cold winds. Harrowing and/or rolling may be done in the spring to break up a soil cap which may have formed and to firm the soil around plants which have been heaved by frost action.

(b) *Spring planted.* A fairly fine seed-bed is required in the spring-time (very fine if grasses and clovers are to be undersown). If the seed-bed is dry or very loose after drilling it should be consolidated by rolling—this is especially important if the crop is undersown and where the soil is stony.

(2) *Root crops*, e.g. sugar-beet, swedes, and carrots: also *kale*. These crops have small seeds and so the seed-bed must be as fine as possible, level, moist and firm. This is very important when precision drills and very low seed rates (to reduce seed and thinning costs) are used. Good, early, ploughing with uniform, well-packed and broken furrow slices will considerably reduce the amount of work required in the spring. If

possible, deep cultivations should be avoided in the springtime so keeping frost mould on top and leaving unweathered soil well below the surface. Roots are usually sown on ridges in the higher rainfall areas; this avoids damage by surface water and also makes singling and harvesting easier.

(3) *Direct seeding of grasses and clovers.* Same as for roots.

(4) *Beans and peas.* Similar to cereals—but tilth need not be so fine. Peas grown on light soils may be drilled into the ploughed surface (or after one stroke of the harrow) if the ploughing has been well done.

(5) *Potatoes.* This crop is usually planted in ridges 60–90 cm wide and so deep cultivations are necessary. The fineness of tilth required depends on how the crop will be managed after planting. A fairly rough, damp seed-bed is usually preferable to a fine, dry tilth which has been worked too much. Some crops are worked many times after planting—such as harrowing down the ridges, ridging-up again, deep cultivations between the ridges and final earthing-up. The main object of these cultivations is to control weeds but the implements often damage the roots of the potato plants.

Most annual weeds can be controlled by spraying the ridges when the potato sprouts start to appear. This can replace most of the inter-row cultivations and reduce the number of clods produced by the rubber-tyred tractor wheels.

Recent developments

Direct Drilling (slit-seeding, zero tillage) and *Minimal Cultivations* techniques are now becoming widely-established alternatives to conventional cultivations on many farms and on a wide range of soil types including difficult clays.

For *direct drilling*, several types of special drills are available, using such developments as heavily weighted discs for cutting slits, strong cultivator tines, or modified rotary cultivators.

The *advantages* of the technique are:

(1) Considerable saving in time (quarter to half the normal).
(2) It may be possible when normal cultivations are not, but this is not to be encouraged—particularly on heavy soils.
(3) Very little loss of soil moisture when compared with ordinary cultivations.
(4) It has a controlling effect on "take-all" in cereal crops and usually gives higher yields of plumper grains.
(5) Better control of wild oats and blackgrass because shed seeds left on the surface are eaten or otherwise destroyed and dormancy is not encouraged by burial of the seeds.
(6) Better soil structure—which also means better drainage: this is especially so when the technique is introduced at a time when the soil has a good structure. Normal cultivations can destroy a good structure built up by a period in grass. When the technique is used for several years, organic matter residues accumulate in the surface layer of the soil where it is of greatest benefit and is not diluted by burial (e.g. ploughing) and mixing with deeper soil layers. Earthworms (and moles) increase.

Some *disadvantages* are:

(1) Does not bury trash, so burning of stubble or destruction of grass swards with chemicals is necessary.
(2) Should not be used where perennial weeds such as couch, field bindweed, coltsfoot, and corn mint are abundant. However, "Roundup" (glyphosate) may be used to destroy these.
(3) Winter wheat crops take longer to establish in the firm soil, so should be sown before the end of October. Also, more likely to get trouble with hares and rabbits on the firmer soil and especially if stubble is left.
(4) Is not a success if soils are badly drained.
(5) If stubble is badly rutted after harvest, this must be remedied by conventional cultivations.
(6) Slug damage can be serious—especially in wet conditions—and slug pellets should be used to control them (these may be placed in the slit with the seed).

(7) Heavy soils may crack along the slits in very dry weather and the seed may not establish properly.

(8) Toxic substances develop in the buried trash (e.g. acetic acid) see page 58.

Possible uses for the technique are:

(a) For cereals in a continuous cereal system or following peas, beans, oil-seed rape, potatoes. If following grass (destroyed by chemicals), trouble may arise from grass seedlings growing in the cereals if a suitable herbicide is not used.

(b) Kale after grass; establishes well, even in dry weather, and the undisturbed turf is much firmer for grazing in autumn.

(c) Root crops, such as sugar-beet, on light soils.

(d) Oil-seed rape after cereals or other crops.

(e) Green forage crops after early-harvested cereals for autumn and early winter grazing.

Minimal cultivations systems are those using various types of cultivators, instead of ploughing, and in such a way that only the minimum depth of soil is moved to allow drilling to take place.

(See references for further reading at end of chapter.)

Tillage implements

The main implements used for tillage are:

Ploughs. Ploughing is the first operation in seed-bed preparation on most farms and is likely to remain so for some time yet, although many farmers are now using rotary cultivators, heavy cultivators with fixed or spring tines, and mechanically-driven digging or pulverizing machines, as alternatives to the plough. Good ploughing is probably the best method of burying weeds and the remains of previous crops; it can also set up the soil so that good frost penetration is possible. Fast ploughing produces a more broken furrow slice than slow steady work. The *mounted* or *semi-mounted plough* is replacing the *trailed type* on most farms because of ease of handling. *General-purpose mouldboards* are com-monly used; the shorter *digger* types (concave mouldboards) break the furrow slices better and are often used on the lighter soils. *Deep digger ploughs* are used where deep ploughing is required, e.g. for roots or potatoes. The *one-way (reversible)* type of plough is fairly popular for crops such as roots and peas: it has right-hand and left-hand mouldboards and no openings or finishes have to be made when ploughing so the seed-bed can be kept level. *Round-and-round* ploughing with the ordinary plough has almost the same effect although this is not a suitable method on all fields.

The proper use of skim and disc coulters and careful setting of the plough for depth, width and pitch can greatly improve the quality of the ploughing. The furrow slice can only be turned over satisfactorily if the depth is less than about two-thirds the width; the usual widths of ordinary plough bodies vary from 20 to 35 cm. If possible, it is desirable to vary the depth of ploughing from year to year to avoid the formation of a plough pan. Very deep ploughing which brings up several centimetres of poorly weathered subsoil must be undertaken with care: the long-term effects will probably be worthwhile but, for a few years afterwards, the soil may be rather sticky and difficult to work. Buried weed seeds, such as wild oats which have fallen down cracks, may be brought to the surface and may spoil the following crops. "*Chisel ploughing*" is a modern term used to describe the work done by a heavy-duty cultivator with special spring or fixed tines—it does not move or invert all the soil as the ordinary plough does. *Disc* ploughs have large saucer-shaped discs instead of shares and mouldboards. Compared with the ordinary mouldboard ploughs they do not cut all the ground or invert the soil so well but they can work in harder and stickier soil conditions. They are more popular in dry countries. Double mouldboard *ridging* ploughs are used for potatoes and some root crops in the wetter areas.

Cultivators. These are tined implements which are used to break up the soil clods (to ploughing depth). Some have tines which are rigid or are held by very strong springs which only give when an obstruction such as a strong tree root is struck. Others have spring tines which are constantly moving according to the resistance of the soil—they have a very good pulverizing effect and can often be pulled at a high speed; they can be useful for dragging the rhizomes of weeds, such as couch, to the surface. The shares on the tines may be of various widths. The pitch of the tines draws the implement into the soil. Depth can be controlled by tractor linkage or wheels. The timing of cultivations is very important if the operation is to be effective.

Harrows. There are many types of harrows: the zigzag type, which has staggered tines, is the commonest. Harrow tines are usually straight, but may vary in length and strength on the heavy and light types. *Drag harrows* have curved ends on the tines.

These implements are often used to complete the work of the cultivator. Besides breaking the soil down to a fine tilth they can have a useful consolidating effect due to shaking the soil about.

"Dutch" harrows have spikes fitted in a heavy wooden frame and are useful for levelling a seed-bed as well as breaking clods.

Some harrows, e.g. the chain type, consist of flexible links joined together to form a rectangle. These follow an uneven surface better, and do not jump about so much on grassland as the zigzag type. Most chain harrows have spikes fitted on one side. They are sometimes used to roll-up weeds such as "couch" grass.

Special types of harrows fitted with knife-like tines are used for improving matted grassland, by tearing out surface trash.

Power-driven reciprocating harrows on which rows of tines are made to move at right angles to the direction of travel, and rotary harrows, can result in much better movement of the soil in one pass. They are most valuable when preparing fine seed-beds for potatoes, sugar-beet and other crops on the heavier soils.

Hoes. These are implements used for controlling weeds between the rows in root crops. Various shaped blades, and discs may be fitted to them. Most types are either front-, mid- or rear-mounted on a tractor. The front- and mid-mounted types are controlled by the steering of

the tractor drive. The rear-mounted types usually require a second person for steering the hoe.

Disc harrows consist of "gangs" of saucer-shaped discs about 30–60 cm in diameter. They have a cutting and consolidating effect on the soil and this is particularly useful when working a seed-bed on ploughed-out grassland, some discs have scalloped edges to improve the cutting effect.

The more the discs are *angled*, the greater will be:

(a) the depth of penetration,
(b) the cutting and breaking effect on the clods,
(c) the draught.

To increase the effect of the operation, the rear gangs should be angled more than the front gangs. Disc harrows are widely used for preparing all kinds of seed-beds but it should be remembered that they are expensive implements to use. They have a heavy draught and lots of wearing parts (discs, bearings and linkages) so should only be used when harrows would not be suitable. They tend to cut the rhizomes of weeds such as "couch" and creeping thistle into short pieces which are easily carried about and so encourages the spread of these weeds. Discing of old grassland before ploughing will usually allow the plough to do better work and a better seed-bed can be made. Heavy discs, and especially those with scalloped edges, are very useful for working in chopped straw after combining.

Rotary cultivators (e.g. rotavator). This type of implement consists of curved blades which rotate round a horizontal shaft set at right angles to the direction of travel. The shaft is driven from the P.T.O. of the tractor; depth is controlled by a land wheel or skid. This implement can produce a good tilth in difficult conditions and in many cases may replace all other implements in seed-bed preparation. A light fluffy tilth is sometimes produced which may "cap" easily if wet weather follows. The fineness of tilth can be controlled by the forward speed of the tractor—fast speed, coarse tilth. It is a very useful implement for mixing into the soil the remains of kale and sprout crops, or straw. The rotating action of the blades helps to drive the implement forward; so extra care must be taken when going down steep slopes. The blades cut up rhizomatous weeds (e.g. couch) but if the implement is used several times (at 10–14-day intervals) in growing weather it can completely destroy the couch either by burying the sprouted pieces or throwing them out on the surface to die off in drying winds. A similar method (working 20–25 cm deep) will control bracken.

In wet heavy soils the rotating action of the blades may have a smearing effect on the soil. This can usually be avoided by having the blades properly angled. "Rotavating" of ploughed or cultivated land when the surface is frozen in winter can produce a good seed-bed for cereals in the spring without any further working or loss of moisture; if there is couch present and the weather is dry, much of it may die off. Narrow rotary cultivator units are available for working between rows of root crops.

A light high-speed rotary cultivator can be used for quick shallow working either for stubble cleaning or seed-bed preparation.

Rolls. These are used to consolidate the top few centimetres of the soil so that plant roots can keep in contact with the soil particles and the soil can hold more moisture. They are also used for crushing clods and breaking surface crusts (caps). Rolls should not be used when the soil is wet—this is especially important on the heavier soils. The two main types of rolls are the *flat* roll which has a smooth surface and the *Cambridge* or *ring* roll which has a ribbed surface and consists of a number of heavy iron wheels or rings (about 7 cm wide) each of which has a ridge about 4 cm high. The rings are free to move independently and this helps to keep the surface clean. The ribbed or corrugated surface left by the Cambridge roll provides an excellent seed-bed on which to sow grass and clover seeds or roots. Also, it is less likely to "cap" than a flat rolled surface.

When rolling a growing crop, e.g. young cereals, tractor wheel-slip must be avoided as this will tear out the seedlings.

Very heavy rolls are sometimes used for levelling fields in the spring prior to taking a white clover seed crop.

A *furrow press* is a special type of very heavy ring roller (usually with three or four wheels) used for compressing the furrow slices after ploughing—it is usually attached to and pulled alongside the plough.

A *subsoiler* is a very strong tine (usually two or more are fitted on a tool-bar) which can be drawn through the soil and subsoil (about 0.5 m deep and 1 m apart) to break pans and produce a heaving and cracking effect. This will only produce satisfactory results when the subsoil is reasonably dry (see STL 114) and drainage is good. The modern types with "wings" fitted near the base of the tine produce a very good shattering effect.

Pans

A *pan* is a hard, cement-like layer in the soil or subsoil which can be very harmful because it prevents surplus water draining away freely and restricts root growth.

Such a layer may be caused by ploughing at the same depth every year. This is a *plough pan* and is partly caused by the base of the plough sliding along the furrow. It is more likely to occur if rubber-tyred tractors are used when the soil is wet and there is some wheelslip which has a smearing effect on the bottom of the furrow. Plough pans are more likely to form on the heavier types of soil. They can be broken up by using a *subsoiler* or deeper ploughing.

Pans may also be formed by the deposition of *iron compounds*, and sometimes *humus*, in layers in the soil or subsoil. These are often called *chemical* or *iron pans* and may be destroyed in the same way as plough pans. *Clay pans* form in some soil formation processes.

Soil loosening

Soil loosening is a new term used to cover many different types of operation (including subsoiling) which aim to improve the structure of the soil and subsoil by breaking pans and having a general loosening effect.

Ordinary subsoilers, fitted with wings, can be used. The depth of working and distance between the tines are very important. Other implements which may be used are:

—the *Paraplow*, the angled mainframe carries three or four sloping tines which have adjustable rear flaps and there are also adjustable sloping discs; the effect is to give a fairly uniform lifting and cracking to about 350 mm.

—the *Shakaerator*, the five strong tines are made to vibrate in the ground by a mechanism driven from the tractor P.T.O.

—*modified rotary cultivators with fixed tines attached.*

The soil loosening process is very necessary in situations where shallow cultivations—up to 150 mm—have been used for many years and the soil below has become consolidated and impermeable to water and plant roots; rainwater collects on this layer and causes plant death by waterlogging and/or soil erosion.

Soil capping

A soil *cap* is a hard crust, often only about 2–3 cm thick, which sometimes forms on the surface of a soil.

It is most likely to form on soils which are low in organic matter. Heavy rain or large droplets of water from rain-guns (see "Irrigation") may cause soil capping. Tractor wheels (especially if slipping), trailers and heavy machinery can also cause capping in wet soils.

Although a soil cap is easily destroyed by weathering (e.g. frost, or wetting and drying) or by cultivations, it may do harm while it lasts by:

(a) preventing water moving into the soil,
(b) preventing air moving into and out of the soil in wet weather,
(c) hindering the development of seedlings from small seeds such as grasses and clovers, roots and vegetables.

Chemicals such as cellulose xanthate can be sprayed on seed-beds to prevent soil capping—they enhance seedling establishment and do not affect herbicidal activity in the soil.

Control of weeds by cultivation

The introduction of chemicals which kill weeds has reduced the importance of cultivation as a means of controlling weeds. The cereal crops are now regarded by many farmers as the *cleaning crops* instead of the roots and potato crops, mainly because chemical spraying of weeds in cereals is very effective.

However, weeds should be tackled in every way possible and there are still occasions when it is worthwhile to use cultivation methods.

Annual weeds can be tackled by:

(a) Working the stubble after harvest (e.g. discing, cultivating or rotavating) to encourage seeds to germinate: these young weeds can later be destroyed by harrowing or ploughing. Unfortunately, this allows wild oats to increase (page 170).
(b) Preparing a "false" seed-bed in spring to allow the weed seeds to germinate—these can be killed by cultivation before sowing root crops.
(c) Inter-row hoeing of root crops which can destroy a lot of annual weeds and some perennials.

Perennial weeds, e.g. couch grass, creeping thistle, docks, field bindweed and coltsfoot, can only be satisfactorily controlled by fallowing (i.e. cultivating the soil periodically through the growing season instead of cropping) but this is expensive. A fair amount of control can be obtained by short-term working in dry weather.

Couch grass is easily killed by drying winds if the rhizomes can be dragged out to the surface free of soil. This may be done in August and September, or in the early spring, using cultivators or shallow ploughing to loosen the soil, followed by drag or spring tined harrows to drag it out on top and shake off the soil; dry weather is necessary to do this properly.

In damp soil conditions, the rotary cultivator can be used three or four times at 2–3-week intervals to chop up and exhaust couch grass rhizomes. It is very important that the first time over should be on firm soil (e.g. stubble), and the tractor moving in low gear. Chopping the rhizomes into short pieces encourages nearly all the buds to send out shoots and so helps to exhaust them.

The deeper rooted bindweed, docks, thistles, and coltsfoot cannot be satisfactorily controlled by these methods, but periodic hoeing and cultivating between the rows of root crops can considerably reduce these weeds by cutting off new shoots and so exhausting them.

Fallowing. The object of a long-term fallow is to dry out the soil by frequent working and so dry out and kill the perennial weeds. On the medium to lighter soils this is done by frequent cultivations. On heavy soil, the field is ploughed when damp in spring to make it dry into hard lumps. These lumps are then moved by cross-ploughing or deep cultivations to help dry them out. During the summer, alternate periods of wetting and drying break the lumps down to a fine tilth and then annual weeds may germinate to be destroyed by further working. Fallowing is not very common nowadays because of the cost of the work and the loss of profit on a crop.

One of the best methods of controlling wild onion is by taking seven spring-sown crops in succession and ploughing each year in November.

Thorough cultivations which provide the most suitable conditions for rapid healthy growth of the crop may result in the crop outgrowing and smothering the weeds.

SUGGESTIONS FOR CLASSWORK

Examine:

(a) Samples of rock, e.g. granite, basalt, chalk, limestone.

(b) Various soil types in the field or in suitable blocks in the classroom, e.g. clay, sand, loam, peat. Handle them when wet and dry. If possible, examine the soil profiles.

(c) Note the crumbling effect on clods of frost action, and wetting and drying.

(d) Visit farms on clay, loam and sandy soils and discuss the management of these soils.

(e) Visit a drainage scheme in progress. Note how levels are taken and used, and the depth and spacing of the drains.

(f) See seed-beds being prepared for cereals, roots and potatoes and make notes on the cultivation work which was carried out.

(g) Visit one or more irrigation schemes and note the sources, cost and availability of the water; also the type of application equipment, how it is used and labour requirements.

FURTHER READING

Davis, Eagle and Finney, *Soil Management*, Farming Press.
Modern Farming and the Soil, MAFF Report, HMSO.
Curtis, Courtney and Trudgill, *Spoils of the British Isles*, Longmans.
Allen, *Direct Drilling and Reduced Cultivations*, Farming Press.
Handbook of Direct Drilling (ICI).
Cereals without Ploughing, MAFF PFE. No. 6.

3

FERTILIZERS AND MANURES

SUPPLYING PLANT NUTRIENTS TO THE SOIL

If good crops are to be continuously removed from a field or a farm, there should be at least as much nutrients returned to the soil as have been removed in the crops. Table 11 gives average figures for nutrients removed by various crops.

When supplying nutrients to the soil it is usual to apply more than enough for the needs of each crop because some nutrients may be lost by drainage (e.g. nitrogen and potash) and some will become "fixed" or unavailable in the soil (e.g. phosphate). Where one crop, e.g. potatoes or roots, has been heavily manured with fertilizers and FYM, it may be possible to reduce the amount of nutrients supplied to the following crop.

TABLE 11   NUTRIENTS REMOVED BY CROPS

Crop (good average yield)		N	P_2O_5	K_2O	
		kg/ha	kg/ha	kg/ha	
Wheat					
grain 5 tonnes/ha		93	43	30	If cereal straw is burnt on the field after combining, the
straw 5 tonnes/ha		17	7	40	potash is not lost but may be unevenly distributed
	Total	110	50	70	
Barley					
grain 4 tonnes/ha		67	33	22	
straw 3 tonnes/ha		17	4	31	
	Total	84	37	53	
Oats					
grain 4 tonnes/ha		67	33	22	
straw 5 tonnes/ha		15	9	74	
	Total	82	42	96	
Potatoes					
tubers 40 tonnes/ha		130	60	240	The response to phosphatic fertilizers is greater than these
dry haulm 3 tonnes/ha		50	6	110	figures suggest
	Total	180	66	350	
Sugar-beet					
roots 40 tonnes/ha		71	39	78	If sugar-beet tops or kale are eaten by stock on the field
fresh tops 35 tonnes/ha		119	39	202	where grown then most of the nutrients may be returned
	Total	190	78	280	to the soil
Kale					
fresh crop 50 tonnes/ha		224	67	202	

47

48

TABLE 12   THE NEED FOR AND EFFECTS OF NITROGEN, PHOSPHORUS AND POTASSIUM

Plant nutrient	Crops which are most likely to suffer from deficiency	Field conditions where deficiency is likely to occur	Deficiency symptoms	Effect on crop growth	Effects of excess	Time and method of application
Nitrogen (N)	All farm crops except legumes (e.g. beans, peas, clover) It is especially important for leafy crops such as grasses, cereals, kales and cabbages	On all soils except peats, and especially where organic matter is low and after continuous cereal crops	Thin, weak, spindly growth; lack of tillers and side shoots; small yellowish-green leaves, sometimes showing "autumn" tints	Increases leaf size, rate of growth and yield Produces darker green leaves	Causes "lodging" of cereal crops Delays ripening Produces soft growth which is more susceptible to disease and frost May spoil crop quality by lowering the starch or sugar content If combine-drilled germination of seed may be damaged	Nitrogen fertilizers applied in seed-bed or top-dressed in spring Anhydrous ammonia may be injected 15–22 cm into the soil at up to 250 kg/ha to supply grass needs for 3–4 months
Phosphorus (P)	Root crops (e.g. sugar-beet, mangolds, swedes, carrots), clovers, lucerne, potatoes and kale	Clay soils; acid soils —especially in high rainfall areas, chalk and limestone soils and peats Poor grassland	Similar to nitrogen except that leaves are a dull, bluish-green colour with purple or bronze tints	Speeds up growth of seedlings and increases root development: hastens leaf growth and maturity Encourages clover development in grassland Improves quality of crops	Might cause crops to ripen too early and so reduce yield if not balanced with nitrogen and potash fertilizers	Phosphorus fertilizers applied in seed-bed for arable crops; "placement" in bands near or with the seed reduces the amount which has to be applied Broadcast on grassland in autumn or early spring
Potassium (K)	Potatoes, carrots, beans, barley, clovers, lucerne, sugar-beet and mangolds	Light sandy soils, chalk soils, peats, badly drained soils, grassland repeatedly cut for hay, silage or "zero" grazing	Growth is squat, and growing points "die-back", e.g. edges and tip of leaves die and appear scorched	Crops are healthy and resist disease and frost better Prolongs growth Improves quality Balances N and P fertilizers	May delay ripening too much May cause magnesium deficiency in fruit and glasshouse crops and "grass-staggers" in grazing animals	K. ferts. broadcast or "placed" in seedbed for arable crops: broadcast on grassland in autumn or mid-summer
Magnesium (Mg)	Cereals, potatoes, sugar-beet, peas, beans, kale; "grass staggers" in cattle and sheep	Light sandy soils, chalk soils, often of a temporary nature due to poor soil structure, excessive potash, etc.	Chlorotic patterns on leaves (short of chlorophyll), e.g. "May Yellows" in barley	Associated with chlorophyll, and phosphorus metabolism	Unlikely, requirements about same as phosphorus, and one-tenth that of nitrogen and potash	Use FYM or slurry; lime with magnesium limestone; Epsom salts; fertilizer containing kieserite

Nitrogen is supplied by fertilizers, organic matter (e.g. FYM), nodule bacteria on legumes (e.g. clovers, peas, beans, lucerne), and nitrogen-fixing micro-organisms in the soil. It is difficult to estimate how much nitrogen is produced by legumes and micro-organisms; clovers in grassland may supply up to 250 kg/ha and micro-organisms about 60 kg/ha per annum.

Phosphates and *potash* are supplied by the soil minerals, organic manures and fertilizers.

The farmer has to decide each year what fertilizers to put on each crop. This is partly a haphazard choice and partly based on the results of experiments and his previous experience on his farm. Soil analysis, as now carried out, gives a poor indication of nitrogen requirements, and is only a very rough guide to the need for phosphates and potash.

Trace elements

The need for trace elements is likely to be greatest on very poor soils, where soil conditions such as a high pH make them unavailable, where intensive farming (with high yields) is practised, and on farms where organic manures such as FYM and slurry, and fertilizers such as basic slag, are not used. The importance of trace elements in plant nutrition is most appreciated when plants are grown in culture solutions circulated past their root systems (hydroponics). This is a form of horticultural crop production which is rapidly increasing especially where the solutions can be automatically replenished with nutrients.

Supplying trace elements to plants can be troublesome and care is required to prevent over-dosing which may damage or kill the crop. To facilitate their application and availability to the plants, trace elements such as calcium, copper, iron, manganese and zinc are now obtainable in a chelated form. These *chelates* are "protected" water-soluble complexes of the trace elements with organic substances such as EDTA. They can be safely applied as: foliar sprays—for quick and efficient action, or, to the soil—for root uptake without wasteful "fixation" in the soil because they do not ionize. They can be safely mixed with many spray chemicals. Trace elements may also be applied as *frits*, which are produced by fusing the elements with silica to form glass which is then broken into small particles for distribution on the soil.

Mixtures containing an assortment of chemicals—some of which may be essential plant nutrients—are available for spraying on crops which may be suffering from deficiency diseases. Very often these are chemical by-products, or derived from seaweed, or just an optimistic cocktail which may or may not be worth using. Occasionally, the results from using one of these sprays can be dramatically beneficial and this can lead to extravagant claims being made by unscrupulous salesmen.

It is very important, whenever possible, to obtain expert advice before using trace elements, as crops which are not growing satisfactorily may be suffering because of factors other than trace-element deficiency, for example, poor drainage, drought, frost or mechanical damage, viral or fungus diseases, major nutrient deficiency, etc.

UNITS OF PLANT FOOD

For many years the recommended rates for plant food applications to crops have been expressed in units—each unit of plant food being 1% of 1 cwt (i.e. 1.12 lb) and approximately 0.5 kg.

With the introduction of metrication, the kg is the unit of plant food and recommendations will be in terms of kilogrammes per hectare (kg/ha). For example, the recommendation for a cereal crop may be to apply

100 kg/ha N, 50 kg/ha P_2O_5 and 50 kg/ha K_2O.

To convert this into numbers of bags of fertilizer per hectare it is necessary to use the per cent analysis figures for the fertilizer to be used; this is clearly stated on each bag, e.g. 20:10:10 means that this particular fertilizer contains 20% N, 10% P_2O_5 and 10% K_2O, always given in that order.

100 kg of (20:10:10) fertilizer contains 20 kg N, 10 kg P_2O_5 and 10 kg K_2O, therefore a 50-kg bag of it contains 10 kg N, 5 kg P_2O_5 and 5 kg K_2O, i.e. the number of kg of plant food in a 50-kg bag of fertilizer is half the per cent figures.

Another example, a bag (50 kg) of (10-25-25) fertilizer contains

    5 kg N,   12.5 kg P_2O_5 and   12.5 kg K_2O

To return to the recommendation above (in kg/ha) of 100 N, 50 P_2O_5, 50 K_2O this would be supplied by 10 bags of 20:10:10 fertilizer. This is an easy one; in more complicated cases the figures may not work out exactly and a compromise has to be accepted. Sometimes a fertilizer is chosen which gives the correct amount of P and K but not enough N, so a top-dressing of N would be given. For example:

	N	P_2O_5	K_2O
Spring cereals recommendation (kg/ha)	100	50	50
Four bags of 10:25:25, combine-drilled, supply	20	50	50
Difference	80	—	—

The 80 kg of N would be top-dressed and may be supplied by approximately $4\frac{1}{2}$ bags per hectare of "Nitram" (34.5%): $80 \div 17.25 = 4.6$.

If bulk fertilizer is used, each tonne contains 10 times the per cent of each plant food: 1 tonne (1000 kg) $= 100 \times 10$ kg,

so 1 tonne of 20:10:10 contains 200 kg N,
  100 kg P_2O_5 and 100 kg K_2O.
and half a tonne supplies 100 kg N,
  50 kg P_2O_5 and 50 kg K_2O.

In the case of liquid fertilizers, 1000 litres weigh approximately 1 tonne, so the amount required can be worked out as above and litres substituted for kilogrammes.

The cost of plant food per kilogramme can be calculated from the cost of fertilizers which contain only one plant food such as nitrogen in "Nitram", P_2O_5 in triple superphosphate, and

K_2O in muriate of potash. If 1 tonne of "Nitram" (34.5% N) costs £138 and contains 345 kg N, then

$$1 \text{ kg of N costs } \frac{13{,}800}{345} \text{ pence, i.e. 40p.}$$

If 1 tonne of triple super (45%) costs £180 and contains 450 kg P_2O_5, then

$$1 \text{ kg of } P_2O_5 \text{ costs } \frac{18{,}000}{450} \text{ pence, i.e. 40p.}$$

If 1 tonne of muriate of potash (60%) costs £120 and contains 600 kg K_2O, then

$$1 \text{ kg of } K_2O \text{ costs } \frac{12{,}000}{600} \text{ pence, i.e. 20p.}$$

One tonne of a (10:25:25) compound contains:

N	P_2O_5	K_2
100 kg	250 kg	250 kg

The value of this, based on the costs of a kg of N, P and K above is:

N	$100 \times 40 =$	4,000	
P	$250 \times 40 =$	10,000	
K	$250 \times 20 =$	5,000	
Total		19,000p.	$= £190$

If the well-mixed, granulated compound costs £205 per tonne then the extra cost compared with "straights" is £15.0 (£205 − 190).

The above keep changing as costs keep on rising but the actual costs at any time could be substituted in the calculations.

It is also possible to compare the values of "straight" fertilizers of different composition on the basis of cost per kg of plant food, e.g. the cost of a kg of N in "Nitram" (34.5%) is

$$(345 \text{ kg N/tonne}) \quad \frac{£138 \text{ per tonne}}{345} = 40p,$$

the cost of a kg of N in "Nitro-chalk" (26% N) is

$$(260 \text{ kg N/tonne}) \quad \frac{£108 \text{ per tonne}}{260} = 41.5p$$

so, apart from less handling costs, the "Nitram" is the lower-priced fertilizer.

(See AF 53, Fertilizer Regulations, 1977.)

STRAIGHT FERTILIZERS

Straight fertilizers supply only one of the major plant foods.

Nitrogen fertilizers (N)

The nitrogen in many straight and compound fertilizers is in the ammonium (NH_4 ions) form but this is quickly changed by bacteria in the soil to the nitrate (NO_3 ions) form. Many crop plants, e.g. cereals, take up and respond to the NO_3 ions quicker than the NH_4 ions but other crops, e.g. grass and potatoes, are equally responsive to NH_4 and NO_3 ions.

Commonly used nitrogen fertilizers are:

(1) *Ammonium nitrate*, $NH_4 \cdot NO_3$ (33.5–34.5% N). This is a very popular fertilizer for top-dressing (half the nitrogen is very readily available) and is marketed in a special prilled or granular form to resist moisture absorption. It has been used by guerillas to make bombs, but is safe if stored in sealed bags and well away from combustible organic matter. Heavy dressings tend to make the soil acid.

(2) *Ammonium nitrate lime* (21–26% N). These granular fertilizers are mixtures of ammonium nitrate and lime, sold under various trade names; they take up moisture readily when the bags are opened and go pasty; they do not cause acidity in the soil.

(3) *Urea* (46% N). This is the most concentrated solid nitrogen fertilizer and is marketed in prilled form; it is sometimes used for aerial top-dressing. In the soil, urea changes to ammonium carbonate which may temporarily cause a harmful local high pH. Nitrogen, as ammonia, may be lost from the surface of chalk or limestone soils, or light sandy soils when urea is applied as a top-dressing; when it is washed or worked into the soil it is as effective as any other nitrogen fertilizer; chemical and bacterial action changes it to ammonium and nitrate forms. If applied close to seeds, urea may reduce germination. It is also used to make plastics, and so it is expensive.

(4) *Sulphate of ammonia* (S/A, $(NH_4)_2 \cdot SO_4$). This was the main source of nitrogen, but is seldom used now. It consists of whitish, needle-like crystals and is produced as a by-product from gasworks or synthetically from atmospheric nitrogen. Bacteria change the ammonium nitrogen to nitrate. It has a greater acidifying action on the soil than other nitrogen fertilizers. Some nitrogen may be lost as ammonia when it is top-dressed on chalk soils.

(5) *Nitrate of soda*, $Na \cdot NO_3$ (16% N) is obtained from natural deposits in Chile and usually marketed as moisture-resistant granules. The nitrogen is readily available and the sodium is of value to some crops such as sugar-beet and mangels. It is expensive and so is not widely used.

(6) *Calcium nitrate* (15.5% N). This is a double salt of calcium nitrate and ammonium nitrate in prilled form. It is mainly used on the Continent.

(7) *Anhydrous ammonia* (82% N). This is ammonia gas liquefied under high pressure, stored in special tanks and injected about 15 cm into the soil from pressurized tanks through tubes fitted at the back of strong tines; strict safety precautions must be observed. The ammonia is rapidly absorbed by the clay and organic matter in the soil and there is very little loss if the soil is in a friable condition and the slit made by the injection tine closes quickly. It is not advisable to use anhydrous ammonia on very wet or very cloddy soils or where there are lots of stones, but it can be injected when crops are growing, for example into winter wheat crops in spring, between rows of Brussels sprouts, and into grassland. The cost of application is much higher than for other fertilizers, but the material is cheap, so the applied cost per unit compares very favourably with other forms of nitrogen. On grassland it is usually applied twice—in spring and again in mid-summer—at up to 250 kg/ha each time. In cold countries it can be applied in late autumn for the following season, but the mild periods in winters in this country usually cause heavy losses by nitrification and leaching.

(8) *Aqueous ammonia* (about 28% N). This is ammonia dissolved in water under slight pres-

sure; it must be injected into the soil (10–12 cm), but the risk of losses is very much less than with the anhydrous ammonia; also, cheaper equipment can be used.

(9) *Aqueous nitrogen solutions* (26–32% N). These are usually solutions of mixtures of ammonium nitrate and urea, and are commonly used on farm crops; they are not under pressure and can be sprayed on the soil. If injected or worked in they are just as effective as other nitrogen fertilizers, but there is a risk of some loss from surface applications to chalk and very sandy soils: they can be sprayed on growing cereals and grass with a slight risk of scorch damage which can be avoided with special jets or by dribbling the solution through flexible polythene tubes (see page 54, Liquid Fertilizers).

(10) *Gas liquor* (1.7% for "10 oz liquor")—a variable by-product from gasworks before North Sea gas was available; the nitrogen is mainly present as ammonium salts. It is likely to scorch growing crops and has been used as a combined top-dressing and selective herbicide to kill weeds such as charlock in kale (it contains phenolic substances) at about 33–35 hl/ha.

Various attempts have been made to produce slow-acting nitrogen fertilizers and reasonable results have been obtained with such products as resin-coated granules of ammonium nitrate (26% N), sulphur-coated urea prills (36% N) (bacteria slowly break down the yellow sulphur in the soil), and urea formaldehydes (30–40% N).

Organic fertilizers such as *Hoof and Horn* (13% N)—ground up hooves and horns of cattle, *Shoddy* (up to 15% N), waste from wool mills, and *Dried Blood* (10–13% N), a soluble quick-acting fertilizer, are usually too expensive for farm crops and are mainly used by horticulturists.

Much of the nitrogen now supplied to farm crops comes from compound fertilizers in which it is usually present as ammonium nitrate, ammonium phosphate, or urea. (See AF 52.)

Phosphate fertilizers (P)

By custom and by law the quality or grade of phosphate fertilizers is expressed as a percentage of phosphorus pentoxide (P_2O_5) equivalent.

(1) *Ground rock phosphate.* The natural rock ground to a fine powder—i.e. 90% should pass through a "100-mesh" very fine sieve (16 holes/mm^2). The best ones contain about 29% P_2O_5 which is insoluble in water. They should only be used on acid soils in high rainfall areas and for grassland and brassica crops (e.g. swedes, turnips, kale).

"*Hyperphosphate*" is a softer rock phosphate obtained from North Africa. It can be ground to a very high degree of fineness, 90% passing through a "300-mesh" sieve—48 holes/mm^2. Because of this it will dissolve more quickly in the soil, although it should still only be used under the same conditions and for the same crops as ordinary ground rock phosphate.

(2) *Superphosphate* (super). This contains 18–21% water-soluble P_2O_5 produced by treating ground rock phosphate with sulphuric acid. It also contains gypsum ($CaSO_4$), which may remain as a white residue in the soil, and a small amount of unchanged rock phosphate. It is suitable for all crops and all soil conditions.

(3) *Triple superphosphate.* This contains about 47% water-soluble P_2O_5 and is produced by treating the rock phosphate with phosphoric acid. 1 bag triple super = $2\frac{1}{2}$ bags ordinary super.

(4) *Basic slags.* These are by-products in the manufacture of steel. The total amount of phosphate present varies between about 8–22% P_2O_5. This is insoluble in water but most of it may be soluble in a 2% citric acid solution (this is a guide to its solubility in the soil). A good slag should have a high percentage P_2O_5; a high proportion of this (80%+) should be *citric soluble* and over 80% should pass through a "100-mesh" sieve—16 holes/mm^2. Slags are not so quick acting as "supers" and give best results on acid soils. Cattle may be poisoned by eating herbage recently treated with slag which has not been washed off the leaves. Rate of application 0.75–1.25 tonnes/ha. Slags contain some trace elements. Supplies are now very limited. (See AL 442.)

Potash fertilizers (K)

The quality or grade of potash fertilizers is expressed as a percentage of potassium oxide (K_2O) equivalent.

(1) *Muriate of potash* M/P (potassium chloride) as now sold usually contains 60% K_2O. It does not store very well and does not spread easily unless specially treated. This is the commonest source of potash for farm use and for the manufacture of compounds containing potash.

(2) *Sulphate of potash* S/P (potassium sulphate). This is made from the muriate and so is more expensive per unit K_2O. It contains 48–50% K_2O and is the best type to use for quality production of crops such as potatoes, tomatoes and other market garden crops.

(3) *Kainit and potash salts.* These are usually a mixture of potassium and sodium salts, and sometimes magnesium salts. They contain about 12–30% K_2O. They have most value for sugar-beet and similar crops for which the sodium is a useful plant food. (See AL 443.)

Salt

Common salt (sodium chloride) is a cheap and useful fertilizer for sugar-beet and similar crops—applied at the rate of 0.5 t/ha.

P and K analyses

An increasing number of research workers, writers and advisers are now expressing amounts of plant nutrients in terms of the elements P (phosphorus) and K (potassium) instead of the commonly used oxide terms P_2O_5 and K_2O respectively. Throughout this book the oxide terms are used but these can be converted to the element terms by using the following factors:

$$P_2O_5 \times 0.43 = P, \text{ e.g. } 100 \text{ kg } P_2O_5 = 43 \text{ kg P,}$$
$$K_2O \times 0.83 = K, \text{ e.g. } 100 \text{ kg } K_2O = 83 \text{ kg K.}$$

COMPOUND OR MIXED FERTILIZERS

These fertilizers supply *two* or *three* of the major plant foods (i.e. nitrogen, phosphorus and potassium). They are produced by mixing such fertilizers as ammonium nitrate, ammonium phosphate, and muriate of potash or by more complex chemical processes.

About 75% of all fertilizers now used in this country are compounds. These are *well mixed* by machinery, are *granulated* and *store well*. This is a great saving in labour at a busy time because fertilizers mixed on the farm do not store well.

The concentration of compounds varies widely; some recently introduced contain 60 units of plant food per cwt, e.g. 10% N, 25% P_2O_5, 25% K_2O. If fertilizers are mixed on the farm it is possible to calculate the analysis of the mixture as follows:

Suppose the following are mixed:

2 parts sulphate of ammonia,
3 parts superphosphate and
1 part muriate of potash.

	N	P	K
200 kg of S/A (21%) contain	42 kg	—	—
300 kg of super (18%) contain	—	54 kg	—
100 kg of M/P (60%) contain	—	—	60 kg
600 kg of the mixture contain	42 kg	54 kg	60 kg
100 kg of the mixture contain	7 kg	9 kg	10 kg

The analysis of the mixture is therefore 7:9:10.

Purchased compound fertilizers usually cost £15.00 or more per tonne than the equivalent in "straights".

PLANT FOOD RATIOS

Fertilizers containing different amounts of plant food may have the same plant food ratios. For example:

	Fertilizer	Ratio	Equivalent rates of application
(a)	12:12:18	$1{:}1{:}1\frac{1}{2}$	5 parts of (a) = 4 parts of (b)
(b)	15:15:23	$1{:}1{:}1\frac{1}{2}$	
(c)	15:10:10	$1\frac{1}{2}{:}1{:}1$	7 parts of (c) = 5 parts of (d)
(d)	21:14:14	$1\frac{1}{2}{:}1{:}1$	
(e)	12:18:12	$1{:}1\frac{1}{2}{:}1$	5 parts of (e) = 6 parts of (f)
(f)	10:15:10	$1{:}1\frac{1}{2}{:}1$	

Some examples of compounds and possible uses are shown in Table 13.

Fertilizers are supplied in various ways:

Solids.

(a) *50-kg polythene bags.* These can be stored outside but should be covered with a polythene sheet. If stored on pallets, the manual work of handling is considerably reduced.
(b) *Bulk*—can be stored in dry, concrete bays covered with polythene sheets. 1 tonne occupies about $1\,\text{m}^3$. It can be moved by tractor hydraulic loaders or augers into trailers or trailer spreaders. It may also be supplied to the field as required in self-unloading 2-tonne bins.

Liquids. In the United Kingdom as a whole the liquid fertilizer share of the fertilizer market is about 8%. However, in the typical arable areas where the system has, until recently, been concentrated, liquid nitrogen has a share in the range of 10–27%, and compounds account for between 5 and 12% of the total tonnage of compound fertilizer used.

Liquid fertilizers are non-pressurized solutions of normal solid fertilizer raw materials. They should be distinguished from pressurized solutions such as aqueous ammonia (see page 51) and, more particularly, anhydrous ammonia (see page 51). Liquids are stored chiefly in steel tanks on the farm although a cheaper form of storage—butyl sheeting supported by a steel shell can be used instead.

Most of the application equipment used by the farmer is hired from the fertilizer company. The broadcasters, which range from a 600-litre tractor-mounted to the 2000-litre capacity trailed applicator, have spray booms of 10–20 m wide. Although the broadcaster is specially designed for fertilizer, by changing the jets it can be used for other agricultural chemicals, and so the liquid system fits in well with tramlining (see page 00) because only two pieces of equipment have to be matched. Equipment is also available

TABLE 13

Compound (N:P:K)	Crop	Rate 50 kg bags/ha	N	P	K
12:12:18	potatoes	25	150	150	225
20:10:10	spring cereal	10	100	50	50
0:20:20	autumn cereal	5	0	50	50
9:25:25	autumn cereal	5	22	60	60

for the placement of fertilizer for the potato and brassica crops, and a combine drill attachment can be used for sowing the liquid and cereal together. For top-dressing, and to minimize scorch, a special jet can be fitted by hand; this produces larger droplets for a better foliar run-off.

At present, liquid compounds are only about 60% as concentrated as the solid compound, although this is not considered to be very important as the material is so easily handled by the pump. It is potash which is the main constraint on the concentration of the compounds because it is the least soluble of the important plant foods considered to be necessary for crop nutrition.

The kilogram cost of the plant food is 5–8% more expensive than that of the solid, although the handling and application costs are considerably lower.

There is no difference in subsequent plant growth following the application of the fertilizer in a liquid or solid form.

APPLICATION OF FERTILIZERS

The main methods used are:

(a) *Broadcast distributors* using various mechanisms such as:

(1) *Pneumatic types.* A wide range of machines are now available which distribute the fertilizer by blowing it through various types of tubing; the spreading width can be selected to suit "tramline" widths.

(2) Other types involving rollers, brushes, chains, etc.

(3) *Spinning disc types.* These may be mounted or semi-mounted on the three-point tractor linkage or may be trailer types. Accuracy of distribution varies considerably, and is very dependent on accurate setting and amount of overlap.

(b) *Combine drills.* Fertilizer and seed (e.g. cereals) from separate hoppers are fed down the same or an adjoining spout. A *star-wheel* feed mechanism is normally used for the fertilizer and this usually produces a "dollop" effect along the rows. In soils low in phosphate and potash

this method of *placement* of the fertilizer is much more efficient than broadcasting.

(c) *Placement drills.* These machines usually place the fertilizer in bands 5–7 cm to the side and 3–5 cm below the rows of seeds. It is more efficient than broadcasting for crops such as peas, and also sugar-beet on some soils, e.g. sandy soils. Other types of placement drills attached to the planter are used for applying fertilizers to the potato crop.

(d) *Broadcast from aircraft.* This is useful for top-dressing of cereals—especially in a wet spring; also for applying fertilizers in inaccessible areas such as hill grazings. Highly concentrated fertilizers should be used, e.g. urea whenever possible.

(e) Liquids injected under pressure into the soil.

(f) Liquids (non-pressurized) broadcast or placed.

(See also MAFF Booklet 2292.)

Machinery maintenance

All machinery for fertilizer application should be thoroughly washed after use and coated with a rust-proofing material during long idle periods, e.g. creosote.

ORGANIC MANURES

Farmyard manure (FYM)

This consists of dung and urine, and the litter used for bedding stock. It is not a standardized product, and its value depends on:

(1) *The kind of animal that makes it.* If animals are fed strictly according to maintenance and production requirements, the quality of dung produced by various classes of stock will be similar. But in practice it is generally found that as cows and young stock utilize much of the nitrogen and phosphate in their food, their dung is poorer than that produced by fattening stock.

(2) *The kind of food fed to the animal that makes the dung.* The richer the food in protein and

mineral, the richer will be the dung. But it is uneconomical to feed a rich diet just to produce a richer dung.

(3) *The amount of straw used.* The less straw used, the more concentrated will be the manure and the more rapidly will it break down to a "short" friable condition.

Straw is the best type of litter available, although bracken, peat moss, sawdust and wood shavings can be used. About 1.5 tonnes of straw per animal is needed in a covered yard for 6 months, and between 2–3 tonnes in a semi-covered or open yard.

(4) *The manner of storage.* There can be considerable losses from FYM because of bad storage. (See STL 67 and 171.)

Dung from cowsheds, cubicles and milking parlours should, if possible, be put into a heap which is protected from the elements to prevent the washing out and dilution of a large percentage of the plant food which it contains. Dung made in yards should preferably remain there until it is spread on the land, and then, to prevent further loss, it is advisable to plough it in immediately.

FYM is important chiefly because of the valuable physical effects on the soil of the humus it contains. It is also a very valuable source of plant foods, particularly nitrogen, phosphate and potash, as well as other elements in smaller amounts. An average dressing of 25 tonnes/ha of well-made FYM will provide about 40 kg N, 50 kg P_2O_5, and 100 kg K_2O in the first year. At least one-third of the nitrogen will be lost before it is ploughed in, although this depends on the time of application, and all the plant food in FYM is less readily available than that in chemical fertilizers.

Application. The application of FYM will be dealt with under the various crops. (See STL 171 and 67.)

Liquid manure, and slurry

With the introduction of more intensive livestock enterprises, the rising cost of straw for bedding, and the need for cheap and effective mechanical methods of dealing with animal excreta, more and more farmers are now dealing with manure in a liquid or semi-solid (slurry) form instead of the traditional solid form as produced in straw-bedded yards. However, in many cases this change has created more problems than it has solved, for example—this slurry must not be allowed to pollute watercourses; also, trouble can arise from the nuisance of smells and possible health hazards over a wide area when the slurry is diluted and distributed by rain-guns as organic irrigation. If slurry is stored in anaerobic conditions, dangerous, obnoxious gases (mainly hydrogen sulphide) are produced and are released when it is being spread. These are not as acceptable to the surrounding community as the smell of FYM and so complaints, etc., may follow its use. The Water Authorities have a duty to prevent the pollution of streams and rivers with farmyard effluents and their advice should be sought on what is allowable and how problems (if any) can be dealt with. This usually means that the farmer has to find a way to utilize his slurry.

Table 14 sets out the average amounts and

TABLE 14

Livestock	Undiluted		Available nutrients in kg/m³ (total in brackets)			
	litres per day	m³ over winter	N	P_2O_5	K_2O	Mg
Dairy cows (500–600 kg)	45	7+	2.5 (5)	1 (2)	4.5 (5)	0.6
Fattening cattle (400–500 kg)	36	5+	2 (4)	1 (2)	4.5 (5)	—
Fattening pigs (75 kg)	4.5	—	4 (6)	2 (4)	2.7 (3)	—

composition of slurry (faeces + urine) produced by some stock.

The figures in the table can only be used as a basis to work from and only apply to slurry collected in an undiluted form, e.g. under slatted floors or passageways where washing and rain water are excluded. It may apply to slurry stored in sealed compounds or modified tower silos, in a few seasons when the amount of rainwater entering it is balanced by evaporation.

The slurry on most farms has varying amounts of water added to it and it is very difficult to be certain of its composition even when samples are analysed. However, from the figures in the table, it should be possible to know the total nutrients entering the sealed storage, and if the total amounts were spread evenly on one or more fields, the amount of nutrients/hectare could be estimated irrespective of the dilution of the material, e.g. with a tower silo storage: 100 cows over winter produce about 700 m^3 of excreta, i.e.

$$700 \times 2.5 = 1750 \text{ kg N}, 700 \times 1 = 700 \text{ kg P}_2O_5$$

and

$$700 \times 4.5 = 3150 \text{ kg K}_2O.$$

This will usually be diluted, but if it is mixed by recirculating it regularly, and spread over, say, 25 hectares in the spring, then the amount of nutrients available per hectare is approximately:

70 kg N, 28 kg P_2O_5 and 126 kg K_2O.

Rota-spreaders and similar machines can deal with fairly solid slurry; modern vacuum tankers and pumps are very efficient (when working properly) in dealing with slurry with less than 10% d.m. (i.e. slightly diluted). For rain-gun application, the slurry has to be diluted 2 parts water to 1 part slurry. In this case the nutrients per m^3 given in the table should be divided by 3; the amount applied at one setting can be varied according to the ability of the soil to absorb it, the crops to be grown, etc. Special care is required when applying slurry to grazing land because of the risk of taint, possible spread of disease organisms in fresh material (e.g. salmonella, brucella, foot and mouth), and possible

killing of some plants by heavy dressings.

A high proportion of the nitrogen in slurry may be lost by volatilization, and when put on fields with poor soil structure (allowing run-off) during the autumn and winter.

Slurry can be applied to any crop. It is usually most abundant on dairy farms and can be very well utilized on the fields used for silage or hay, or, applied to fields to be sown with kale or maize. Several weeks must elapse after application to grazing land before cows will readily eat the grass.

The injection of slurry into the soil using strong tines fitted behind the slurry tanker can reduce (but not eliminate) wastage and offensive odours, but it is expensive and can spoil the surface of a grass field where there are stones present.

The increasingly high price of fertilizers makes it highly desirable that the valuable nutrients in organic manures should be utilized as fully as possible. This is best achieved if the slurry can be applied at a time when growing crops can utilize it, and so storage is usually necessary.

Pig slurry is a useful source of copper for deficient soils, but toxicity can occur if very large amounts are used.

TABLE 15 NUTRIENT VALUE OF UNDILUTED SLURRY

| Source | m^3/ha | kg/ha | | | Worth |
		N	P_2O_5	K_2O	£
Dairy cows	40	100	40	180	54
Fattening pigs	25	100	50	70	44

Assuming 1 kg N is worth 22p, P_2O_5 (27p), K_2O (12p).

New techniques for separating slurry into solid and liquid fractions by screening or centrifugal action are now being used. The solid part (18–30% dm) can be handled like FYM, and the liquid portion is much less likely to cause tainting of pastures, and the disease risk is reduced. Table 16 is an example of separated cow slurry.

TABLE 16

1000 kg slurry dm, unseparated			Mechanically separated slurry					
			720 kg liquid at 14% dm			280 kg fibre at 26% dm		
N	P_2O_5	K_2O	N	P_2O_5	K_2O	N	P_2O_5	K_2O
4.4	0.6	2.5	3.2	0.5	2.1	1.1	0.1	0.4

(See also *MAFF Bulletin* No. 210, Booklet 2081, and STL nos. 67, 172 and 185.)

Cereal straws

Straw has some value as a source of organic matter, but its plant food content is low.

Two tonnes straw (approximately that produced per hectare) contains about 30 kg N, 5 kg P_2O_5 and 45 kg K_2O, so that the soil should gain some plant food value when straw is ploughed in. However, under naturally low nitrogen conditions the sudden influx of carbohydrate in the form of straw widens the carbon:nitrogen ratio (it should be 10:1) and so, unless extra nitrogen is added (25 kg/ha), the following crop will be starved of the plant food.

The other advantages of putting the straw back into the soil is that it can help to improve the structure of poor, sandy soils.

It is important that the straw should be spread evenly prior to ploughing. Ideally it should be chopped, and/or the straw spreader attachment can be used. As far as possible, and it is not always possible, straw should be shallow-ploughed when it will break down more quickly than if deep-ploughed. However, it is normally deep-ploughed and this, as well as being more expensive, dilutes the organic matter from the vital top few centimetres where it is most needed by the plant's roots. It has also been found that as straw decomposes under anaerobic conditions it produces toxic substances, notably acetic acid, which can prevent the development of seedlings.

Good stubble hygiene is almost impossible to carry out when the straw is left on the surface, and there is definite evidence of an increase in weeds (grass weeds and others) and plant diseases following several years of ploughing-in the straw and stubble.

Although it is unpopular with the general public, it is impossible to ignore the advantages of burning the straw and stubble. Table 17 summarizes three years' results from five of the Ministry's Experimental Husbandry Farms, and it shows that burning is markedly superior to any other straw-disposal method when it is a question of the yield of the crop.

The reasons for the yield advantage are not difficult to understand. Apart from hygiene, the straw can itself obstruct the development of the seedling plant. With direct-drilling especially, it is not unusual for the seed to lie on the top of the straw (which has been pushed down into the coulter slit) out of contact with the soil.

The yield, and cost advantages, do not take into account the additional expense involved in disposing of the straw compared with burning, as in Table 18.

TABLE 17   THREE-YEAR AVERAGE PERCENTAGE YIELD OF WINTER WHEAT FOLLOWING DIFFERENT STRAW DISPOSAL METHODS

Basic cultivation prior to drilling	Burnt	Previous straw chopped	Baled
Ploughed	100	92	91
Cultivated	100	89	87
Direct-drilled	100	78	78
Mean	100%	86%	85%

TABLE 18

Method of straw disposal	Index of cost
Baling and transporting	100
Chopping and cultivating-in or use of chisel plough	76
Ploughing-in (straw spreader back of combine)	68
Burning	4

This table gives the order of magnitude rather than the accuracy of detail, and it emphasizes the very substantial costs that would have to be borne by the farmer if straw burning were no longer practised.

The ley

When a field is put down to grass, fertility and soil structure are improved, and this increased fertility can be utilized by a succession of arable crops before the field is put back again to grass. This is the Alternate Husbandry System of Farming, and it has for many years formed the basis of mixed arable and livestock farming in this country.

Work at Experimental Stations and some of the Ministry Experimental Husbandry Farms (i.e. on different soil types) questions the validity of this system of farming as trials over the last 20 years show that increase in arable crop yields following grass is very slight compared with the same number of years' continuous arable where disease, pests, and weeds are not limiting factors.

Green manuring

This is the practice of growing and ploughing in green crops to increase the organic matter content of the soil. It is normally only carried out on light sandy soils.

White mustard is the most commonly grown crop. Sown at 9–17 kg/ha it can produce a crop ready for ploughing within 6–8 weeks. Fodder radish is becoming more popular for green manuring, and, like mustard, it can also provide useful cover for pheasants.

However, Rothamsted has shown that a short ley has very little benefit in the way of building up organic matter in the soil. There must, therefore, be even less from quick-growing crops (which break down equally quickly in the soil) such as mentioned above.

Seaweed

Seaweed is often used instead of FYM for crops such as early potatoes in coastal areas, e.g. Ayrshire, Cornwall and the Channel Islands. Ten tonnes contain about 50 kg of N, 10 kg P_2O_5 and 140 kg K_2O. It also contains about 164 kg of salt. The organic matter in seaweed breaks down rapidly.

Poultry manure

The annual production of poultry manure (about 5 million tonnes) is sufficient to put a tonne on every hectare of arable land in England and Wales. However, the implied possibilities do little to solve the problem of utilization because most of it is produced in very large units and transport creates problems. 1000 layers in cages produce about 1 tonne/week of fresh manure which is relatively low in organic matter and potash; about half the total production comes from battery cages and an increasing amount of this is being dried in special very high-temperature driers (or out-of-season grass driers).

Smell and cost are two problems.

The dried, sterilized, product weighs about 0.5 tonne/m^3, compared with 1 tonne/m^3 for fresh manure and granular fertilizers. The cost is considerably higher as a fertilizer than granular fertilizers but there are possible outlets for use in gardens; it is also likely to be used in increasing quantities as protein supplement in animal feeding stuffs. Deep-litter layers produce about 4% of total manure. Broilers on deep litter (four or five lots per year) produce more than one-third of the total, and because of the heat in the houses the manure is fairly dry and rich in nitrogen (high protein feeds).

TABLE 19

Source	Average annual production per 1000 birds in tonnes	% dry matter	Total nutrients kg/tonne of product available nutrients in brackets		
			N	P₂O₅	K₂O
Battery layers (fresh)	55	30	17 (11)	14 (7)	7 (5)
„ „ (dried)	14	86	50 (27)	46 (22)	18 (22)
Broilers	10	71	24 (16)	22 (11)	14 (10)
Deep-litter	60	68	17 (11)	18 (9)	13 (10)

(See also AL 320.) The above are average figures.

Fresh poultry manure usually has a high copper and zinc content, and there is a possible risk of causing toxicity to crops if it is used at rates above 70 tonnes/ha. It can be used at reasonable rates for most crops.

Waste organic materials

Various waste products are used for market garden crops—partly as a source of organic matter and partly because they release nitrogen slowly to the crop. They are usually too expensive for ordinary farm crops.

Shoddy (waste wool and cotton) contains 50–150 kg of nitrogen per tonne. Waste wool is best, and is applied at 2.5–5 tonnes/ha.

Dried blood, ground *hoof and horn* and *meat and bone meal* are also used; the nitrogen composition is variable.

(See *MAFF Bulletin* No. 210, *Organic Manures*.)

RESIDUAL VALUES OF FERTILIZERS AND MANURES

The nutrients in most manures and fertilizers are not used up completely in the year of application. The amount likely to remain for use in the following years is taken into account when compensating outgoing farm tenants.

All the *nitrogen* in *soluble* nitrogen fertilizers (e.g. ammonium nitrate and some compounds, and in dried blood) is used in the first year.

For *nitrogen* in bones, hoof and horn, meat and bone meal:
Allow $\frac{1}{2}$ after one crop and $\frac{1}{4}$ after two crops.

Phosphate in *soluble* form, e.g. super, basic slag, compounds:
Allow $\frac{2}{3}$ after one crop, $\frac{1}{3}$ after two and $\frac{1}{6}$ after three crops.

Phosphate in *insoluble* form, e.g. bones, ground rock phosphate:
Allow $\frac{1}{3}$ after one crop, $\frac{1}{6}$ after two and $1/12$ after three.

Potash, e.g. muriate or sulphate of potash, compounds:
Allow $\frac{1}{2}$ after one crop and $\frac{1}{4}$ after two.

Lime: one-eighth of the cost is subtracted each year after application.

ORGANIC FARMING

This is a term used to describe farming systems which do not involve the use of so-called artificial fertilizers and crop-protection chemicals. To the advocates of such systems, chemicals are synonymous with poisons and no arguments, however sensible and logical, will convince them otherwise.

It is bewildering that fertilizers are condemned as being artificial when, in fact, phosphate and potash fertilizers are derived from natural rocks and most nitrogen fertilizers are made from atmospheric nitrogen. Basic slag—the most artificial of all our fertilizers—appears to be an exception and is widely approved and used in many organic farming systems!

Organic material, whether as farmyard manure, slurry, composts, grass turf, straw, or other crop residues, can have a beneficial effect on most soils by improving the soil structure and moisture-holding capacity, and, also, by supplying a wide range of plant nutrients in unpredictable amounts. Most farmers appreciate the value of organic materials and will utilize these, wherever possible, and when there is little risk of spreading diseases and weeds (cereal straw can cause problems in this way and so is often burnt).

However, it is quite unrealistic to rely entirely on organic material to supply the nutrients required to produce the crop yields which are needed to meet the ever-increasing world demand for food. Individual farmers may appear to be growing satisfactory crops without resort to "artificial" or "bag" fertilizers, but in these cases they are either buying in nutrients in large quantities of feeding-stuffs to feed to intensive livestock enterprises, or are depleting the nutrient reserves in their soils. It is perhaps too obvious that all (or many) farmers could not follow such systems.

Wild, irresponsible statements, to the effect that the soils of many countries have been systematically poisoned for years by chemical fertilizers with disastrous results to the health of plants, animals, and human beings, are often put out by ill-informed people. These persons are usually well educated, often in good positions, and have the ability to express themselves in a very influential manner. Many of the organic farming enthusiasts are very sincere in their activities, but, because they are ill-informed, or acting deliberately, they mislead the general public and the *full* facts of their production methods are not always disclosed. The lower crop yields which many organic farmers obtain are usually compensated by the very inflated prices, paid by gullible people, for food grown with "organics". There is no clear-cut evidence to show that food produced by organic farmers is in any way superior to that produced by farmers using properly balanced "artificial" fertilizers.

Another bewildering aspect of organic farming is the acceptance by its advocates that all crops are equally good provided no "artificial" fertilizers are used, i.e. they are grown in natural conditions. However, it is well known that the amounts of plant nutrients available to crops varies very widely not only between farms and fields, but also in the same field, and so the crops growing in these various situations will be different.

Proper use of fertilizers enables the farmer to make up soil deficiencies when producing good crops and also to raise the general level of nutrients in circulation on the farm.

More and more crop-protection chemicals are becoming available for dealing with insect pests, diseases and weeds, and, provided they are used properly, they can be very useful and safe. Special care is required when mixtures of chemicals are used. It is very important that all such chemicals are fully tested for safety, to operators when applying them, to wild life, and also for possible harmful residues in foodstuffs. Some insects and fungi develop resistance to specific chemicals after a period of usage, and so different chemicals have to be used to counter this resistance; in general, it is best to use these chemicals only when necessary.

SUGGESTIONS FOR CLASSWORK

1. Examine samples of all the commonly-used "straight" fertilizers and a few compounds and note differences between them.
2. Visit experimental farms or demonstration plots to see the differences in crop growth due to deficient, excess and correct supply of nutrients.
3. If possible, visit farms to see bulk and sack storage of fertilizers, methods of handling and application.
4. When visiting farms note the various methods used for handling FYM and slurry.

FURTHER READING

Fertilizer Recommendations, MAFF Bulletin No. 210, HMSO.
Organic Manures, MAFF Bulletin No. 210, HMSO.
Cooke, *Fertilizing for Maximum Yields*, C.L.S.
Cooke, *Control of Soil Fertility*, Crosby Lockwood.
Lime and Fertilizer Recommendations, MAFF Booklet 2191.
Fertility Facts, Fertilizer Manufacturer's Association.

4

CROPPING

CLIMATE AND WEATHER AND THEIR EFFECTS ON CROPPING

Climate has an important influence on the type of crops which can be grown satisfactorily nearly every year. It may be defined as a seasonal average of the many *weather* conditions.

Weather is the state of the atmosphere at any time—it is the combined effect of such conditions as heat or cold, wetness or dryness, wind or calm, clearness or cloudiness, pressure and the electric state of the air.

The daily, monthly and yearly changes of temperature and rainfall give a fairly good indication of the conditions likely to be found.

Average yearly figures such as 1000 mm rainfall and temperature 10°C are of very limited value.

The climate of this country is mainly influenced by:

(1) its distance from the equator (50–60°N latitude),
(2) the warm Gulf Stream which flows along the western coasts,
(3) the prevailing south-west winds,
(4) the numerous "lows" or "depressions" which cross from west to east and bring most of the rainfall,
(5) the distribution of highland and lowland—most of the hilly and mountainous areas are on the west side,
(6) its nearness to the continent of Europe; from there hot winds in summer and very cold winds in winter can affect the weather in the southern and eastern areas.

Local variations are caused by *altitude, aspect* and *slope*.

Altitude (height above sea level) can affect climate in many ways. The temperature drops about 0.5°C for every 90 m rise above sea level. Every 15 m rise in height usually shortens the growing season by 2 days (one in spring and one in autumn) and it may check the rate of growth during the year. High land is more likely to be buffeted by strong winds and is likely to receive more rain from the moisture-laden prevailing winds which are cooled as they rise upwards.

Aspect (the direction in which land faces) can affect the amount of sunshine (heat) absorbed by the soil. In this country the temperature of north-facing slopes may be 1°C lower than on similar slopes facing south.

Slope. When air cools down it becomes heavier and will move down a slope and force warmer air upwards. This is why frost often occurs on the lowest ground on clear still nights whereas the upper slopes may remain free of frost. "Frost pockets" occur where cold air collects in hollows or alongside obstructing banks, walls, hedges, etc. (see Fig. 35). Frost-susceptible crops such as early potatoes, maize and fruit should not be grown in such places.

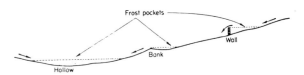

Fig. 35. Diagram to show how frost pockets are formed as cold air flows down a slope.

Rainfall in the British Isles

This comes mainly from the moist south-westerly winds and from the many "lows" or "depressions" which cross from west to east.

Western areas receive much more rain than eastern areas—partly because of the west to east movement of the rain-bearing air and partly because most of the high land is along the western side of the country.

The average annual rainfall on lowland areas in the west is about 900 mm (35 in.) and in the east is about 600 mm (24 in.). It is much greater on higher land.

Temperature of the British Isles

The temperature changes are mainly due to:

(1) the seasonal changes in length of day and intensity of sunlight;
(2) the source of the wind, e.g. whether it is a mild south-westerly, or whether it is cold polar air from the north or from the continent in winter;
(3) local variations in altitude and aspect;
(4) night temperatures are usually higher when there is cloud cover which prevents too much heat escaping into the upper atmosphere.

The soil temperature may also be affected by colour—dark soils absorb more heat than light-coloured soils. Also, damp soils can absorb more heat than dry soils.

The average January temperature in lowland areas along the west side of the country is about 6°C and about 4°C along the east side.

The average July temperatures in lowland areas in the southern counties is about 17°C but this drops to about 13°C in the north of Scotland.

Cropping in the British Isles

Grass grows well in the wetter, western areas and so dairying, stock rearing and fattening can be successfully carried on in these areas.

The drier areas in the east are best suited to arable crops which require fairly dry weather for easy harvesting.

The mild, frost-free areas in the south-west of England and Wales (i.e. parts of Devon, Cornwall, and Pembroke) are suitable for early crops of potatoes, broccoli, flowers, etc. The Isle of Thanet (Kent) and the Ayrshire coast are also early areas free from late frosts.

The more exposed hill and mountain areas are unsuitable for intensive production because of the lower temperatures, very high rainfall, inaccessibility and steep slopes. These are mainly rough grazings used for extensive cattle and sheep rearing. Large areas are now forestry plantations.

Most of the chalk and oolitic limestone areas (in the south and east) are now used for large-scale cereal production—particularly barley. The leys grown in these areas are mainly used by dairy cattle or sheep, or for herbage seed production. Cereals are grown on all types of soil. Maincrop potatoes and root crops, such as sugar-beet, are grown on the deeper loamy soils of the midlands and eastern counties. Carrots are grown on some light soils in the eastern counties.

Mixed farming (i.e. both crop and stock enterprises on the same farm) is found on most lowland farms. The proportion of grass (and so stock) to arable crops usually varies according to soil type and rainfall. The heavier soils and high rainfall areas usually have more grass than arable crops.

ROTATIONS

A *rotation* is a cropping system in which two or more crops are grown in a fixed sequence. If the rotation includes a period in grass (a ley), which is used for grazing and conservation, the system is sometimes called "alternate husbandry" or mixed farming. The term "ley farming" describes a system where a farm or group of fields is cropped entirely with leys which are reseeded at regular intervals; some people describe any cropping system which includes leys as "ley farming".

Farm crops may be grouped as follows:

(a) *Cereals (wheat, barley, oats* and *rye).* These are *exhaustive* crops because they are removed from the field and usually sold off the farm (i.e. they are cash crops). They encourage weeds— especially grass weeds such as couch. If grown continuously on the same field, fungous diseases such as take-all and eyespot, or pests such as eelworms can seriously reduce yields. Continuous spring barley crops are least likely to suffer losses. Cereals have peak demand for labour in autumn (ploughing and some sowing), spring (sowing) and late summer (harvesting). Large-scale mechanization has greatly simplified cereal production.

(b) *Potatoes* and *root crops* such as *sugar-beet, mangolds, carrots,* etc. These are mainly high-value cash crops and require deep soils. They have heavy demands for plant nutrients but allow the farmer to use large amounts of fertilizers and FYM and so build up fertility. Timely cultivations before sowing and during the early growth period can control most troublesome weeds—hence the reason for regarding this group as "cleaning" crops. This is expensive and chemical weed control is now widely used. It is very risky to grow any of these crops continuously—mainly because of eelworms. This group has a high labour demand—especially for harvesting of potatoes and singling of sugar-beet. However, mechanization has solved many of the problems.

(c) *Pulse crops,* e.g. *peas* and *beans.* In many ways these crops resemble cereals but they can build up nitrogen in the nodules on their roots. They should not be grown continuously because of build-up of fungous diseases (e.g. clover and bean stem rot) and pests (e.g. pea root eelworm). They can provide a break from continuous cereal growing.

(d) *Restorative crops,* i.e. the crops which are usually fed off on the fields and so return nutrients and organic matter to the soil.

The best examples are *leys, kale* and *roots for folding off.*

A good crop rotation would include several crops because this would:

(1) reduce the financial risk if one crop yielded or sold badly,
(2) spread the labour requirements more evenly over the year,
(3) reduce the risk of diseases and pests associated with single cropping (mono-culture),
(4) probably give better control of weeds,
(5) provide more interest for the farmer.

However, most of these objectives could be obtained without having a rigid system of cropping. The present tendency is to break away from traditional systems and to simplify the cropping programmes as much as possible.

This approach has been encouraged by:

(1) the need to economize on labour and capital expenditure,
(2) better machinery for growing and harvesting crops,
(3) much better control of pests and diseases— mainly by chemicals and resistant varieties,
(4) chemical weed control,
(5) guaranteed price systems for most crops.

Many well-tried rotations have been practised in various parts of the country. One of the earliest and best known was the *Norfolk Four-Course* rotation which was well suited to arable areas in eastern England.

It started as:

Turnips or *swedes.* Folded off with sheep in winter. Roots.
Spring barley (undersown). Cash crop. Cereal.
Red Clover. Grazed in spring and summer. Ley.
Winter wheat. Cash crop. Cereal.

This was a well-balanced rotation for:

(1) building up and maintenance of soil fertility,
(2) control of weeds and pests,
(3) employment of labour throughout the year,
(4) providing a reasonable profit.

However, considerable changes have occurred over the years mainly due to:

(1) the introduction of fertilizers, other crops and better machinery,

(2) greater freedom of cropping for tenant farmers,

(3) the need for increased profits.

Some of the *changes* which have occurred are:

(1) *Sugar-beet, potatoes, mangolds*, and *carrots* have replaced all or part of the folded roots.

(2) *Beans* and *peas* have replaced red clover in some areas or alternatively a 2- or 3-year *ley* has been introduced.

(3) Two or three successive cereal crops have replaced the barley and wheat crops.

An example of a wide variation is:

Winter wheat	⎱ Replacing winter
1 or 2 crops of spring barley	⎰ wheat.
Sugar-beet or potatoes	Root break.
1–3 cereal crops	Replacing spring barley.
2–4-year ley	Replacing 1-year red clover ley.

Where there is a big difference in the types of soil on a farm it may be advisable to have one rotation for the heavy soils and another for the light soils.

The most suitable rotation or cropping programme for a farm must be based on the management plan for the farm. It should provide grazing and other foods for the livestock and also the maximum possible area of cash crops. The cash crops grown will partly depend on the amount of labour available throughout the year.

CONTINUOUS CEREAL PRODUCTION AND "BREAK" CROPS

Continuous cereal production has become an increasingly common system of farming on a wide range of soil types and climatic conditions. In many cases the whole farm is producing cereals whilst on other farms only certain fields are continuously in corn.

The main reasons for this are:

(a) simplification, and relatively low labour and capital requirements compared with root crops and livestock enterprises;

(b) improved methods of harvesting, drying and storing grain;

(c) development of chemicals which give good weed control;

(d) development of fungicides to control diseases;

(e) reasonably profitable system.

Barley is the commonest cereal used for continuous cropping—mainly on loams and lighter types of soil—whilst continuous winter wheat has been increasing on heavy soils. Oat crops are not grown continuously because of their susceptibility to cyst (root) eelworm.

Many farmers are now justifiably worried about the long-term prospects of continuous cereal production because:

(1) Yields are not increasing sufficiently to maintain profits as costs rise, where low-cost streamlined systems are used (modern, intensive systems are better).

(2) Fungous diseases such as take-all and eyespot (see Table 54) and pests such as cyst (root) eelworm have been building up in the soil and seriously affecting yields in many cases. Take-all and cyst eelworms attack the roots of the cereal whereas eyespot attacks the stems a few centimetres above the soil. The general effect is to interfere with the normal absorption or translocation of water and nutrients, especially nitrogen and magnesium, with disastrous effects on the grain yield of individual plants. Also, they cause early ripening, or death, of the infected plants, and these are easily spotted in the crop as the season advances. After 4 or 5 years the damaging effects of take-all are much reduced.

(3) Grass weeds are not being adequately controlled and these reduce yields by competition and as hosts of take-all and eyespot.

(4) Leaf diseases such as *yellow rust* of wheat, *rhynchosporium, net blotch* and *brown rust* of barley, *crown rust* of oats and *mildew* of wheat, barley and oats can seriously reduce

yields by interfering with photosynthesis in the upper (flag) leaf and the ear where most of the carbohydrate in the grain is produced. These fungous diseases are spread by air-borne spores from neighbouring crops, from over-wintering crops or plants which have developed from shed grain at harvest time; they are increasing in importance due to the large increase in the area of cereals on all ploughable land, and carelessness in allowing self-sown plants to grow unhindered. Fungicides now available can control most of the cereal diseases.

Where there is a need to increase the profitability of a farm and adequate capital is available then high-value cash crops such as sugar-beet, potatoes or vining peas and beans may be introduced (if the soil depth and texture is suitable) or leys may be grown for stocking with dairy or beef cattle or sheep. In many cases adequate capital is not available for such enterprises and so combinable crops are grown which give a "break" from continuous corn but can be grown, harvested and dried with the cereal machinery, e.g. *beans, threshed peas, oil-seed rape, linseed, maize* and *seed crops of grass, clover* and *sugar-beet*.

A good "break" crop should provide at least as good an income as the cereal area it replaces: it should not allow weeds to spread; and, it should allow a crop of winter wheat to follow so that the farm income can be improved.

The crop of wheat following a "break" crop usually yields very well but the following one or two crops usually yield less than the average before the "break" crop and this raises doubts about the value of a "break" crop unless it gives a better profit than the cereal it replaces.

In a continuous barley system, eyespot is the most likely cause of trouble because barley is fairly resistant to take-all and cyst eelworm. Consequently, oats, which are resistant to eyespot, or an eyespot-resistant variety of wheat (see NIAB list) may be introduced occasionally—say once every 5 or 6 years—to increase the farm income because such oat or wheat crops usually yield over 5 tonnes/ha where the average barley yield is about 3.5 tonnes/ha.

Fungicides are now available for the control of eyespot.

CEREALS

The *cereal* (*corn, grain*) crops grown in this country are *wheat, barley, oats, rye* and *maize*; wheat and barley are the most important.

Cereal production has considerably expanded and improved with the introduction of better methods of sowing, combine-harvesters, driers, bulk handling and chemical aids such as herbicides, fungicides, insecticides and growth regulators. Modern intensive methods (see page 72) are not simple but can be very rewarding.

The cereals are easily recognized by their well-known grains (see Fig. 36a and b), ears or flowering heads (see Fig. 37a–e), and in the early leafy stages as shown in Fig. 38.

It is now common practice to use weight measures for seed rates (kg) and yields (tonne) of cereals; volume measures such as m^3 are used for bulk storage and hectolitre (100 litres) for specific weights.

Table 20 gives some average figures.

TABLE 20   (OF SOME HISTORICAL INTEREST)

	Wheat	Barley	Oats	Rye
Peck (2 gal) 9 litres	7 kg	6.25 kg	4.75 kg	6.25 kg
Bushel (4 pecks)	28 kg	25 kg	19 kg	25 kg
Sack or coomb (4 bushels)	112 kg	100 kg	75 kg	100 kg
Quarter (2 sacks)	224 kg	200 kg	150 kg	200 kg
Weight per m³	750 kg	700 kg	500 kg	700 kg
Cubic metres per tonne	1.25	1.4	2.0	1.4
Kilogrammes/hectolitres	75	70	50	70

The main facts about growing cereals are given for each separate crop later; the following apply to cereals in general.

Harvesting. Threshing is the separation of the grains from the ears and straw. In *wheat* and *rye* the chaff is easily removed from the grain. In *barley*, only the awns are removed from the grain—the husk remains firmly attached to the kernel. In *oats* each grain kernel is surrounded by a husk which is fairly easily removed by a rolling process—as in the production of oatmeal; the chaff enclosing the grains in each spikelet threshes off. Most cereal crops in all parts of the country are now harvested by combine. The grain is bulk handled on most of the larger farms and stored in silos or loose on barn floors. Storage in sacks is still used on some farms.

Methods used for *drying* are set out below with usual moisture extraction rates in brackets:

(1) Various types of *continuous-flow driers*: hot air takes out the excess moisture (6% per hour) and ambient air then cools the grain but this may not be sufficient cooling in very hot weather.

(2) *Batch driers*: drying similar to (1) but the grain is held in batches in special containers during the drying process (6% per hour in small types; 6% per day in silo types).

(3) *Ventilated silos or bins*: cold or slightly heated air is blown through the grain in the silo—this can be a slow process, especially in damp weather ($\frac{1}{3}$–1% per day).

(4) *Sack driers*: heated air is blown through the sacks of grain laid over holes in a platform or stacked to form a tunnel (1% per hour).

(5) *Floor drying*: a large volume of cold or slightly heated air is blown through the grain to remove excess moisture. The air may enter the grain in several ways, e.g.

(a) from ducts about a metre apart on or in the floor.

(b) from a single duct in the centre of a large heap (Rainthorpe system),

(c) through a perforated floor which may also be used to blow the grain to an outlet conveyor when emptying.

Floor drying is becoming increasingly popular because it can be done in a general-purpose building. It is a cheap method and requires very little labour when filling. Rate of drying $\frac{1}{3}$–1% per day.

Grain intended for the following purposes should not be heated above the following *maximum* temperatures:

For seed, malting barley up to 24% m.c.	49°C
For seed, malting barley above 24% m.c.	43°C
Milling for human consumption	
e.g. wheat flour	66°C
For stock feeding	82°C

Safe moisture contents for storage of all grains:

In bulk (e.g. silos, loose on floor)	for long period	14% or less
In bulk	up to one month	14–16%
In sacks	for long period	16–18%
In sacks	a few weeks	18–20%

Only fully ripe grain in a very dry period is likely to be harvested in this country at 14% moisture. In a wet season, the moisture content may be over 30% and the grain may have to be dried in two or three stages if a continuous-flow drier is used.

Damp grains, above the limits set out above, will heat and may become useless. This heating is mainly due to the growth of moulds and respiration of the grain. Moulds, beetles and weevils may damage grain which is stored at a high temperature, e.g. grain not cooled properly after drying; or grain from the combine on a very hot day. Ideally, the grain should be cooled to 18°C—this is difficult or impossible in hot weather.

Heating and destruction of dry grain in store several months after harvest may be caused by grain weevils and beetles. Insecticides can be used to fumigate silos and grain stores before harvest or applied to the grain when it is being stored. Special formulations of the insecticide *malathion* are commonly used for this purpose. (See page 183.)

An established method is the storage of damp grain, straight from the combine, in sealed silos. Fungi, grain respiration and insects use up the oxygen in the air spaces and give out carbon dioxide, and the activity ceases when the oxygen is used up. The grain dies but the feeding value does not deteriorate whilst it remains in the silo. This method is best for damp grain of 18–24% moisture, but grain up to 30% or more may also be stored in this way although it is more likely to cause trouble when removing it from the silo, e.g. "bridging" above an auger. The damp grain is taken out of the silo as required for feeding. This method cannot be used for seed corn, malting barley, or wheat for flour milling (see STL 33).

Another recent development is the storage of damp grain by cooling it. Chilled air is blown through the grain and the higher the moisture content of the grain the lower the temperature must be, e.g.

Moisture content of grain, %	16	18	20	22
Temperature of grain (approx,°C)	13	7	4	2

This method is cheaper than drying and the grain stores well, i.e. the germination is not affected and mould growth does not develop and so it can be used for seed, malting or milling. Because the moisture content is higher than in dried grain it is very suitable for rolling for cattle food (see STL 42).

Another method of storing damp grain safely and economically is by sterilizing it with a slightly volatile acid such as *propionic acid*. The acid is sprayed onto the grain from a special applicator as it passes into the auger conveying it to the storage heap; 5–9 litres per tonne of acid is required. Grain stored in this way is not suitable for milling for human consumption or for seed but it is very satisfactory for animal feeding and after rolling or crushing it remains in a fresh condition for a long time because the acid continues to have a preservative effect (see STL 100).

The latest development is the use of alkali which not only preserves damp grain but also increases its digestibility, and the grains can be fed whole to livestock.

Grain quality in cereals. Good-quality grain is clean, attractive in appearance, and free of mould growth and bad odours. Unsprouted, plump (well-filled) grains, with thin husk or skin, are desirable because they give a high extraction rate of the valuable endosperm. Good quality is indicated by high 1000 grain weight, and high specific weight (kg/hl)—provided there are not many small or broken grains which are wasteful because they are lost in the cleaning processes, old cracked grains go mouldy on the malting floor. Live pests such as weevils, beetles and mites must not be present. For long-term storage the moisture content should be about 14%—very dry grain of about 10% is not wanted because it has to be wetted before milling, rolling or malting.

Grain which is over-heated when being dried, or in storage, is spoiled for seed and malting because in these cases the germination should be well over 90%; over-heating also damages the protein which produces gluten in wheat flour.

A low nitrogen (protein) content is required for malting barley, but for other purposes a high protein content is preferable.

Wheat is the only cereal which contains a protein which produces gluten when mixed with water and so is suitable for bread-making. For seed purposes, the germination must be high (over 90%), the grains should be plump if they have to grow in adverse conditions, and should be as vigorous as possible.

Various quality standards may be required when selling grain—these are set out in Table 21.

Cereal straw is a problem on many farms. It can be baled to go for:

litter—less important nowadays where stock are kept in cubicles;

feeding—low-quality bulk food, unless treated with alkali to improve its digestibility but this is expensive;

industrial uses, e.g. paper, boards, insulating materials, fuel, etc.

Straw can be useful for improving poor soil structures if it is chopped and/or spread and mixed into the soil as soon as possible after harvest.

TABLE 21   GRAIN-QUALITY STANDARDS—A SUMMARY COMPARISON
Minimum quantity for intervention buying in the U.K. 100 tonnes (wheat and barley)

	Wheat				Barley
	Intervention		HGGA classification scheme	Denaturing quality used as a trade guide	Intervention
	Bread	Feed			
Moisture content (maximum %)	15	15	16	18	15
Specific weight (minimum kg/hl)	72	68	72	70	63
Misc. impurities (maximum %)	3	3	2.5	3	3
Sprouted grains (maximum %)	6	8	not applicable	8	8
Total impurities (maximum %) of which:	10	12			12
Broken grains (max. %)	5	5			5
Grain impurities, do.	5	12			12

Note: the above may be changed from time to time.
See also the Home Grown Cereals Authority publications.

Burning is the cheapest and simplest method of disposal of straw and a good burn helps to control diseases such as net blotch, leaf-blotch, septoria and eyespot; if this is done carelessly, serious damage can be caused to other crops, hedges, trees, buildings and wild life. The NFU Code of Practice for straw-burning should be strictly followed. The banning of straw-burning could lead to greater damage, because those who are determined to burn their straw will "accidentally" drop matches and there will be no indication of safety precautions having been taken.

(See also the ADAS reports on Straw Utilization conferences, and Staniforth *"Straw for fuel, feed and fertilizer"*. Farming Press Ltd.)

Cereal grains

Wheat, Rye, and *Maize* grains consist of the seed enclosed in a fruit coat—the pericarp—and are referred to as "naked" caryopses (kernels).

In *Barley, Oats,* and *Rice,* the kernels are enclosed in husks formed by the fusing of the glumes (palea and lemma) and are referred to as "covered" caryopses (kernels).

The following enlarged drawings show the relative sizes and shapes of the six main cereals:

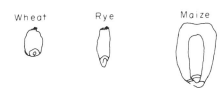

FIG. 36a. "Naked" kernels.

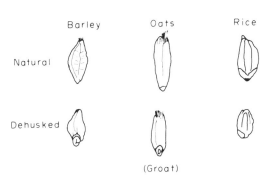

FIG. 36b. "Covered" kernels.

Cereal ears

See drawings of ears (heads) of wheat, barley, oats, and rye.

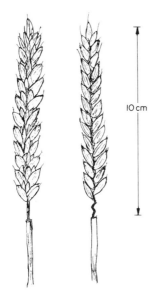

FIG. 37a. Wheat spikelets alternate on opposite sides of the rachis. 1–5 grains develop in each spikelet. A few varieties have long awns.

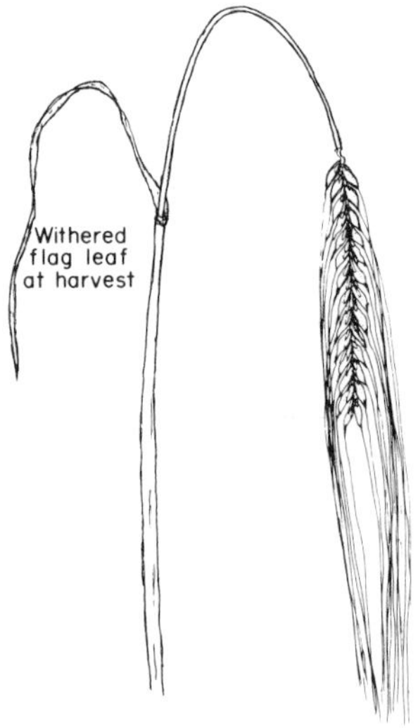

FIG. 37c. 2-row Barley heads hang down when ripe; each grain has a long awn; the small infertile flowers are found on each side of the grains.

FIG. 37b. 6-row Barley, all three flowers on each spikelet are fertile. Awns are attached to the grains.

FIG. 37d. Grain easily seen in the spikelets of rye.

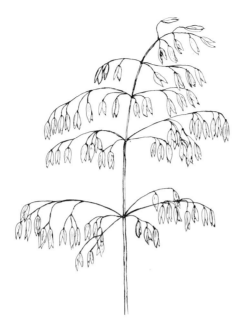

Fig. 37e. Oats.

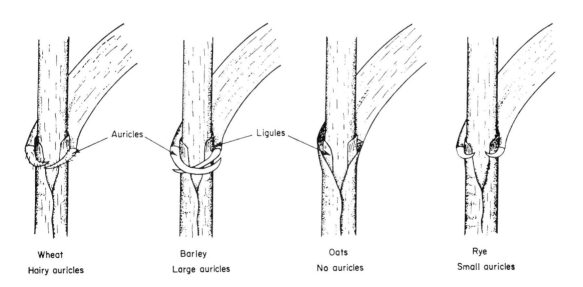

| Wheat | Barley | Oats | Rye |
| Hairy auricles | Large auricles | No auricles | Small auricles |

Fig. 38. Diagrams showing method of recognizing cereals in the leafy (vegetative) stage.

CEREAL YIELDS (ABOVE THE AVERAGE)

The high cereal yields obtained with systems—such as those of: Professor Laloux with winter wheat in Belgium, the Schleswig-Holstein method of growing winter wheat in northern Germany, Peter Lippiatt's methods of producing very high yields of winter and spring barley and winter wheat on 340 hectares of Cotswold land on a continuous cereal system, and many well-tried consultancy methods of producing very high cereal yields on the farms supervised—have stimulated many farmers to look carefully at the growth factors which are important in determining yield, and also to consider how they might apply the principles involved on their own farms: assuming drainage, soil structure and pH are satisfactory, this usually means modifications in seed-bed preparations, seed rates and drilling methods, fertilizer rates and times of application, and the correct usage of many of the herbicides, fungicides, insecticides, growth regulators, trace elements, etc., which are now available.

It is now common practice to spray cereal crops many times during the growing season and this is best achieved by following the same wheelways each time—these "tramlines" can be introduced when drilling by many simple and some complicated methods, or, by repeatedly using the wheel marks made by the first spraying. The loss of yield caused by wheelways is small and is more than compensated-for by the ease and accuracy of spraying—provided the lines are properly positioned and carefully followed each time (they also allow motorized crop inspection!). (See STL 189.)

The actual yield of a cereal crop is determined by the contributions made by the three components of yield:

> number of ears per hectare,
> number of grains per ear, and
> weight (size) of the grains.

These components are interrelated; for example, by increasing the number of ears (e.g. by denser plant populations or more tillering), the number of grains per ear may be reduced and, also, the size of the grains. There are many different opinions on what the ideal numbers or size of each component should be, and, of course, it will vary according to the type and variety of cereal, the soil and climatic conditions, and the possible occurrence of uncontrollable diseases and pests.

The main features of the systems mentioned above are as follows:

Professor Laloux's system for growing winter wheat is one which can be followed fairly easily by the ordinary farmer and does not aim at maximum yields. Assuming good soil structure, removal of previous crop residues, and effective early weed control, the objective is an established plant population of about $200/m^2$, and about 400–500 ears/m^2. Phosphate and potash applications must be adequate and will vary according to the level of soil fertility; similarly for nitrogen, but an important part of the system is the application of nitrogen at precise growth stages. A rational use of fungicides is recommended.

Full details of the Laloux system are set out in a booklet published by the Kenneth Wilson Group.

Schleswig-Holstein system—this rich farming area in north Germany has been producing very high average yields of winter wheat for many years (8–10 tonnes/ha on many farms), mainly because the crop is very carefully managed at all stages. High seed rates (200–220 kg/ha) are used hopefully to produce about 750 ears/m^2: high phosphate and potash levels are maintained by heavy annual dressings. Lime, magnesium and trace elements are used as required. Nitrogen (about 160 kg/ha) is applied as split dressings of 80 in early spring, 60 at ear emergence and the remainder at an intermediate stage. "Cycocel" is applied twice for improved straw strength; the straw is chopped and ploughed-in (burning is banned).

Conventional cultivations and drilling are carried out but are very carefully done and "tramlines" are used on most farms.

Every crop is visited every day in the growing season, and very considerable attention paid to controlling diseases by spraying.

The wheat crops are usually grown in a simple rotation with winter barley and oil-seed rape.

Peter Lippiatt's system—an average overall yield of over 7 tonnes/ha is obtained from 340 hectares of cereals grown on a continuous system for up to 30 years (about 100 winter barley, 180 spring barley and 60 winter wheat). The keynote to this success is very careful attention to detail at every stage of production, and especially to really effective preventative measures to control diseases and pests (every field is visited at least twice a week and immediate action taken if diseases or pests are found). The straw is usually burnt to help control diseases and conventional cultivations (well done) are used on this brashy soil. The seed is sown very carefully with a narrow-spaced (12 cm) grain drill, and seed rates (kg/ha) are above average—winter wheat (200), winter barley (180), spring barley (150), with the object of obtaining over 750 wheat ears/m^2, and over 900 barley ears/m^2. Phosphate and potash levels are kept high by generous annual dressings. Nitrogen (kg/ha) is applied at intervals as required, so that the crops are never short of N; up to 180 for winter cereals and 140 for spring barley.

Chlormequat is applied twice to the wheat to strengthen the straw and also to improve the action of fungicides on eyespot and foot-rot diseases.

Spraying costs are high, but bonus payments are obtained for seed crops and malting barley (Golden Promise).

The increasing complexities of modern cereal production has created a need for specialized advice and demonstrations. The Cereal Unit (at the National Agricultural Centre), established in 1977, has been giving a lead by organizing national demonstrations, conferences and courses, publications, and by various CDIU groups (Cereal Development and Information Unit), e.g. the Cotswold group which started by developing modern methods for growing winter barley in conjunction with D. M. Barling's studies of ear development at the Royal Agricultural College. This was followed by studies on winter and spring wheats. The Cotswold Cereal Centre (established in 1979) is sponsored by about 200 farmers and 20 trade organizations to study current problems of cereal production in detail and to pass on the results and advice to members as soon as possible. Free advice and publications are available to farmers from ADAS and various chemical firms, and an increasing number of paid consultants provide on-farm specialist advice on all aspects of cereal production.

Terms such as "blueprint", "planned approaches", and "integrated systems" are sometimes used to describe the present increasing attempts to maximize cereal profits.

Accurate spacing and uniform depth of planting of seed in well prepared seed-beds are very important if maximum yields are to be obtained. This usually involves doing seed counts/kg and setting the drill carefully: narrow (10–12 cm) rows are preferable. Broadcasting seed can be successful provided the seed is uniformly distributed and covered. Stony soils and wet cloddy conditions give a crop a poor start.

Seed rates should be chosen with the object of establishing a desirable plant population. The factors to be taken into account are:

(a) seed size, e.g. increase rate for large seed (low seed counts/kg) and decrease the rates for small seed (high seed counts/kg);
(b) tillering capacity—some varieties tiller more freely than others, and so their seed rate may be reduced;
(c) seed-bed conditions—the seed rate should be increased in cloddy and stony conditions because of the poor conditions for seedling growth; vigorous seed establishes better in poor conditions;
(d) time of sowing—for late autumn sowings and very early spring sowings the seed rate should be increased;
(e) the possibility of seedling losses by pests.

The most desirable plant population to aim for is somewhat debatable and must be related to the potential yielding capacity of the field; in general, it is better to go for a high population than to depend too much on tillering to make up for a rather low plant establishment and the risk of uneven ripening.

Varieties. There are very many cereal varieties now on the market and new ones are introduced every year. However, there are only a few outstanding varieties of each cereal and these are described in the annual *Recommended Lists* from the National Institute of Agricultural Botany (NIAB) for England and Wales, and from the Agricultural Colleges in Scotland and the Department of Agriculture for Northern Ireland.

Seed-dressing. Cereal seed should be dressed with an organo-mercurial or other seed-dressing to control fungous diseases such as smuts and leaf-stripe; seed treatment chemicals such as "Baytan" will control a wide range of cereal diseases; insecticides, e.g. gamma HCH, may also be included to control wheat bulb fly and wireworm. (See Tables 53 and 54 and leaflet 816.)

Fertilizers. These are usually combine-drilled with the seed, although the star-wheel mechanism used on most drills gives a very uneven distribution of fertilizer. A recent trend is to broadcast the fertilizer and use narrow row widths (100–125 mm) when sowing the seed; this reduces the labour costs and gives good results on fertile soils. All the fertilizer could be applied in the seed-bed with spring crops. With winter crops most of the nitrogen is top-dressed in the spring. Nitrogen rates are the most difficult to decide on—especially after grassland and where organic manures have been used. Approximately 10 kg of P_2O_5 and K_2O are removed by each tonne of cereal harvested, e.g. 7 t/ha wheat removes 70 kg P_2O_5 per hectare.

Average fertilizer requirements;

Phosphate (P_2O_5) 55–75 kg/ha.
Potash (K_2O) 55–75 kg/ha.
Nitrogen (N) Autumn 0–25 kg/ha.
    Spring—in dry, arable areas 75–100 kg/ha.
    Spring—high rainfall and very fertile areas
      25–60 kg/ha.

Weed Control—see Chapter 6, page 166.

Wheat

The present production of wheat in the United Kingdom is about 8.5 million tonnes per year but less than half of this is suitable for flour production because the varieties are not suitable, the weather spoils some, and it is not always marketed in suitable quantities. Of the remainder, some is used for seed, breakfast foods and malting but most of it goes into compound feeding-stuffs (wheat should not make up a high proportion of a stock ration because of its pasty nature). In future, higher premiums may encourage more flour-quality wheats to be grown so that a higher proportion of home-grown flour can be included in our bread flour.

In the production and use of flour, quality requirements can be divided into two groups: *milling* and *baking* qualities (see *Farmers Guide to Wheat Quality*, Cereal Unit, NAC).

Milling quality. This refers to the ease of separation of white flour from the germ and bran (pericarp and outer layers of the seed); it is a varietal characteristic and can be improved by breeding. In the milling process, the grain passes between fluted rollers which expose the endosperm and scrape off the bran before sieving. The endosperm is ground into flour by smooth rolls and in this process the good-quality endosperms (hard wheats) break along the cell walls into smooth-faced particles which slide over each other without difficulty, and so can be easily sieved, but the poor-quality endosperms (soft wheats) produce broken cells with jagged edges which cling together in clumps and so make sieving slow and difficult, also, some of the cell contents are lost. "Milling value" is a measure of the yield, grade and colour of flour obtained from sound wheat.

Baking quality. Wheat is the only cereal which produces flour which is suitable for *bread-making* because the dough produced has elastic properties; this is due to the gluten (hydrated insoluble protein) present. The amount and quality of the gluten is a varietal characteristic but can also be affected by soil fertility and climatic conditions.

Bread-making quality. To provide large, soft,

finely textured loaves, the baker requires flour (and dough) with a large amount of good-quality gluten (at least 11.6% on dry matter basis in the grain). This is produced by strong-textured wheats. A good dough will produce a loaf of about twice its volume. The small amounts of alpha-amylase enzyme present in sound wheat is desirable for changing some starch to sugars for feeding the yeast used in bread-making, but wet and germinated wheat contains excessive amounts of alpha-amylase and so excessive amounts of sugars and dextrines are produced, resulting in loaves with very sticky texture and dark-brown crusts (alpha-amylase activity can be measured by the Hagberg falling number test). Mini bread-making tests are now being used to evaluate wheat samples.

The protein in overheated (over 60°C) wheat grain is spoiled (denatured) for bread-making.

The amount of water-absorption by flour is also important in bread-making (it is a varietal characteristic and depends on the protein content and the amount of starch granules damaged by milling); high water uptake—by the damaged granules of hard wheats—means more loaves, which keep fresh longer, from each sack of flour.

Biscuit quality. A very elastic type of gluten is not required for biscuits which should remain about the same size as the dough; *weak* textured wheat varieties are suitable for biscuit-making flour.

The Flour Milling and Baking Research Association produces tables showing four quality grades for each of the following varietal characteristics: milling value, bread-making value, biscuit value, and water-absorption.

The following are some examples of varietal quality:

	Winter	Spring
Hard and Strong varieties:	*Avalon*	*Sicco, Timno*
Hard and Weak varieties:	*Hustler, Huntsman, Virtue*	
Soft and Strong varieties:	*Flanders*	
Soft and Weak varieties:	*Aquila, Fenman, Guardian, Norman, Rapier*	

The poorer-quality grain and the by-products from white flour production, i.e. bran (skin of grain) and various inseparable mixtures of bran and flour (e.g. weatings), are fed to pigs, poultry and other stock.

Wheat straw is used mainly for bedding (litter), but some varieties produce good-quality thatching straw which is very valuable; some is still used for covering potato clamps.

Yield (tonnes/ha)	Average	Good	Excellent
Grain	4	6	12
Straw	3	4	4

Soils and climate (pH should be higher than 5.5). Wheat is a deep-rooted plant which grows well on rich and heavy soils and in the sunnier eastern and southern parts of this country. Winter wheat can withstand most of the frosty

TABLE 22   FLOUR REQUIREMENTS IN THE U.K. AND SOURCES OF SUPPLY (MILLION TONNES)

Type of flour and quantity	Wheat equivalent	Homegrown	Imported Strong	Weak
Bread	2.44	3.3	1.5	1.3	0.5
Biscuit	0.41	0.55	0.5	—	0.05
Household	0.83	1.15	0.4	0.2	0.55
Total	3.68	5.00	2.4	1.5	1.10

conditions of this country but is easily killed by waterlogged soil conditions.

Place in rotation. Wheat is the best cereal to grow when the soil fertility is high because of its resistance to lodging and its high yields and good prices. It is commonly taken for 1 or 2 years after grassland, potatoes, beans or oilseed rape but it is now being grown continuously on many farms because eyespot can now be controlled by fungicides and after 4 or 5 years the so-called "take-all barrier" is passed and yields remain fairly constant.

Seed-beds. A fairly rough autumn seed-bed prevents "soil-capping" in a mild, wet winter and protects the base of the plants from cold frosty winds. In a difficult autumn, winter wheat may be successfully planted in a wet sticky seed-bed and usually produces a good crop. Spring wheat should only be planted in a good seed-bed. Soil-acting herbicides require fine seed-bed conditions.

Time of sowing. Winter wheat: late September–early February (early sowing preferable). Spring wheat: February–early May (March best).

Methods of sowing—depth should be about 3–5 cm.

(1) Drilling (a) combine (seed and fertilizer) drill 15–18 cm.
    (b) seed drills 10–18 cm (10 cm preferable).
(2) Broadcasting, e.g. by hand, fiddle, aeroplane, fertilizer spinner.

Seed-rate:
Winter wheat 150–250 kg/ha.
Spring wheat 170–220 kg/ha.
} Sow at least 400 seeds/m² to obtain at least 300 plants/m²

Varieties of wheat. Winter wheats usually yield better than spring varieties. Most of the modern varieties have red (brown) grain; the white (cream) grained varieties sprout too readily in a damp season. Most winter wheats must pass through a period of low temperatures and short days if they are to yield well—hence they should

only be sown in the spring in exceptional circumstances (see NIAB list).

Recommended varieties:

	Winter	Spring
Good-quality milling and bread	Avalon	Sicco, Timno
Biscuit quality	Brigand, Fenman, Guardian	—
Others	Hustler, Virtue	—

See also NIAB recommended list.

Breeding work is now going on in this country to improve yields of wheat by producing dwarf types and special hybrids which are suitable for British conditions: types having nodular nitrogen-fixing bacteria (like legumes) may be produced by modern plant breeding methods—in the distant future.

TABLE 23   FERTILIZERS

	kg/ha		
	N	P_2O_5	K_2O
Winter wheat—autumn	0–25	50–90	50–90
Winter wheat—spring	50–200	—	—
Spring wheat	50–150	50–60	50–60
With 30 t/ha FYM or equiv. slurry	30–50	—	—

The lower figures for nitrogen shown in Table 23 would be adequate if the wheat were taken after grassland containing clovers or lucerne.

Spring grazing of winter wheat. If the crop is well forward in the spring and the soil is dry, it can provide useful grazing for sheep or cattle about late March. It should only be grazed once and should be top-dressed with nitrogen afterwards unless the reason for grazing was to reduce the risk of lodging. Yields are likely to be reduced by grazing.

Harvesting. Winter wheat ripens before spring wheat; the crop is harvested in August and September. Indications of ripeness for harvesting:

(a) *Binder* (mainly of historical interest) Straw: yellowish, all greeness gone.
Grain: in cheesy condition, firm but not hard.

(b) *Combine* (7–10 days later) Straw: turning whitish; nodes shrivelled.
Grain: easily rubbed from ears, hard and dry.

Tillering of wheat (and other cereals). The production of side shoots (i.e. tillering) is a very important characteristic of cereals. Where the plants are thinly spaced more side shoots are likely to be produced than where they are close together, and so a crop which is uneven in the seedling stage can even up considerably before harvest. All side shoots do not produce ears.

Lodging in wheat (and other cereals). Lodging (laying flat) is usually caused by wind and rain. It is mostly likely to occur when:

(a) the field is in an exposed situation;
(b) the variety is weak-strawed;
(c) excessive amounts of nitrogen are present in the soil, resulting in long weak straw and delayed ripening—this is likely to occur after a clovery ley or where too much nitrogen-fertilizer has been applied. A wet season makes matters worse;
(d) the straw is elongated and weakened due to shading when the plants are too dense or shaded by trees;
(e) the crop is attacked by the eyespot fungus or foot-rot.

If the crop lodges when the straw is still green and growing the stems can bend at the nodes and grow upright again, but yield will be reduced.

Lodging near maturity may not affect yield provided the pick-up reel mechanism on the combine is set properly when harvesting. It does, however, increase the cost of harvesting and grain quality may be spoiled.

The growth regulating (anti-gibberelin) chemical chlormequat (CCC, Cycocel), is sprayed on wheat crops to reduce the risk of lodging, and consequent yield loss, by strengthening and shortening the straw (shorter internodes and thicker stem walls). This treatment allows optimum rates of nitrogen to be used; eyespot attacks are not so severe, but the shorter, denser crops may be attacked earlier by leaf diseases, such as yellow rust, unless controlled by fungicides.

Barley

Barley is a very important arable crop in the U.K. (30% of arable area); the grain is used mainly for: *feeding*—to pigs (ground), cows and intensively fed beef (rolled)— and *malting* (about 2 M tonnes; the best-quality grain is usually sold for malting, i.e. plump, sound grains, high germination (about 90%), but low nitrogen (protein) % (1.3–1.6 N); some varieties are more suitable than others for malting (see table, page 78 and NIAB list).

In the malting process the grain is soaked, then sprouted on a floor to produce an enzyme (diastase, maltase), dried on a kiln and the rootlets removed to leave the malt grains.

In the brewing process, the malt is crushed (grist), soaked in warm water in mash tuns where the enzyme changes the starch into sugars which then dissolve in the liquor before being drained off. The remainder of the malt (brewers grains) is a valuable cattle food.

The sugary liquor (wort) is boiled with hops to give it a bitter flavour and keeping quality, and to destroy the enzymes. The strained "hopped" wort is then fermented with yeast which converts the sugars to alcohol and so produces beer.

To produce whiskey, hops are not added, and the fermented wort is distilled to produce a more concentrated alcoholic liquid called malt whiskey.

Some malt is used to produce malt vinegar.

A substantial premium, which varies from year to year, is paid for good malting barley.

Barley straw is mainly used for litter, but some is fed to cattle.

Yield (tonnes/ha)	Average	Good	Excellent
Grain	4	5.5	9
Straw	3	4	5

Soils and climate. Barley can be grown on arable land throughout the U.K., provided the pH of the soil is about 6.5; it is a shallow-rooted crop which grows better than other cereals on thin chalk and limestone soils, but will grow on a wide range of soils provided they are well drained; on organic and very fertile soils, barley is likely to lodge—especially in a wet season—and the grain is not likely to be of malting quality.

Place in rotation. Barley is usually taken at a stage when the fertility is not very high (wheat taken then), although modern varieties stand very well. On many farms it is grown continuously on the same fields and produces reasonable yields.

Seed-beds. Similar or finer than wheat; best to sow shallow (3–5 cm).

Time of sowing. Winter barley: early September (e.g. Scotland) to early November (e.g. South Devon), aim to have plants well-established before winter. Spring barley: February–early April (early sowing best).

Methods of sowing. Same as for wheat.

Seed-rates (kg/ha): winter barley 150–180, spring barley 125–180, aim for 300 plants established/m² (sow 300–400 seeds/m²).

Winter versus spring barley. Modern two-row varieties of winter barley are frost-hardy, of reasonably good quality, and yield better than spring barley—especially in areas which suffer from drought in summer. To obtain best results with winter barley, the crop must be sown early enough to develop a good root system and become well-tillered before winter, and protected (if necessary) from leaf diseases, e.g. mildew and rhynchosporium, using fungicides; an aphicide e.g. "Ambush", may be required

where BYDV is expected. Autumn weed control is also desirable for both annual broad-leaved and grass weeds using suitable herbicides. Long runs of winter cereals tend to encourage grass weeds such a blackgrass and sterile brome, and control can be very expensive. The very early ripening time of winter barley (a) makes it easier to control perennial weeds by pre-harvest "Roundup" spray—especially "onion" couch which senesces earlier than ordinary couch, and (b) allows for earlier establishment of leys, stubble turnips and oilseed rape. The fungicides now available make it possible to grow good crops of spring barley in areas where drought is not a problem.

Popular varieties:

	Winter	Spring
Malting quality	*Maris Otter, Tipper*	*Triumph, Kym, Atem, Golden Promise*
Feeding quality (2-row) (6-row)	*Igri, Sonja, Video Gerbel, Athene, Pirate*	*Goldmaker, Patty* —
Cyst-eelworm resistant	—	*Tyra*

See also NIAB recommended list.

Fertilizers. General requirements are similar to wheat except that the amount of nitrogen applied in spring to winter barley should be 60–90 kg/ha and to malting barley 40–70 kg/ha. Up to 150 kg/ha of nitrogen are required for continuous barley crops. FYM is seldom applied before planting barley but many crops would benefit from it or slurry.

Spring grazing of winter barley. It can be grazed like winter wheat but this is seldom done.

Harvesting. Winter barley ready July–early August. Spring barley ready August–early September.

Barley is ready for harvesting by binder or combine when the straw has turned whitish and the ears are hanging downwards parallel to the straw. In a crop containing late tillers, harvesting should start when most of the crop is ready.

Malting crops are usually left to become as dead ripe as possible before harvesting. Harvesting of feeding barley is often started before the ideal stage—especially if a large area has to be harvested and the weather is uncertain. The ears of over-ripe crops and some varieties break off easily; this can result in serious losses.

The pick-up reel now fitted to most combines is very useful for picking up laid crops.

Oats

Oats are mainly used for feeding to livestock—they are particularly good for horses and are also valuable for cattle and sheep but are not very suitable for pigs because of their high husk (fibre) content. The best-quality oats may be sold for making oatmeal which is used for breadmaking, oatcakes, porridge, breakfast foods and for feeding chickens.

Barley has largely replaced oats in cattle rations in recent years and the area in oats has been declining for many years. Recently there has been some renewed interest in the crop as a possible "break" from continuous barley because oats are resistant to several of the diseases which affect barley—especially eyespot and some types of take-all. There is also a renewed interest in the "naked" oats species which has a very high feeding value (better than maize) because the husk threshes off the grain. The varieties of this species are very low yielding for various reasons such as being more susceptible to disease, lodging and shedding before harvest. A new, higher-yielding variety (Nuprime) has been bred in France and breeding work has now started in this country to produce better varieties for grain production for possible replacement of imported maize.

Oat straw is very variable in quality. The best quality from leafy varieties is similar in feeding value to medium-quality hay. Some short, stiff-strawed varieties are no better than barley straw when grown in the warmer, drier parts of this country.

Yield	Average	Good
Grain	4 tonnes/ha	5 tonnes/ha
Straw	3 tonnes/ha	5 tonnes/ha

Soils and climate. pH should be about 5 or over—if too much lime is present, manganese deficiency (grey-leaf) may reduce yields.

Oats do best in the cooler and wetter northern and western parts of this country, but even in these areas they have been replaced by barley on many farms. They will grow on most types of soil and can withstand moderately acid conditions where wheat and barley would fail.

Place in rotation. Oats can be taken at almost any stage in a rotation of crops. If grown too often cyst eelworms may cause a crop failure.

Seed-bed and *methods of sowing:* similar to wheat and barley.

Time of sowing: winter oats, late September–October; *spring oats,* February–March.

Seed rate: 190–250 kg/ha.

Varieties. Winter varieties are not so frost-hardy as winter wheat or barley; they usually yield better—especially in the drier districts—and are less likely to be damaged by *frit fly* than spring oats.

Popular varieties:

Winter:	*Pennal, Peniarth* Eelworm resistant: *Panema*
Spring:	Mainly grain: *Leanda, Cabana, Pinto, Perona* Grain and straw: *Karin, Ayr Commando, Maelor* Cyst eelworm resistant: *Trafalgar*

See also recommended lists.

Fertilizers: similar to wheat, but less nitrogen.

$$\text{N } 50\text{–}90 \text{ kg/ha} \qquad P_2O_5 \ 50\text{–}60 \text{ kg/ha}$$
$$K_2O \ 50\text{–}60 \text{ kg/ha}$$

Spring grazing. Similar to wheat; may be desirable if there is a risk of lodging because the grazing results in shorter straw.

Harvesting. There is still a small proportion of the oat crop cut with the binder—usually when the straw is still green or just turning yellow; this early cutting gives better-quality straw and there is less shedding of grain.

If combined, the crop must be left until it is

fully ripe; there is then a serious risk of shedding by high winds or if bad weather holds up the work. Swathing reduces this risk.

Rye

Rye is grown on a small scale in this country for grain or very early grazing. The grain is used mainly for making rye crispbread; it is not in demand for feeding to livestock.

The long, tough straw is very good for thatching and bedding but is no good for feeding.

	Average	Good
Grain	3 tonnes/ha	4 tonnes/ha
Straw	4 tonnes/ha	5 tonnes/ha

Soils and climate. Rye will grow on poor, light acid soils and in dry districts where other cereals would fail. It is mainly grown in such conditions for grain because, on good soils, although the yields may be higher, it does not yield or sell so well as other cereals.

Rye is extremely frost-hardy and will withstand much colder conditions than the other cereals.

Place in rotation. Rye can replace cereals in a rotation—especially where the fertility is not too high. It can be grown continuously on poor soils with occasional break of carrots, sugar-beet or leys. Grazing rye is usually taken as a catch crop before kale or roots.

Seed-beds and methods of sowing: similar to wheat.

Time of sowing: winter, September–October; *spring,* February–March. Grazing rye should, if possible, be sown in late August or early September.

Seed rate: for grain, 190 kg/ha. Early sowing for grazing 125–190 kg/ha.

Varieties. Most of the rye varieties grown now are winter types. Rye, unlike the other cereals, is cross-fertilized so varieties are difficult to maintain true to type and new seed should be bought in each year.

Popular varieties:

| Winter (grain): *Amino, Ashill Pearl, Dominant* |
| Grazing: *Lovaszpatonia, Greenfold, Rheidol* |

Fertilizers. On light soils, for grain.

N 50–75 kg/ha
P_2O_5 40–50 kg/ha
K_2O 40–50 kg/ha

Same amounts used for grazing rye on better soils. The nitrogen applied as a top-dressing in February.

Spring grazing. The special varieties for early spring grazing (late February–March) can be grazed at least twice if the grazing is started before the plants develop hollow stems (see page 143). Once-grazed crops can be left to harvest as grain which is usually sold for seed.

Harvesting. Rye is normally the first of the cereal crops to ripen. It is cut with the binder or combine near the dead ripe stage when the grain is hard and dry and the straw turning from a greyish to whitish appearance. The ears sprout very readily in a wet harvest season. If the crop is grown for its high-quality straw it should be cut with the binder before the grain develops and so the straw is not damaged by threshing.

Mixed corn crops

Mixtures of cereals (*dredge corn*) are grown in some areas—particularly in the south-west. The commonest type is a mixture of barley and oats in various proportions with a total seed rate of about 220 kg/ha. The yield of grain is usually better than if either crop were grown alone. Varieties must be chosen which ripen at the same time.

Sometimes cereals and peas or beans are mixed (*mashlum*)—this type can be used for silage or grain.

Winter and spring mixtures are used.

Growing and harvesting: grain—similar to oats; silage—forage harvester.

Triticale

This is a new crop produced by crossing a tetraploid wheat with rye and treating seedlings of the sterile F1 plants obtained so that their chromosome number is doubled and they become reasonably fertile. It is intermediate between wheat and rye in most of its characteristics. A lot of enthusiasm has been shown for it by a few people, but in spite of the claims made for it, triticale has not been a success as a grain crop so far. Some people are tempted to buy it in small packets at ridiculous prices for germinating and eating as a health food. It has a possible future as a feeding grain to replace wheat in livestock areas, and on light soils where droughts occur. Some new varieties are promising.

Maize

Maize was introduced into this country about 200 years ago as a forage crop, but never really became established until recently with the introduction of much more suitable varieties, improved cultural and harvesting methods, and much enthusiasm.

It is a tall annual grass plant with a strong, solid stem carrying large narrow leaves. The male flowers are produced on a tassel at the top of the plant and the female some distance away on one or more spikes in the axils of the leaves. (This separation simplifies the production of hybrid seed.) After wind pollination of the filament-like styles (silks) the grain develops in rows on the female spike (cob) to produce the maize ear in its surrounding husk leaves.

The main uses for the crop in this country are for making *silage* (cut when the grain on the cob is still soft or cheesy) or for harvesting as ripened grain; some is grazed or cut and fed as a forage crop, and there is a limited, but profitable, market for "corn-on-the-cob" or sweet corn as a vegetable—this latter is a special type of maize in which some of the sugar produced is not converted into starch and is harvested when the grain is in the milky stage. (Leaflet 297.)

Climatic requirements. Maize growing in Britain is limited because it will not grow until the temperature is above 10°C—most other crops and weeds start growth at 4–5°C. Good yields are very much dependent on plenty of sunshine; exposed (windy) situations are not suitable. Maize for grain can only be grown successfully in lowland areas (below 120 m) in the south-eastern counties; for silage, the area can be extended to sheltered parts of the midland and south-western counties. Wet autumns make grain harvesting very difficult.

Suitable soils. Ideally, a rich, deep, well-drained loam is best. Light soils are reasonable if they do not dry out. Thin chalk soils should be avoided and heavy soils are usually very slow to warm up in spring. A soil pH of 6.0 or over is desirable.

Place in rotation. When maize became established as a forage and/or grain crop about 10 years ago, it was considered possible to grow it continuously on convenient fields without any problems. It is now realized that it can carry take-all on its roots, so it is not a good "break" crop, and Fusarium stalk-rot and smut are more troublesome in fields where maize is grown in close succession. However, it is not affected by cereal cyst eelworm, eyespot and the usual leaf diseases of other cereal crops. When grown as a grain crop, it may be harvested too late to allow winter wheat to be planted.

Seed-bed. The seed should be planted about 4–6 cm deep in a level, moist, friable seed-bed. This can usually be produced by good ploughing, about 20–25 cm deep, in autumn, leaving a fairly level surface for frost action and avoiding deep working in spring which might produce a cloddy tilth. Soil pans should be broken up by subsoiling.

Manuring. Yields can be improved by applying FYM or slurry, if available.
Nitrogen—up to 125 kg/ha applied in seed-bed. Phosphate and potash—about 60 kg/ha P_2O_5 and K_2O ploughed in during the autumn, plus an extra 50–60 kg/ha in the seed-bed where one or both nutrients are deficient.

Sowing. Maize should be drilled at the end of April or early May with a precision drill fitted with the correct belt, cell wheel or plate to ensure an even and correct distribution of undamaged seed. A row width of 75 cm is satisfactory for most conditions. A 20-row headland should be used to facilitate harvesting.

TABLE 24 Plant populations

	Grain	Silage
Plants per acre	35,000	45,000
Plants per hectare	90,000	110,000
Distance between plants in 75-cm rows	15 cm	11 cm
Seed rates in kg/ha	24–30	30–40

Varieties. Most of the hybrid varieties grown in Britain are Flint × Dent crosses. Some examples of these are shown in Table 25.

TABLE 25

	Grain	Silage	Sweet corn
Early	Dekalb 202 LG 11	LG 11 Brutus, Passat	John Innes Hybrid Kelvedon Sweetheart
Late	Anjou 196		October Gold

See NIAB recommended list.

Pests and diseases. Seedling blight—use thiram seed dressing.

Foot-rot or stalk-rot—choose resistant varieties.

Frit fly and wireworm—use phorate granules in seedbed.

Leatherjackets and wireworm—use HCH.

Birds (rooks, pigeons and pheasants) can cause very serious damage at the emergence stage—especially on small areas; various controls can be used as dawn and dusk patrols, bangers, dead cats, and black nylon thread about waist height between canes or sticks at 10–40-metre spacings.

Weed control. Weeds can ruin a maize crop because of their faster growth rate in May and June. However, most annual broad-leaved weeds and blackgrass can be controlled by the herbicide atrazine applied at planting time; much heavier doses applied in two stages (half worked into seedbed in early spring and half at planting will kill couch grass, but because of

residues a second crop of maize must follow). Couch is also controlled by EPTC *plus* antidote ("Eradicane"). "Fortrol" (Cyazine) at drilling time and 2,4-D post-emergence may be used for broad-leaved weeds and "Suffix" for wild oats.

Harvesting (a) For silage—see page 155.

(b) For grain—usually ready between mid-October and mid-November. Frosts will kill off the foliage and this facilitates combining and helps to dry the grain, which in the yellow hard condition has a moisture content of 35–45% at harvest. Various types of machines are available for harvesting—some are ordinary combines with header attachments which only remove the ears from the standing crop and then thresh off the grain; others thresh the whole crop and some deliver the complete or dehusked ears into a trailer to be shelled later.

Drying the grain can be a problem and is very expensive. It usually has to be dried in two or three stages in a continuous drier from 40% to 15% moisture content. Floor-drying cannot be used for threshed grain, but the whole ears can be dried in this way.

Wet storage is possible in air-tight tower or butyl silos, but it can be difficult to unload and deteriorates rapidly when taken out of the silo.

Propionic acid, applied at 18–20 litres/tonne through a special applicator, will preserve the wet grain very satisfactorily in heaps in existing buildings.

Straw-choppers or a rotavator may have to be used to help dispose of the maize trash after harvest.

Yield. Silage (very variable)   25–60 tonnes/ha
Grain   4–6 tonnes/ha
Vegetable cobs   up to 75,000/ha
(See Maize Development Association literature and STL 93 and AL 297.)

PULSE CROPS

Pulse crops are *legumes* which have edible seeds; the main ones grown in this country are the various types of *beans* and *peas*. Bacteria on the roots of these crops can fix nitrogen, so, normally, they do not require nitrogen fertilizers and the following crops benefit from nitrogen left in the soil: about 60 kg/ha.

Beans and peas are useful, protein-rich grains for blending with cereals for feeding farm stock, but yields may be disappointing and harvesting troublesome. At present, peas are mainly grown for human consumption in the canned, quick-frozen, or dried state; beans for human consumption (canned, frozen and dried) are increasing in importance as a farm crop now that they can be mechanically harvested, e.g. dwarf stringless beans.

The smaller (tick) beans and maple peas are popular for pigeon feeding, and high prices are paid for good-quality grain.

Beans and peas can be used in mixtures with cereals for ensilage.

Stock-feeding beans usually grow well on heavy soils and loams, whereas peas, French and runner beans prefer the medium and lighter soils.

Beans or peas provide a useful break between cereal crops but should not be grown in successive years because of fungous diseases and pests—although climbing French (kidney) beans are sometimes grown on the same site for several years where fixed support wires have been erected.

The introduction of improved varieties, subsidized prices, chemical control of weeds and pests such as aphids, may result in a greater area being grown for stock feeding.

Irrigation is beneficial on light soils and in dry seasons—especially when the pods are setting and filling.

Beans

There are several types of beans grown on a field scale in this country; the main ones, and approximate per cent of total area (63,000 ha), are:

(1) *Vicia faba*, which includes the *field bean* (73%)—grain used for stockfeeding; and the *broad beans* (4%)—immature seed for human consumption, either fresh or in a processed form (canned and frozen).

(2) *Phaseolus vulgaris* (*French beans*) grown for sale fresh (3%) or processed from (a) dwarf "green" beans (18%) harvested as immature pods to be canned, frozen or dried; (b) dwarf "navy" beans (1%) harvested as a dried seed crop and later canned in tomato sauce (baked beans).

(3) *Phaseolus coccineus* (*runner beans*)—garden and market garden crop—almost entirely for the fresh trade (2%) (see HE 3).

Field beans

This crop has been grown in this country for centuries and has fluctuated in popularity. It is most consistently grown in heavy-land arable areas such as Essex.

Field beans grow well on clays and heavy loams which have a good structure, are well-drained and pH over 6.5 (nodule bacteria work better); organic soils and too much nitrogen usually results in too much straw, lodging and low grain yields.

The varieties grown can be grouped as:

(a) *Winter beans* (large seeds), e.g. *Throws MS* (a 4-line synthetic variety), *M. Beagle* and *Bulldog* (stiff-strawed), *M. Beaver* (suits E. Anglia), *Banner*.

(b) *Spring beans* (1) *tick beans* (small seeds) *M. Bead* and *Herra* are very small and suitable for the racing pigeon trade; *Blaze* and *Danas* are high-yielding and suitable for livestock feeding. (2) *horse beans* (large seeds and later ripening) e.g. *Stella Spring* (see also NIAB Farmers leaflet no. 15).

In general, compared with spring beans, winter beans yield at least 0.2 tonne/ha more, ripen up to a month earlier, are more susceptible to chocolate spot and lower in protein (23%) compared with 27% for spring beans.

Seed-beds. Deeper, but otherwise similar to those for cereals.

Sowing. Winter beans do best if sown in early October; spring beans in early February to early March. The seed rates are: winter beans, about 200 kg/ha (they branch better than spring beans), spring tick types—200 kg/ha, but might range up to 250 for horse types to establish 30–40 plants/m².

The seed should be dressed with thiram or captan.

The methods of sowing can vary; in difficult autumn conditions they may be ploughed in (7–10 cm deep), or broadcast on ploughing and covered by harrowing. If inter-row cultivations are likely, the rows will be 45–60 cm apart. Normally, the crop is sown with the ordinary corn drill (15–18-cm rows) but care is required with the large seeded varieties to avoid "bridging" or damage to the seed. If simazine is used to control weeds, the seed must be planted at least 8 cm deep, which might reduce yield in some soil conditions.

Fertilizers. 0–75 P_2O_5, 0–110 K_2O—depending on the soil fertility; an average of 50 of each may be combine-drilled; no nitrogen is required, but beans usually benefit from well-rotted FYM or slurry.

Weed control. Perennial weeds, e.g. couch, thistles, field bindweed, must be destroyed before the crop is sown. Simazine has been used for many years to control most annual weeds except cleavers, polygonum species, wild oats and cereals are not usually controlled satisfactorily. This cheap soil-acting herbicide is depth-selective for beans, provided they are planted at least 8 cm deep—if not, the bean leaves become blackened round the edges and the plants may die. Other herbicides are now used, e.g. propyzamide ("Kerb") and "Opogard"—pre-emergence winter beans and weeds ("Carbetamex" pre- and post-emergence); also, dinoseb acetate "Ivosit" post-emergence. Wild oats can be controlled by "Kerb" and "Carbetamex", or tri-allate "Avadex") worked into seed-bed, or "Suffix" post-emergence (fuller details in Booklet 2373).

Pollination. Honey bees (2 or 3 hives/ha) are usually necessary, especially with seed crops, to obtain good pollination of the beans. This results in a quick set (fertilization) of the maximum number of pods which each plant can fill, and so more even ripening, an earlier harvest, and better yields. Some new hybrid varieties being developed are self-fertilizing and so are less dependant on bees than the common open-pollinated varieties. Systematic aphicides which may be required to control blackfly should, if possible, be applied before flowering (June) to give better control with less risk of damage to the bees. These chemicals, which are normally only required for spring beans, may be applied as sprays (e.g. pirimicarb) or granules (e.g. phorate or disulfton).

Diseases. Several fungous diseases can damage beans; ·*chocolate spot* may ruin a winter bean crop, and is encouraged by dense foliage and high humidity conditions; early "Benlate" sprays may be helpful (see page 201). *Leaf spot*—somewhat similar, but with grey centres to the brownish spots and tiny black dots; encouraged by cool wet conditions; use tested disease-free seed, and kill volunteer bean plants in neighbouring cereal crops (MCPA). *Stem rot* (bean sickness)—control by wide rotation 1 year in 5 for beans/red clover crops (see Booklet 2373).

Pests. The following are the main pests and numbers of the advisory leaflet in brackets: black bean aphid (blackfly)—mainly on spring beans (see Table 53 and AL 54), pea and bean weevil (AL 61), stem eelworm (AL 178), pea cyst eelworm (AL 462) (see page 188).

Harvesting. Winter beans are usually ready for harvesting in August/September; spring beans in September/October. They ripen unevenly, lower pods first. Most crops are now harvested by combine when the leaves have withered and nearly all the pods are ripe and the seeds ready to shatter. It is best to combine in dull weather or in mornings and evenings to reduce losses at the cutter-bar; desiccation of green growth with diquat can make combining easier. Sheep will pick up and eat the shed seeds.

Drying and storage. Combined beans may require drying before storage, these large seeds must be dried carefully in continuous driers—preferably in two stages if the moisture content is over 20%. They can be dried successfully on floor driers, but the air may escape too easily and some beans near the floor may not be dried if the ducts are widely spaced. Store at 14% m.c. 1.25 m³/tonne—the same as wheat.

Yield. Winter beans 3–4.5 tonnes/ha. Spring beans 2.5–4 tonnes/ha.

Broad beans

Broad beans are very similar to field beans and much of the growing process is the same (see table on processed bean crops, page 86).

Weed control—tri-allate, simazine and dinoseb acetate as for field beans, also, chlorpropham plus diuron or fenuron (pre-emergence).

Varieties (only white-flowered types suitable for canning) e.g. *Triple White* (white-seeded), *Beryl* (very small seeds and late). *Feligreen* and *Staygreen* are green-seeded varieties.

Harvesting. The crop is harvested by cutting it with a pea-cutter when the beans are still soft (as from garden); tenderometer reading 120–140 for freezing; 130 and over for canning. It is allowed to wilt in the windrows and then vined with a pea viner. The beans must be rushed to the processing plant immediately after vining, to be canned or frozen.

In the processing plants, broad beans are fitted in between the pea and green bean crops so there is only limited room for expansion.

Yield. 2–4 tonnes/ha.

Green beans

This crop has expanded very fast over the past 20 years from a market garden crop sold as immature pods, to an important farm crop which is mainly grown for processing. This development has been encouraged by the expanding market for convenience foods, and also by the introduction of harvesters which flick the bean pods and leaves off the plant and separate the leaves in an air stream. The early machines worked on single, widely spaced rows, but the latest machines can deal with narrow rows and can go in any direction through the crop. The freshly harvested pods are then taken to the processing plant to be canned, frozen, or dried. They are not so perishable as peas or broad beans. Efficient weed control has also helped to popularize the crop; "Treflan" into the seed-bed, and/or "Basagran" post-emergence will deal with the annual weeds; perennials should be destroyed before planting. If weeds are not controlled, the crop may be unsaleable. "Ivorin" (dinoseb acetate) is also approved for pre-emergence annual weed control. EPTC worked into the seed-bed at least 3 weeks before planting will control couch and most wild oats.

Green beans prefer loams and lighter types of soil. It is a late-sown, short-season crop and so requires a deep, free-draining, moisture-retentive soil, and a very good seed-bed for fast, steady growth. It is also susceptible to frost and cold weather and so most crops are grown in the eastern counties from the Fens southwards (see Table 26).

TABLE 26

Crop	N	P_2O_5	K_2O
Broad beans	0–25	40	100–150
Dwarf green French beans	130–170	100–120	50–60
Dwarf dried "navy" beans	100–130	50–90	30–40
Runner beans	120–180	50–90	120–180

Varieties. There are four main groups of green beans: (a) *Tendercrop* types (pods about 125 mm), e.g. *Cascade*, *Provider*, white-seeded *Tendercrop*; (b) short pods (about 75 mm), e.g. *Chicobel*; (c) wax pods (golden-yellow colour), e.g. *Sunglo*; (d) flat pods (like runner beans, but not stringy), e.g. *Bina*.

Harvesting. The harvested bean pods must be whole, undamaged, separated (not in clusters) and free of stems, leaves, soil and stones, etc. As the crop matures, the pods lengthen rapidly, then enlarge as the seeds develop rapidly; the average length of the largest seed in each of ten randomly-selected pods is used as a guide to the time to harvest the crop. In practice, the length of the ten seeds is measured and the crop is ready for freezing when the total length is 100 mm for large-seeded varieties (80 mm for small-seeded types). The canning stage is later and the respective lengths are 120 mm and 100 mm.

Dried beans ("navy")

This crop is a relatively new one to this country. It is grown in the same way as green beans, but it requires better harvest weather and is more cold-sensitive, so is grown only in the south-east of England.

Varieties. *Purley King* (Seafarer) and *Gratiot* are small-seeded "navy" varieties; *Limelight* is a butter bean.

Harvesting. The crop is ready for harvesting when all the leaves have fallen off the plants, and when the moisture content of the seed is 20–25% (if under 18% the seeds crack, and if over 30% they are crushed in the combine); a low drum speed must be used. Desiccation with diquat will remove green rubbish before combining. Dry carefully—as for dried peas.

Yield. Very variable, 1–3 tonnes/ha.

Fertilizers for beans. French beans do not normally develop nitrogen-fixing nodules as other legumes do in this country, and so nitrogen has to be applied. Research work is in progress to try to overcome this by isolating a suitable strain of Rhizobium bacteria. Table 26 shows the approximate needs of the crop in kg/ha of plant food.

Seeding. The seed should be precision-drilled (or carefully with other drills) at a seed rate calculated according to factors such as: the number of seeds/kg, the plant population required, the germination %, time of sowing and an estimate of the likely seed-bed losses, and possibly using a formula:

$$\text{seed rate (kg/ha)} = \frac{10{,}000 \times \text{no. plants/m}^2}{\text{no. seeds/kg}}$$

$$\times \frac{100}{\text{germination \%}}$$

$$\times \frac{100}{100 \text{ less \% field loss}},$$

e.g. a seed sample has 3500 seeds/kg, germination 90%, target population is 30 plants/m², and estimated field loss is 10%, then:

$$\text{seed rate (kg/ha)} = \frac{10{,}000 \times 30}{3500}$$

$$\times \frac{100}{90} \times \frac{100}{100 - 10}$$

$$= 105.8 \text{ kg/ha}.$$

Table 27a presents a general guide.

Dates of sowing and harvesting green and broad beans have to be carefully worked out to give an even flow of produce to the factory. The green bean season lasts about 6 weeks, broad beans usually less.

Diseases and pests. Many pests and diseases attack these bean crops; halo blight and anthracnose can cause a lot of trouble in some seasons; all these problems are fully dealt with in the PGRO handbook.

TABLE 27a

Crop	Row width (cm)	Plant/m² (nos.)	Seed rate (kg/ha)
Broad beans	20–45	15 (12–20)	140+
Green beans	15 (10–20)	27 (25–30)	130 (90–180)
Dried beans ("navy")	40	30 (20 *Limelight*)	80

TABLE 27b

Crop	Dates of sowing	Dates of harvest	Yield (tonnes/ha)
Broad beans	Feb.–early May	mid-July–end Aug.	3 (2–4)
Green beans	mid-May–end of June	mid-Aug.–late Sept.	7 (4–9)
Dried beans ("navy")	about mid-May	September	1.5 (1–3)

Irrigation. This can be very beneficial in a dry season—especially when the pods are developing—but care is required because it may spread some fungous diseases.

(For full details about bean crops for processing, see the *PGRO Pea and Bean Growing Handbook*, Vol. 2, *Beans*; also Booklet 2373, *Field Beans*.)

Peas

Nearly all the peas grown in this country are for human consumption. The purple-flowered, brown and yellow mottled maple peas are only grown to a very limited extent for racing pigeons and stock (yield up to 4 tonnes/ha), or for forage (alone or in mixtures); new forage peas (*Rosakrone, Krupp, Lioness* and *Minerva*) may revive an interest in growing peas for stock (sown in February at 125 kg/ha). (Leaflet 801.)

The main types for human consumption (all white-flowered dwarfs) are:

1. *Vining peas* (two-thirds total market)—used for canning fresh ("garden peas"), quick-freezing, or artificial drying. These are normally grown under contract or by co-operating groups. There is a wide range of early-, medium- and late-ripening varieties; the dark-skinned types are suitable for all processes, but the light (pale) skinned types are only suitable for canning (may be dyed). The seed is supplied by the contracting firms, and time of sowing (February–May) is carefully decided on a "heat unit" basis, as well as variety, to spread the harvest evenly. Yield 5 (3–8) tonnes/ha (contract price).
2. *Threshed (dry) peas* (one-third total market)—sold dry in packets or loose, or canned (processed). The four main groups are: (a) Marrowfats (wrinkled)—best quality, e.g. *Maro, Progreta.* (b) Large blues (round), e.g. *Rondo, Allround*—for fertile soils. (c) Small blues (round), e.g. *Vedette*, susceptible to herbicides. (d) White-seeded, e.g. *Frimas* (has over-wintered well), *Birte.* These peas should be sown in early March, if possible. Yield 3–4.5 tonnes/ha (contract or free-market price).
3. *Pulling peas* (less than 5% of market)—sold as fresh peas in pods. A wide range of varieties grown mainly as a market garden crop. Yield 8–12 tonnes/ha (market price).

New types of peas with no (or very small) leaves and/or stipules (leafless and semi-leafless types) have been bred by the John Innes Institute. They are mostly white-seeded peas and the variety, *Filby*, is promising. The tare-leaved mutant variety *Progreta* (selected from Maro) has very small stipules and leaves (some changed to tendrils). Both these new types of pea yield well (even with less leaves) and have a promising future because the crop is more open and easier to harvest—especially as a dried crop, but weed control is more difficult. Some of these leafless types and other varieties are being sown in October to give an earlier harvest and better yields in a dry season, e.g. *Frogel* and *Frimas*. However, there are problems with frost heaving, wet soils, thinned crops (pigeon damage and pests) leading to uneven ripening and greater weed problems.

Soils and climate. Peas grow best on loams and lighter types of soil; seed-bed preparation and harvesting can be difficult on heavy soils; all soils must be well drained and have a pH over 5.5.

The threshed peas require good weather for harvesting (less than 50 mm rain in July and in August)—especially where the crop is dried on four poles or racks; desiccated direct-combined crops are less likely to suffer damage.

The vining and pulling peas are grown in most of the arable farming areas of the country within easy range of a processing plant or market.

Seed-bed. It should be possible to drill the peas about 5 cm deep in a fairly loose tilth; on light easy-working soils it is possible to sow into well-ploughed land with the minimum or no previous cultivation; rolling may be necessary to push stones into the surface and to firm very loose soils, but it should not be done on wet, heavy land.

Sowing—normally in narrow rows (15–20 cm) and herbicides used: inter-row cultivations require 30–40-cm rows and yield is lower. The seed rate should be carefully calculated according to seed size (nos./kg), vigour and germination per cent, quality of the seed-bed, time of sowing, and cost of seed. To produce good profitable crops, the plant population is very important, e.g. for vining peas the aim should be 100 plants/m^2 (i.e. 5 cm apart in 20-cm rows); for marrowfats 70–80/m^2; large blues 65/m^2, and small blues 100/m^2. Average seed rate 230 kg/ha (150–300). All seed should be protected from seedling diseases by a seed dressing such as thiram or drazoxolon, or metalaxyl/captan (downy mildew).

Fertilizers. P$_2$O$_5$ 0–50 kg/ha K$_2$O 0–150 kg/ha, depending on the level of fertility; nitrogen is not normally required. The fertilizer is normally broadcast in the seed-bed, but "placing" in bands near the seed is more efficient for wide rows. If manganese deficiency is expected or diagnosed the crop can be sprayed with 5–10 kg/ha of manganese sulphate.

Weed control. It is most important that weeds are controlled in the pea crop because they can cause serious yield losses, encourage disease, make harvesting difficult, and spoil the produce, e.g. the seed heads of mayweeds, thistles and poppies, or volunteer potato and black nightshade berries in vining peas. Peas are a fairly open crop—especially the new leafless types— and offer little competition, so herbicides must be very efficient. Inter-row cultivations to control weeds are seldom used now because, by changing to narrow rows and chemical control, yields have nearly doubled. If possible, couch and other perennial weeds should be killed pre-harvest in the previous cereal crop with "Roundup": post-emergence couch control in the growing crop may be possible with "Clout" or "Fusilade". Wild oats—propham or tri-allate, worked into soil before sowing; "Hoegrass" or barban post-emergence. Broad-leaved annuals —these are mainly controlled by soil-acting residual chemicals sprayed on after sowing; a fine, moist seed-bed is necessary for good results; some examples are: cyanazine ("Fortrol"), prometryne ("Gesagard"), trietazine + simazine ("Remtal" and "Aventox"), chlorpropham + diuron ("Residuron extra"), dacthal

+ methazole ("Delozin S"). Additionally, post-emergence treatment with: dinoseb(amine or acetate), Basagran MCPB, cyanazine + MCPB + MCPA; MCPB for creeping thistles and fathen. Most soil-acting herbicides are affected by soil type (see Booklet 2262 and product literature for fuller details on herbicides).

Irrigation. Peas are deep-rooting plants and benefit most from irrigation at flowering time (increases numbers of peas) and at pod-swelling stage (increases size of peas).

Pests and diseases. Peas are affected by many pests and diseases, many of which are controlled by seed dressings, resistant varieties, wide rotation (1 year in 5), fungicides and insecticides. These are fully dealt with in the PGRO handbook. (See pages 14, 189.)

Harvesting. The first operation in harvesting most pea crops is to cut the crop with special peacutters, fitted with lifting fingers, and leave it in windrows. The stage of growth at the time of cutting depends on the type of crop.

Vining peas are ready when the crop is just starting to lose its green colour and the peas are still soft. The firmness of the peas is tested daily (near harvest) with a tenderometer and this gives a guide to the best time to start cutting (the reading for freezing peas is about 100, and for canning, they can be a little firmer, at 120). When ready, the crop must be cut as soon as possible. The windrowed vines and pods are picked up and put through a special vining machine which separates the peas from the pods These machines were originally at fixed sites and the whole crop had to be carted to them. Nowadays, most crops are harvested by large mobile viners which can work 24 hours a day in all weathers, and they normally pick up the crop from the windrow and separate the peas as they travel round the field. The latest machines (pea pickers) go straight into the crop (windrowing is not required) and pick the pods only (similar to the green-bean harvester), and then remove the peas from the pods; this is a considerable saving in staff requirements—especially in difficult harvesting conditions. The shelled peas are rushed to the processing plant for freezing, canning or drying. The haulms (vines) left by some of the harvesters may be made into silage or hay, or, more usually, worked into the soil to make humus. The *threshed* peas are not windrowed until the vines and pods have turned a yellow or light brown colour and before the seed starts to shed. After partly drying on the ground the crop can be put on tripods, four-poles, or racks to complete drying over a period of weeks; this can produce a good sample of dried peas. Alternatively, the standing crop may be desiccated by a chemical, such as diquat, and combined direct 1–2 weeks later. If the peas require drying it must be done slowly and at a low temperature (43°C); a floor drier is suitable. The storage moisture content is the same as for cereals.

Pulling peas are harvested by removing the pods when the peas are in a fresh, sweet condition; this requires a large gang of casual workers, or very expensive machinery, so the area grown is rapidly declining. An increasing area is being grown for the pick-your-own sales (see also Vol. 1, *Pea and Bean Growing Handbook* from PGRO).

Vetches

Vetches are legumes which are used mainly for forage purposes. They used to be common (sown in autumn or spring) for folding off with arable sheep flocks. Nowadays they are sown (autumn or spring) in arable silage mixtures with oats, and cut with a forage harvester at various stages (depending on bulk or quality requirements) up to the milky stage of the oats, e.g. a mixture of 50-kg vetches with 150-kg leafy oats (sometimes they are sown alone or at reduced rates in fields where wild oats are abundant). Autumn-sown crops yield better than spring-sown in dry areas (see page 155).

Lupins

Research workers, plant breeders and a few enthusiastic farmers are exploring the possibilities for growing annual lupins in the U.K. Lupins have never been an important crop in this country except where it has been used for green manuring on acid, sandy soils, and to a limited extent for sheep folding. The main interest now is centred on grain production. The natural lupin seed has 30–40% protein (dehusked grain is as good as soya bean) and 10–12% edible oil. Modern varieties have been bred for low alkaloid content (i.e. low toxicity) and are called "sweet" lupins. The most interesting species are the blue-, yellow- and white-flowered types: the *blue* lupin can yield well on deep, fertile soils, but it is susceptible to Fusarium wilt, and is very late ripening, so is not suitable for this country (it is a very important crop in Western Australia). The *yellow* lupin is grown for seed in north-east Europe, but its best use is for green manuring when reclaiming very acid, sandy soils. The *white* lupin is the most promising for grain production in this country. The *pearl* lupin has possibilities because of its disease resistance and high protein and oil contents, but it is late, and toxic (this could be bred out).

Production possibilities for white lupins. The high yielding potential (5 tonnes/ha) and disease resistance of the Russian variety, *Kievskij*, is very promising. It is a "sweet" type, early ripening (120–140 days growing period) and has pods which do not shatter too readily before harvest—a problem with all types. White lupins will grow on acid, sandy soils, but do better on deep, well-limed, free-draining loams in an open situation. They should be drilled about the end of March into a good seed-bed (probably 35-cm rows) at about 60 kg/ha to give about 30 plants/m² (more research is needed on this).

Fertilizers. Same phosphate and potash as for peas or field beans; no nitrogen is required, but seed may require to be inoculated for some soils.

Weed control. Weeds must be controlled; simazine and other herbicides are showing promise.

Pests. Rabbits are very fond of sweet lupins!

Diseases. Fusarium wilt, foot rot and mosaic virus are problems, and are likely to increase as more lupins are grown.

Harvesting. The big problem!, most lupins ripen very unevenly (usually in September). The large fleshy pods are slow to dry out, and shatter far too easily. The crop must be combined very carefully to reduce losses. The seed should be dried to 14% m.c. (as for threshed peas).

Yield. Possibly 2 tonnes/ha, but usually less.

There is a lot of scope for improvement before lupins become an established grain crop in the southern half of the U.K.

Soya beans

Soya bean seed contains about 40% protein which is used for food manufacture, industry and feeding stuffs, and also 17–20% oil which is also used for human consumption and industry (it is the most important vegetable oil in the world).

The varieties now available are not really suitable for this country and the possibilities of it becoming a farm crop are only of academic interest at present. Types originating in Japan, which are insensitive to day length, and grown in a climate similar to the south of England, may prove to be useful.

OIL-SEED CROPS

Oil-seed rape

Oil-seed rape is now established as an important and profitable crop in the U.K. and, as was predicted, pest and disease problems have increased as the area being grown has increased. However, yields are increasing as new varieties are introduced and more becomes known about the needs of the crop and how to control the pests and diseases. Most crops are grown on contract and expert advice is available at all stages.

The small black seed contains about 42% of oil which is extracted by crushing and used for the manufacture of margarine, cooking fats, and special lubricants for jet engines. The residual cake is used for stock feeding.

Soils and climate. The crop will grow in a wide range of soil and climatic conditions provided the land is well drained, pH over 6, and the soil and subsoil structure is good.

Rotation. Ideally, rape and other brassica crops should not be grown more than 1 year in 5, so as to avoid a build-up of diseases (e.g. club root) and pests. It is an alternative host for sugar-beet eelworm and this could affect the place taken by sugar-beet in a rotation.

Varieties. There are many varieties of the swede and turnip types of oilseed rape. The amount of erucic acid in the seed has been a problem in oil used for human consumption, but varieties now being grown have very small amounts—less than 2%. The main variety in recent years has been *Jet Neuf*, but *Rafal*, *Elvira* and *Norli* are also very good disease-resistant varieties which yield well and ripen early—about the end of July. Winter crops (usually sown in August) can be severely damaged by pigeons—especially in spring; autumn stubble cleaning is not possible before sowing, but preharvest "Roundup" in the previous cereal crop is very good for perennial weed control. Spring varieties (less than 3% of total) e.g. *Brutor*, *Cresor* and *Willie* should be sown in late March, if possible, and are ready for harvesting about early September. The work involved suits some farms better than winter rape, but yields are usually much lower—especially in a dry summer (see NIAB Farmers leaflet no. 9).

Seed-beds. The seed is very small, so it should have very fine, moist conditions in the seed-bed. Direct-drilling can be very successful on soils with good structure. Subsoiling may be necessary to break-up pans but this may damage the surface tilth unless done some time earlier. Rape develops a deep tap-root in suitable conditions.

Sowing. 6–10 kg/ha of seed drilled shallow (2 cm) in 12–18-cm rows; this may have to be mixed with fertilizer or other material in some types of corn drill. A crop of winter rape should have about 60–80 plants/m² to produce the best yield of good plump seed (if too dense the important secondary inflorescences do not develop fully). Ideally, there should be 6–8 medium-sized leaves (10–15 cm high) on each plant before winter sets in. If sown too early, or given too much nitrogen in autumn, too much leaf and flower buds develop in the autumn and this will reduce the yield. For spring rape, a plant population of 100 plants/m^2 is required for a good crop. Sow winter rape about 20 August.

Fertilizers. Good crops will take up large amounts of nitrogen, phosphate and potash, but generally, the response to P and K is low—especially on soils in a good state of fertility. The following are general recommendations (kg/ha):

TABLE 28

	N	P$_2$O$_5$	K$_2$O
Winter rape or seed-bed (Aug./Sept.)	40–60	40–75	40–75
top-dressing (Feb.)	150–225		
Spring rape or seed-bed	100–200	40–75	40–75

Where sulphur is required, use sulphate of ammonia or super-phosphate fertilizer. (For further details see *Oilseed Rape Book*, Cambridge Agricultural Publications.)

Weed control. Several herbicides (and expert advice) are available to deal with most problems; the main sprays used are:

TCA in seed-bed) for wild oats, volunteer cereals and grasses.

"Trifluralin" (into seed-bed) and "Propachlor" (pre-emergence) can be used on winter and spring crops for a limited range of annual weeds.

"Butisan S" (metazachlor)—pre-emergence residual for a wide range of common annual broad-leaved and grass weeds.

"Pradone Plus" (mid-Oct. to Feb.) slowly kills nearly all annual grasses and broad-leaved weeds; acts mainly through roots.

"Kerb" (Oct./Nov.) mainly for annual grass weeds, chickweed and a few others (not mayweeds or marigold); very persistent: "Matrikerb", similar plus mayweeds.

"Benazalox" (Oct./Nov.) specially for chickweed, mayweed, cleavers, charlock.

Dalapon (autumn) for cereals/wild oats; checks crop; deters pigeons.

For fuller details see Booklet 2068, and product literature.

Pests and diseases. The main pests are pollen (blossom) beetle, seed weevil, bladder pod midge, cabbage-stem flea beetle, cabbage-stem weevil, cabbage aphis (see Table 53, and L. 780). The main diseases are club root (good rotation, AL 276), canker, downy and powdery mildews, leaf spot, see Table 54 and Booklet 2278.

Growth stages. The Berkankamp Scale is now being used by some advisers to assist in describing growth stages in the oil-seed rape crop so that precise advice can be given to growers. Briefly, the growth of the crop is divided into four stages:

Stage 1 Leaf development
1.0 cotyledon, 1.1 first leaf, 1.2 second leaf, etc.
Stage 2 Bud development
2.0–2.9
Stage 3 Flower development
3.0–3.9
Stage 4 Seed development
4.0—seeds translucent in lower pods
4.5—seeds in upper pods all brown
5.0—brown pods brittle, stems dry.

Harvesting. The winter crop is usually swathed to allow more even ripening and reduce drying costs, and is ready when most of the middle pods turn yellow and the seed is a chocolate colour. The swathing must be well done, including a vertical knife for heavy and tangled crops. The cut crop should be left on a 25–30-cm stubble to dry and ripen, and combined 7–10 days later. Spring and light winter crops may be combined direct when the seed is black; at this stage losses can readily occur by the ripe pods shattering. Where ripening is uneven, shattering from these ripe pods can be reduced by desiccating the crop with diquat about 3 days after the swathing stage (seed chocolate-black). This also desiccates any green weed growth before combining 4–7 days later.

Drying and storage. The moisture content of the seed at harvest time is likely to be in the range 8–15% for swathed crops and 10–25% for direct-combined crops. The contract price is based on 9% so it should be dried to slightly less than this. Damp rape seed must be dried as soon as possible either by a continuous drier (not above

80°C, and cooled) or by bulk drying on a floor or in bins. The undried grain should not be piled more than 1 m deep, and the drying ducts covered with hessian. The bulk capacity of the seed is 1.4 m³ per tonne (same as barley). Normally, dry cold air is used for drying, but some heat may be required to lower the relative humidity to 70% so that the seed dries to 8% m.c.

Yield. Winter rape 2–4 tonnes/ha, spring rape 2–3 tonnes/ha. When rape seed is being conveyed in trailers or lorries, great care should be taken to block all holes through which it might escape. The straw can be chopped with a flail harvester and ploughed in, but this can encourage slugs in a following winter wheat crop; otherwise, it can be pushed into heaps with a buckrake and burnt (see also *Oilseed Rape Book*, Cambridge Agricultural Publishing).

FLAX AND LINSEED

Flax and linseed are varieties of the same plant which have increased and decreased in importance many times over the past centuries in this country.

Flax is grown for the fibres in the stems, which are used for making linen. At harvest time, the crop is pulled out of the ground and retted (partly rotted) in dams or tanks so that, after drying, the bundles of fibres can be easily removed by a scutching process. The seed yield of flax is low when compared with the linseed varieties and it has not been of any importance as a crop since the period during and shortly after World War II. Recent research work on spraying the crop with "Roundup" before harvest and allowing it to ret as a standing crop promises to simplify the production of this crop.

Linseed has much shorter straw than flax (see Fig. 39) and produces a good yield of seed which contains up to 40% oil and about 24% crude protein. The drying oil produced from linseed is mainly used for making paints, putty, varnishes, oil cloth, linoleum, printers' ink, etc., but these uses are much less important now since the introduction of plastics and latex paints and the use of other oils. It cannot be used in edible

products because of its high linolenic acid content, but the by-product—linseed cake—is a highly-valued, protein-rich, feeding-stuff. (About 12,000 hectares could replace all the present imports.)

Modern varieties of linseed, such as *Antares* and *Linott*, are very much better than the older varieties—especially with regard to yield, shorter and less fibrous straw, earlier and more uniform ripening, less shattering of seed, and disease resistance.

Suitable soils and climate. Linseed can be grown in any of the lowland arable areas of the U.K. but hot weather is desirable for ripening and easy harvesting in August or September. It can be grown on a wide range of soils provided the pH is about 6.5, good fine seed-beds can be pre-

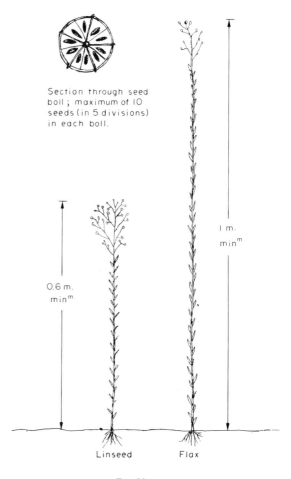

Section through seed boll; maximum of 10 seeds (in 5 divisions) in each boll.

0.6 m. minᵐ

1 m. minᵐ

Linseed   Flax

FIG. 39.

pared, and they do not dry out easily—especially when the seed is forming.

Rotation. This should not be grown more often than 1 year in 5 to avoid carry-over of disease on debris.

Sowing. Ideally, it should be drilled shallow in narrow (12-cm) rows in a fine, firm (but not compacted), level, moist seed-bed, from mid-March to early April. Seed rate—75–90 kg/ha (depending on the seed count and germination) should produce up to 900 plants/m^2.

Fertilizers. Treat as spring barley, i.e. about 50 kg/ha of P_2O_5 and K_2O and 100 nitrogen. Zinc deficiency may occur on over-limed soils—checking growth and showing white spots on the leaves—spray zinc sulphate.

Weed control. Linseed cannot compete with weeds, so they must be controlled, otherwise yields will be very seriously reduced and harvesting very difficult. Herbicides have not been specially developed for linseed, so many different types have been and are being tried. Grass-type weeds can be treated with tri-allate, TCA, barban, dalapon or asulam. Broad-leaved weeds can be treated with linuron, lenacil, trifluralin, MCPA and "Buctril M" (fuller details of weed control given in MAFF booklet STL 183).

Disease control. For rust and wilt, use resistant varieties. For seedling blight, foot and stem rots, and others, use seed dressings.

Pests. Flea beetle—use seed dressing or DDT or HCH sprays. Leatherjackets—use seed dressing or DDT or HCH sprays. Pigeons and rabbits can do much damage in the spring on young crops.

Flowering—most varieties have pale blue flowers which appear in early July—usually early in the morning and only lasting a few hours. The flowering period for the crop may last several weeks. It is normally self-pollinated.

Harvesting. The crop is ready for harvesting when the whole plant is dry, stems yellow-brown and leaves fallen from the base. The seeds should rattle in the bolls (seed capsules) and be plump and brown and, for direct combining, showing the first signs of shattering. Linseed straw is very tough and wiry—so the knife must be kept sharp. The crop may be desiccated with

diquat and/or swathed (80% ripe seeds) before combining—especially where there are weeds present.

Drying and storage—similar to oil-seed rape. The seed must be carefully dried to about 9% moisture content for safe storage (comes in from the field at about 12–16%) (fuller details in STL 183).

Linseed is a small slippery seed which can easily fall through tiny holes so this must be carefully watched at all stages—sowing, combining, in trailers, and in the drier and storage buildings.

Yield. A good yield would be about 2.5–3 tonnes/hectare. Price per tonne is about £130.00 plus about £75/ha deficiency payment.

SUNFLOWERS

Sunflower plants grow successfully in gardens and coverts for game birds in the south of England, but when grown as a seed crop, birds are likely to cause serious losses in the seed-bed and by eating the ripening seeds. Grey mould (Botrytis) can cause serious damage to the leaves, stems (lodging) and heads in a cool, damp season. As yet, there are no satisfactory methods of dealing with these problems, but resistant varieties and fungicides may solve the Botrytis problem, and ways of controlling the birds may be found—probably by increasing the area grown! A few farmers in East Anglia have grown reasonable crops on a small scale, but harvesting and handling the seed afterwards tests their ingenuity.

Modern sunflower varieties are potentially high-yielding; short-stemmed hybrid plants are produced in a similar way to maize hybrids, by using a male sterility and restoring genes technique.

The decorticated (dehusked) seed contains over 40% of valuable, high-quality, edible oil which is used in food manufacture, margarine and cooking oils. Earlier ripening varieties are needed for this country.

Briefly, the production requirements are: well-drained soil, pH over 5.5, good seed-bed—fine and moist; not grown more than 1 year in 4 in the same field; *fertilizer*—75–100 N, 100–125 P_2O_5, 100–125 K_2O, too much nitrogen or very rich soils can cause lodging; *sowing*—about mid-April, about 5 kg/ha seed precision-drilled 15–20 cm apart in 75-cm rows; *weed control*—inter-row cultivations; "Nortron" and "Treflan" are promising; *harvesting*—ready when all the leaves turn yellow (late Sept./Oct.); the heads dry out very slowly—desiccation could help; special lifting fingers direct the heads into the combine; the seed and rubbish may be 30% m.c.—to be dried to 9%, this material is very bulky and augers may not take it; *yield*—2.5 to 3.5 tonnes/ha is possible if all goes well.

POTATOES

Potatoes are tubers in which starch is stored and, depending mainly on the variety, they vary in: shape (e.g. round, oval, kidney, irregular), skin colour (mainly cream and/or red), flesh colour (mainly cream or lemon), depth of eyes (where sprouts develop) and resistance to diseases and discolorations.

Uses. In this country, potatoes are grown on about one farm in ten and are used mainly for human consumption, but in a glut year, some subsidized lots may be used for stock-feeding. In several EEC countries, some of the crop is used for producing starch, alcohol, etc. The consumption of potatoes in the U.K. is fairly constant from year to year (75–90 kg/head), and so the prices obtained vary considerably according to the supply. They are eaten as boiled or baked potatoes, chips, crisps, etc., and an increasing proportion of the crop (about 20%) is processed as crisps, frozen chips, dehydrated instant mash, and canned "new" potatoes.

Good quality ware potato tubers:

(a) are not damaged, frosted, or diseased;
(b) are free of: greening (an indication that a toxic substance—solanine—is present); secondary-growth irregularities such as knobs, cracks and glassiness; and sprout growth which spoils quality and increases preparation costs;

(c) are of reasonable size and shape;
(d) have smooth clean skins with shallow eyes and so easily peeled;
(e) have not been damaged by pests such as wireworms, slugs and cutworms;
(f) have flesh which does not blacken before or after cooking;
(g) do not break down when being boiled.

Potatoes must have special qualities to be suitable for processing, e.g.

for crisps, chips and dehydration, the tubers must have a high dry matter (starch) content and a low sugar content (too much sugar produces dark brown crisps and chips); minimum sizes (mm) crisps 35, chips 45, dehydration 25;

for canning, small (20–40 mm), waxy fleshed, low dry matter tubers are required which do not break down in the cans;

for baking, attractive uniform-shaped tubers weighing about 250 g are preferred and they usually have to be hand-picked off the grading lines.

Varieties. See the NIAB recommended list and Potato Varieties (PMB/NIAB).

In the U.K., varieties are classified as first and second earlies, and early and late main-crops, and each variety can be identified by its foliage (and flowers, if any). In general, early varieties are small bushy plants, early maincrop varieties are of medium height, whilst late varieties have tall upright foliage. The size, number and positioning of the primary and secondary leaflets, the shade of green, the dullness or glossiness of the leaves, the amount of pink or purple colouring in various places, and the wing development on the stems, may all be used as identification aids.

The differences in times of maturity between varieties is mainly because the earlies are "long-day" potatoes, i.e. they can produce a crop during the long days of early and mid-summer, whereas the late varieties are neutral or "short-day" types and will not produce full crops unless allowed to grow on until the shorter days in early autumn. The very dry years—1975 and 1976—

showed up very considerable differences in drought-resistance between varieties, due to a great variation in the rooting depth.

Almost one-fifth of the potatoes grown in the U.K. are earlies. The following are the more important (or promising) varieties:

First Earlies: *Home Guard, M. Bard, Manna, Pentland Javelin*(r), *Ulster Sceptre*(d).
Second Earlies: *Estima*(d), *Maris Peer, Wilja*.
Early Maincrop: *Desirée*(d), *King Edward, M. Piper*(r), *P. Dell, P. Squire*(d), *Romano, Record* (crisping).
Late Maincrop: *Cara*(r and d), *Pentland Crown*(d), *Golden Wonder* (Scotland).

(d)—deep-rooting (drought-resistant) variety.
(r)—resistant to pathotype Ro 1 (formerly A) cyst nematode (eelworm).
New varieties resistant to all types of nematodes will soon be available.

All healthy plants of the same variety are alike because they are reproduced vegetatively, and seed potatoes, as we know them, are not true seed. New varieties are bred by growing (and selecting) plants from the true seeds found in the green tomato-like fruits which develop from flowers on some varieties (usually after crossing two varieties). Varieties which do not normally flower can be induced to do so by grafting them onto tomato plants, and in other ways. It normally takes about 14 years to breed a new variety up to commercial scale production. True botanical seed is now being used for producing potato crops in some countries and the crops are remarkably uniform: virus disease are not carried in the true seed, which is dried to about 4% m.c., is very small and easily transported in small packets instead of heavy sacks of tubers.

Yields can vary a lot depending on climatic and soil fertility conditions.

It is very difficult to decide on the most profitable seed rates for maincrop because factors such as cost of the seed and expected price for the crop, variety and size of the seed must be carefully considered. The figures given in Table 29 are for normal-sized seed (30–60 mm), but if healthy small seed (20–30 mm) is used the rate could be reduced by at least 25%. Large seed can produce more sprouts per tuber than small seed, but small seed produces more sprouts per tonne than large seed; the larger tubers will produce more vigorous early growth. In some countries it is common practice to cut large seed into pieces by hand or machine before planting; the cut pieces are more likely to rot than whole seed but may be treated with "Storite" (thiabendazole). The sliced tubers can be planted immediately, but losses (10% or more in dry seed-beds) may be reduced by "curing" the seed for a short time before planting. The extra machinery and labour costs involved can only be justified by substantially lower prices for the large tubers compared with ordinary seed prices.

Sources of seed (only certified seed can be bought or sold). Growers can plant uncertified home-grown seed, but this is a risky practice if the crop is from poor stock and is not protected by suitable insecticides against aphids, which spread leaf-roll and mosaic virus diseases. Some varieties such as *P. Crown* and *P. Javelin* have a high resistance to virus diseases and so are very much better suited for growing-on as home-grown seed. The ADAS tuber-testing service should be used to check any doubtful stocks of seed before planting. The certified grades of

TABLE 29

	Yield of tubers (tonnes/ha)	Seed rate (tonnes/ha)	Time of planting	Time of harvesting
Earlies	10–30	4–5	Feb.–Mar.	late May–Aug.
Maincrop	25–60	3–4	April	Sept.–Oct.
Seed	20–40	3–5	April	Sept.–Oct.
Canning	10–20	5–7	Mar.–June	June–Oct.
"Blueprint"	60–90	8–9	April	Oct.

seed are: VTSC, FS, SS, AA and CC—details of these are given on page 117.

Sprouting (chitting) tubers before planting. About half of the potato crops in the U.K. are grown from sprouted seed. This is a necessity with earlies to obtain high early market prices, and should be started in the autumn before planting so that single sprouts develop on each tuber (apical dominance effect). It is desirable for maincrops because, on average, it increases yield by 3-5 tonnes/ha, allows more flexibility of planting time, and reduces the risk of yield losses if blight occurs early. Seed crops also benefit from sprouting—rogues (different sprouts) may be removed before planting; the crop bulks earlier, and so the haulms can be destroyed early to check the spread of virus diseases by aphids. To obtain high yields from maincrops and seed crops, several sprouts should be encouraged to grow on each tuber (multi-sprouting).

The physiological age of seed tubers can have a marked effect on how most varieties develop. Physiologically mature seed is produced from early-planted seed crops which are lifted early in warm conditions; such seed produces earlier crops with fewer tubers but they are more susceptible to drought and mature early. Physiologically young seed—produced and stored in cooler conditions—is better for maincrops which grow on to high yields.

The rate of growth of the sprouts is controlled by temperature and is usually fastest about 16°C but varieties differ considerably (see NIAB leaflet). The size of the sprout is controlled by light. Short, sturdy sprouts are formed in well-lighted buildings such as glasshouses or barns fitted with warm-white fluorescent tubes hanging between the rows of seed containers and switched on for 8–12 hours/day after sprouting commences. Storage buildings must be frost-proof and well ventilated to avoid high humidity and condensation; sprouting outside under clear polythene sheeting can be done in sheltered places.

Stackable wooden trays (75 × 45 × 15 cm), each holding about 15 kg (60–80 per tonne), are still the most popular containers for sprouting (especially for earlies), but are being replaced by higher-capacity white plastic trays, or wire crates and pallet boxes holding at least 500 kg in order to speed-up the loading of high-output maincrop planters.

Well-developed sprouts (2–3 cm) give maximum yield advantage but are easily damaged by mechanical planting and so mini-chitted seed (sprouts less than 5 mm) is now used with automatic machines on at least one-third of sprouted maincrops. Mini-chitting requires accurate temperature control, i.e. held about 4–5°C until 3 weeks before planting and then raised to 10–15°C; unused fruit stores can be quite useful for this.

If well-sprouted tubers are taken from a warm store and planted into cold ground, some varieties are likely to emerge very late due to "coiled sprout" development, or "little potato" may develop when no shoots appear and the old tuber is converted into a few new tubers.

If seed is not sprouted before planting, problems can arise in storage. If a sprout suppressant ("Fusarex") is used, adequate ventilation must be allowed before planting (at least 6 weeks). It may be possible to plant in light soils in March before sprouting in store becomes a serious problem, and so overcome the difficulty (see Booklet 2400).

Suitable soils and climate. Earlies do best on light, well-drained soils in areas free of late frosts, but some frost protection is possible with irrigation (see page 38) or with clear polythene sheeting.

Maincrops are best suited to deep, fertile, loam soils because high yields are very important—especially if prices are low.

Seed crops are best suited to the healthy northern and hill areas where aphids are scarce, otherwise efficient aphid control is necessary. Potatoes will grow in acid soils (over pH 5.5); common scab can cause problems on high pH soils in a dry season.

Place in rotation. Potatoes can be taken at various stages, but usually after cereals. After grass, heavy soils may be easier to work, but wireworms and leatherjackets must be controlled. To control cyst nematodes (eelworms), maincrops should not be grown more often than 2 years in 8, seed crops one in 5–7 years, but earlies may be grown continuously if lifted before the nematodes develop viable cysts (i.e. before 21st June).

Seed-bed preparation. Except on light soils, early ploughing is necessary to allow for frost action. In spring, deep cultivations, harrowing, discing and/or power-driven rotary cultivators are used, as required, to produce a fine deep tilth without losing too much moisture. Potatoes will grow in rough, cloddy conditions, but this type of soil structure should be avoided where soil-acting herbicides are used (most farms), and where mechanical harvesters are used (too costly to separate the clods).

Some soils have lots of stones which are likely to cause harvesting problems. These can be collected mechanically and removed, or, as is more usual, they can be mechanically separated into windrows between the ridges so that the potatoes are grown in and harvested faster from stone-free ridges. An alternative is to crush the stones into small pieces—this is better suited to soft stones such as sandstones and Oolitic limestone—because they break fairly easily into rounded pieces. When hard, flinty stones are crushed, many very sharp fragments are produced which can cause a lot of damage to the tubers, machinery and tyres (see STL 178).

On silty clay soils the "Dutch" method of planting may be used. The aim is to build up ridges of clod-free soil over the tubers so that mechanical harvesting is facilitated. Autumn cultivations consist of subsoiling, if necessary, to break soil pans—this should only be done when the soil is reasonably dry and a good cracking effect is possible. Ploughing should be fast, not more than 20–25 cm deep and furrow slices about 25 cm wide. In spring, the soil should be worked about 5–7 cm deep—ideally in one pass, and, if necessary, with power-driven harrows. When planting, the shallow layer of loose soil is used to make low ridges just covering the tubers (sets). The ridges are built up to normal size in several stages over a period of weeks by shallow inter-row cultivations and ridging (weed seedlings are killed in the pro-cess). Building up the ridges in several stages also allows the soil to warm up faster, and this method can also hasten the growth of earlies.

Planting. Nearly all potato crops in the U.K. are planted mechanically with 2–7-row machines—these are either fully automatic or require workers on the machines to place the tubers in cups or tubes. Hand-planting is still practised on some farms (mainly in Scotland) and involves making ridges, possibly applying FYM and fertilizers, planting the tubers from boxes or buckets, and then splitting the ridges to form new ridges over the tubers. Whatever method is used for planting, the ridges must be well formed to protect the new tubers from blight spores, and light which would cause greening.

Spacing of tubers (sets). Row width for earlies and seed (?), 60–70 cm: for maincrop, the commonest width is 75 cm, but 90-cm rows are used on many farms because of the shorter travelling distance (over 2 km/ha), and the higher speed which is possible when planting and harvesting; also, there is no yield loss and fewer problems in producing good ridges, keeping them free of clods, and fewer "greened" and blighted tubers. The spacing between the sets will depend on the seed rate and the size of the seed. This usually involves doing seed counts and using tables or other devices to decide on the machine setting. Automatic planters usually work better with close-graded seed such as 30–40 mm, 40–50 mm, and 50–60 mm, than with the normal 30–60-mm seed size—this may necessitate regrading the seed when it arrives on the farm (see L 653).

Manures and fertilizers. Potato crops benefit from organic manures such as FYM and slurry, and seaweed for earlies near the coast, not only because of the nutrient value (often ignored by farmers), but also because of its beneficial effects on soil structure and moisture retention in a dry season. It can be liberally applied when available—usually 30–60 tonne/ha of FYM or slurry equivalent. Normally this is ploughed-in during the autumn, but it could be applied to hand-planted crops in the spring.

Farmers, in general, apply liberal dressings of fertilizers to their potato crops and defend this by the justifiable claim that any surplus will benefit the following crops. It is either broadcast during the spring cultivations or, more efficiently, "placed" in the ridges by the planters.

The most profitable rates depend on numerous factors such as climate, soil type and state of fertility, previous cropping, variety, method of application, etc., so expert advice can sometimes be helpful—especially regarding the need for magnesium and trace elements.

Too much nitrogen applied to the seed-bed usually delays tuber development and so can reduce yields in some seasons. Nitrogen and potash increase yield by increasing tuber size; phosphate increases the number of tubers. A headache is the only certain result that is likely to follow attempts to put into practice all the varied recommendations on fertilizer (and seed) rates which are in circulation at present. Table 30 is intended as a general guide.

Weed control. Weeds, by tradition, have been controlled in good weather by cultivations and ridging, after planting. The frequent passage of rubber-tyred tractors tends to produce clods in the ridges and this hinders mechanical harvesting; cultivations also damage the potato roots and stolons, and allow moisture to escape from the soil, and so consequently, chemical weed control—mainly by contact and soil-acting residual herbicides—has become normal practice on most farms (see page 175 and Booklet 2260).

Blight. This is the worst fungous disease which attacks the potato crop; it can seriously reduce yield by killing the foliage early, and during periods of heavy rain the spores of the fungus can be washed into the soil and onto the tubers and so cause rotting of the tubers before or during storage. Resistance has been bred in some varieties, but this soon breaks down when new strains of the fungus appear. The disease spreads rapidly in warm, moist conditions if not prevented by effective spraying with fungicides such as maneb, mancozeb and "Patafol". To reduce the risk of spores spreading from the leaves to the tubers, the haulms should be destroyed when about 70% have been killed by blight—this is especially important if heavy rain is expected (see page 203).

Aphids. The considerable increase in physical damage and virus-spread by aphids in recent hot summers has necessitated the use of soil-acting granular insecticides (phorate or disulfoton), or sprays, on more than half the potato crops in the U.K. (see page 192).

Irrigation. In dry seasons, potatoes give very profitable returns from irrigation. Yield increases are usually about 1 tonne/ha per cm of water applied correctly. Quality may be improved by the tubers being more uniform in size and having less common scab, but the dry matter per cent may be lowered if excessive amounts of water are used. Timing is also very important; varieties, such as *King Edward*, which produce lots of tubers should not be irrigated until the tubers are about 1 cm long; if water is applied too late to a crop which has almost died

TABLE 30

| | Kilogrammes per hectare | | |
	N	P$_2$O$_5$	K$_2$O
Nationwide range of conditions	75–225	125–315	125–315
PMB survey average	175	180	250
Earlies	125	125	125
Normal good conditions—fertilizer only	150	200	250
Normal good conditions + 40 t/ha FYM	100	150	150
"Blueprint" (ADAS)—autumn	—	375	250
—spring	125	125	125
—top-dressed	125	—	—
—total	250	500	375
Nutrients removed by a 50 tonnes/ha crop (incl. haulm), but this is only a rough guide to the total uptake by plants	220	80	400

due to drought, then secondary growth may develop as knobs and cracks and spoil the crop. Earlies may be protected from frost damage by keeping the soil moist on warm sunny days which are followed by radiation frosts at night, or in more severe cases, by spraying the crop with about 2–3 mm of water per hour during the frosty period (the latent heat given out, as icicles form on the crop, prevents damage to the leaves). Irrigation may also be used to assist in breaking clods when preparing spring seed-beds in a dry time.

Harvesting. Earlies are harvested when the crop is still growing. To make lifting (and later cultivations) easier, the tops should be destroyed with a flail-type machine. Earlies must be treated gently because the skins are very soft. The time of harvesting (late May–August) is determined by yield and likely price. They are sold in 25-kg paperbags. Maincrops—especially if they are to be stored—should not be harvested until the tuber skins have hardened. This is usually about 3 weeks after the tops have died, or have been desiccated with diquat (not in dry soil conditions), dinoseb in oil, sulphuric acid or metoxuron.

About one-third of the total U.K. crop is picked by hand—usually by gangs of casual workers—into baskets and so into trailers or pallet boxes. About 150 person-hours per hectare are required, but this will vary with the skill of the pickers, the type of basket, the soil conditions, the yield and numbers of tubers, and how the work is organized. The "breadth" system (an area is lifted ahead of the pickers) is more efficient than the single-row "stint" system but it is risky in showery weather because potatoes which are rained-on will not store well. Clods and stones are less troublesome with hand-lifting than with harvesters.

One-row and two-row harvesters of various designs and capabilities are available to lift and load the crop into trailers. The ordinary one-row type can lift about 0.1 ha/hour and the two-row about 0.2 ha/hour, but soil conditions can affect the speed of working and the number of workers required on the machine to separate clods and stones from tubers. About 1000 tonnes/ha of

soil, clods, stones and tubers has to pass through the machine; X-ray controlled "fingers" and high-power suction-fan separators are being used to replace hand work. Very high rates of working are possible with some machines in reasonably dry conditions in stone- and clod-free conditions, e.g. the John Green/Johnson 202 outfit consisting of three two-row elevator diggers, a loader, four trailers and grader/clamping equipment, can lift and store up to 8 ha/day with nine or ten skilled staff.

Damage to tubers by harvesting operations results in serious losses on too many farms due to bad design and/or faulty operation of the machines. Damage can be very serious in very dry soil conditions when the tubers are somewhat dehydrated and more susceptible to damage, and the clods are very hard.

Storage. A high proportion of the maincrop has to be stored—some for only a short period, but some until May/June when the earlies start again. Badly damaged, diseased and rain-wet tubers will not store satisfactorily and should be sold at harvest time.

Potatoes can be stored in good condition in clamps (pits, graves) on a well-drained, sheltered site near a roadway, and covered with a good layer of straw and a 15–30-cm layer of soil; however, riddling and removing the potatoes in winter is usually a miserable job. Some outdoor clamps are more elaborate, e.g. the "Dickie" pie, with straw-bale walls, are more comfortable to work at in winter.

Nowadays, most crops are stored indoors—mostly in 150–500-tonne bulk stores—but also in pallet boxes (these are expensive, but ventilation can be very good and the tubers retain a good appearance). The boxes are usually stored indoors but may be outside on a concrete area and protected with straw bales, plastic sheets, etc. Care is required when filling bulk stores to ensure an even distribution over the floor and avoidance of "soil-cones" which can prevent air circulation. If the heap is more than 2 m deep, some form of duct ventilation will be required to prevent over-heating—this may be convectional or forced-draught ventilation, or, in the more sophisticated buildings, very accurate temperature control is possible by using refrigeration and recirculation ventilation. In most ordinary stores it is usual to put a deep (30–60 cm) layer of straw or nylon quilt on top of the heap to prevent greening and frost damage, and to collect condensation moisture—so keeping the tubers dry. Good ventilation, and allowing the heap to warm up to 15°C for 7–10 days after filling, helps to dry the tubers and to heal wounds; some farmers apply "Storite" to prevent rotting in store. For late storage, sprout growth has to be controlled in spring. This can be done by keeping the heap cool (4–7°C) by cold air ventilation, but this method is not satisfactory for crisping and chipping potatoes, because some starch changes to sugars (some reversal of this process is possible by warming the tubers before sale). Alternatively, chemicals such as technazene ("Fusarex") can be applied when the crop is going into store and is effective for about 5 months when ventilation is restricted. Alternatively CIPC can be introduced through the ventilation system and this prevents sprout growth for 3–5 months.

(See B 2099 and 2290, PMB publications, and *MAFF Bulletin* No. 173.)

Grading. The stored tubers are riddled (graded or sorted) during the winter or early spring when the best potatoes (ware) are separated from the chats (small tubers), diseased, damaged, over-sized, and misshapen tubers, and usually sold in 25-kg paper bags.

A wide range of equipment is available to do this. On ordinary farms it is not always possible to do this work efficiently—especially when selling in pre-packs and for special markets and also finding suitable outlets for the various grades—and so central grading stations, run by groups of co-operating farmers, are operating in some areas. Farm-gate sales are increasing and usually increase the net returns for the crop very considerably.

Yields before harvest may be estimated by digging random samples (see Appendix 12).

Specialist production requirements

Earlies. Early, frost-free area; well-developed single sprouts on fairly large seed—to give early, vigorous growth; close spacing (high seed rate) to make up for the low yield per plant at harvest. Earlies, in gardens, grow well under a covering of black polythene and are easily harvested.

Seed. See page 117.

Processing. Crisps—normally an advantage to grow on contract; high dry matter varieties, such as *Record*, are preferred, but must be handled and stored carefully to avoid damage and high sugar content.

Canned new potatoes. Small (20–30 mm), firm, waxy-fleshed tubers are required; high seed rate of multi-sprouted tubers (*M. Peer* is popular) and harvested in an immature state (see PMB booklet, *The Production of Potatoes for Canning*); best left to specialists.

"Blueprint" crops. In rich soil conditions potato crops in the U.K. have a potential yield of nearly 90 tonnes/ha; ADAS (at Stockbridge EHF) and others have demonstrated that this is possible (and very profitable) on a farm scale by doing everything exceptionally well, e.g. good seedbeds, early-April planting of well-sprouted and cooled-off, large, healthy seed at very high seed rates (over 8 tonnes/ha) and using a high yielding variety such as *P. Crown*; also very high phosphate and potash fertilizer rates (see page 98) and every attempt made to control aphids and blight by frequent sprayings; growth is maintained until late September; irrigation if required. This system has created a lot of interest and has stimulated many growers to adapt it to their own farms to improve their yields.

Ground-keeper potato weeds

Left-over potatoes, growing in the following crops, especially after a mild winter, are a serious problem because they are resistant to most of the ordinary herbicides used in other crops; however, "Roundup" (glyphosate) sprayed on cereal crops 1–3 weeks before harvest will kill them.

SUGAR-BEET

The sugar which is extracted from the crop

supplies this country with about 40% of its total sugar requirements.

It is grown on contract for the British Sugar Corporation (BSC) which has thirteen factories in the British Isles. A contract price per tonne of washed beet containing a standard percentage (usually 16%) of sugar is determined annually by the EEC. Sugar percentage, transport allowance, early and late delivery allowances and other levies, all go to make up the price which the farmer will receive.

There are, in fact, three possible prices which are based on a quota system, viz. A Quota which is the highest price received by the grower and which is paid up to an agreed amount of sugar produced by British growers in a contract year—1st July/30th June. If growers exceed the A Quota amount they are paid for the excess at a B Quota price, which is about 25 percent less than the A Quota. The "C Sugar" price is that paid for sugar produced in the contract year extra to the sum of the A and B Quotas.

Its price depends on the world's supply. It can be very low indeed.

The average sugar content is about 17%.

Apart from its sugar, sugar-beet has two useful by-products:

(1) Beet tops—a very succulent food, but which must be fed wilted.
(2) Beet pulp—the residue of the roots after the sugar has been extracted; an excellent feed for stock.

Yield. Good average yield of washed roots. 45 tonnes/hectare giving about 7.5 tonnes/ha of sugar. Also, 40 tonnes/ha of tops. The yield of tops varies with the variety and growing conditions. Stock farmers may prefer the large-topped varieties.

Seed rate. This depends on seed spacing and row width:

pelleted seed—5–18 kg/ha.
unpelleted seed (now unusual)—1–3 kg/ha.

Time of sowing. Mid-March–mid-April.

Time of harvesting. End of September to December.

Varieties

The NIAB recommended list, and the factory field officer will help in deciding upon the variety to be grown. There are many different varieties and they can be grouped according to the yield of roots, the sugar percentage, and their resistance to bolting, i.e. running to seed in the year of sowing. This is influenced by a cold spell in the three- to four-leaf stage of the plant. Bolting is highly undesirable, as the roots become woody with a low sugar content. Varieties with a high resistance to bolting should be used for early sowing and in colder districts.

Soils and climate. Sugar-beet can be grown on most soils, except heavy clays, which are usually too wet and sticky, thin chalk soils which do not retain sufficient moisture and on which harvesting is difficult, and stony soils on which cultivations and harvesting can be very difficult.

Sugar-beet is a sun-loving crop which will not grow well when there is too much rain and cloud.

Sugar is produced by the conversion of the energy of sunlight into the energy of the plant—sucrose in sugar-beet. Consequently, sunlight and the amount the crop leaf area which can absorb it, is an important factor in determining the amount of sugar produced by the crop. The grower should see that as much sunlight as possible is intercepted and utilized by a well-developed crop (and here irrigation may help) in the long days of May and June. This underlines the importance of early drilling and establishment of the crop as well as a uniform distribution of the plants (see page 102).

Place in rotation. As long as the land is reasonably clean of weeds, the place in rotation is not so important these days. As a condition of the contract with the grower and the BSC there must be at least a 2-year interval between sugar-beet and any other excluded crop, i.e. fodder beet, mangold, red beet, spinach and any brassica. All these crops can perpetuate the beet eelworm.

Seed-bed. The importance of a good seed-bed for sugar-beet cannot be over-emphasized. The success or failure of the crop can, to a great extent, depend on the seed-bed. It must be deep yet firm, fine and level (see page 40).

Manuring

(1) *FYM.* 25 tonnes of well-rotted manure per hectare applied in the autumn.

(2) *Lime.* A soil pH of between 6.5–7.0 is necessary.

(3) *Salt.* $1/2$ tonne/ha broadcast a few weeks before sowing. This should give nearly $1/2$ tonne extra sugar per hectare.

(4) *Magnesium.* Magnesium deficiency has become more evident in recent years, particularly on light, sandy soils. 400 kg/ha Keiserite (magnesium sulphate) should be applied if necessary.

(5) *Boron.* May also be necessary particularly (but not always) on well-limed sandy soils where boron could be in short supply. A boronated compound fertilizer is often recommended in this situation (see Tables 2 and 54).

Other plant foods required can be summarized as in Table 31.

TABLE 31

	N	P_2O_5	K_2O
	kg/ha	kg/ha	kg/ha
With FYM and salt	100	50	88
Without FYM but including salt	125	63	125

Kainit can be used at 500–600 kg/ha instead of potash, salt and magnesium (except in severe cases of magnesium deficiency where Keiserite is still necessary). It should be applied some weeks before sowing the seed.

The phosphate and potash can be broadcast and worked into the seedbed at any time over the preceding winter months.

Except after a very wet autumn and winter which may leave nitrogen reserves on the low side, nitrogen in excess of 125 kg/ha is unnecessary. It may, in fact, depress the sugar percentage without increasing the overall sugar yield.

Nitrogen must be applied in the spring. Band application between the rows after drilling is ideal, but otherwise half the application should be broadcast straight after drilling (not 14 days before!) with the balance after the beet have emerged.

The seed and sowing. The so-called beet seed is really a cluster of seeds fused together, and it usually produces more than one plant when it germinates. This "natural seed" is no longer used.

In order to make it possible to "drill to a stand" or easier to single the crop this "seed" has been separated out from its natural cluster to produce what is known as processed multigerm diploid and/or polyploid seed. The seed then produce a higher percentage of single plants on germination than the original natural seed. However, it has now been superceded by genetical monogerm seed which contains a single embryo from which only one plant will grow. Under laboratory conditions, monogerm varieties produce a minimum of 90% single plants.

Virtually all the crop sown now is with pelleted seed, i.e. it is coated with clay to produce pellets of uniform size and density, 3.50–4.75 mm. This is important for monogerm seed which is rather lens shaped and as such it cannot be handled so well in the precision drill. Some varieties have manganous oxide incorporated in the pellets as an insurance against manganese deficiency.

Drilling to a stand. 90% of the sugar-beet crop is now drilled-to-a-stand, i.e. the individual seeds are placed separately in the position required for the plant, using the precision drill. A very good seedbed and efficient pest and weed control (with herbicides) are essential for the system to be successful.

The main aim of drilling-to-a-stand is to have a reasonably-spaced, yet adequate, plant population. Ideally, this should average 80,000 plants/ha, and when it is less than 65,000 there is usually a serious reduction in final yield.

The number of plants which make up a crop of sugar beet is one of the most important factors in determining, not only the yield, but also the sugar content of the roots.

The seed spacing should be decided on the basis of the germination percentage of the seed (obtainable from the factory which supplies the seed), the possible losses before and after plant emergence (spray damage, inter-row cultiva-tion, wireworm and other pests) and the row width. As far as is consistent with any inter-row work and the harvesting of the crop, the row width should be as narrow as possible, but at present it is unlikely to be less than 46 cm. If, as is more usually the case, a 50-cm row width is being used and the seed is spaced, for example at 15 cm apart, with a plant establishment of 70%, this will give a plant population of 87,000/ha. With the probable losses mentioned, this will bring the population down to a very satisfactory figure. The seed spacing will, in fact, normally vary, depending on conditions, from 12 to 17 cm.

Apart from the saving in labour costs, another important advantage of drilling-to-a-stand is that, compared with what was known as conven-tional growing, the need with larger areas to space out the actual drilling (to make singling easier) no longer arises. Thus, the seed can be sown under the best possible conditions.

Evidence shows that, provided it is success-fully carried out, and this means one mature plant for every two seeds sown, there is no difference in final yield comparing drilling-to-a-stand with other methods of growing, assuming that there is a high level of crop protection.

Shallow-sowing will aid germination: 18–25 mm is ideal with a well-prepared seed-bed. A press wheel fitted immediately behind the seed coulter effectively compresses the soil over the seed and in dry weather this will help germination.

Treatment during the growing period. Band spraying with herbicides at the time of drilling is still the preferred method of controlling weeds *within* the row. It is cheaper than conventional overall spraying and, in fact, some growers like to leave weeds for a time *between* the rows. It is said that they provide alternative food for pests, and also on light soils these weeds help to reduce wind damage.

However, new techniques are being developed for post-emergence weed control in the sugar-beet crop, viz. repeat doses of lower than normal rates of herbicides in a low volume of water at high pressure, and/or repeat doses of low volume full strength herbicide at normal

pressure. These developments do open up the very real possibility of cutting out all inter-row work. But, at present, inter-row cultivation is the main way of keeping the crop clear of weeds.

The crop should emerge in weed-free condi-tions using chemical weed control (see page 175). But as soon as the seedlings can be seen as continuous lines across the field, steerage hoeing should be carefully carried out, usually twice, to assist in weed control. Further hoeing, after the beet has reached the 5–7 leaf stage, should be undertaken, but only as much as is necessary. Weeds have to be kept in check, but any inter-row cultivation does mean a loss of moisture.

Singling (thinning). Where this is still the prac-tice the seed will normally be precision-drilled at a 7.5 cm spacing for subsequent thinning at the 4–7 leaf stage. It is done by hand (mechanical singling has never been very successful, although it could now have better possibilities starting with a thinner row) usually at piecework rates. The final average distance apart of the plants in the row must depend on row width, i.e. rows of 50 cm—plants thinned to 22 cm apart, or rows of 55 cm—plants thinned to 20 cm apart. This will give, at any rate in theory, a plant population approaching the ideal 75,000–80,000/ha.

Virus yellows. It may be necessary to spray the crop to kill the aphids which spread virus yellows (see Table 54), or to apply the insecticide as granules spread on the soil.

Weed beet

Weed beet is now a very big problem for sugar-beet growers. About 50,000 hectares of the total sugar beet area (200,000 hectares) is affected. In 1977, although the British Sugar Corporation were warning growers about it, the situation then was not considered to be serious.

Weed beet are any unwanted beet within and between rows of sown beet and, of course, other crops. Unlike true beet which is a biennial, it produces seed in one year. It originates from several sources, but mainly from naturally-occurring bolters in the commercial root crop, from contamination of seed (by cross pollination from rogue plants), from beet ground keepers either growing in crops following beet, or from old clamp and loading sites, and from seed shed from weed beet itself.

Once weed beet becomes established it is self-perpetuating (although it can remain dormant for many years) and on average it produces 2000 seeds in the year. However, only about 50% survive, but nevertheless a light infestation (suggested 1000/ha) can mean one million weed beet/ha the following year.

Control measures. It is very important to prevent any weed beet seed returning to the soil and so, to start with, inter-row work should be carried out to clear any seedlings which are between the rows (overall spraying is out; at present there are no selective herbicides to control the weed in sugar beet). Following this, as far as possible, all bolters should be removed from the beet field, no later than the middle of July, i.e. before the seed has set. If roguing is not possible, the bolters can be mechanically cut down—usually twice—the first time 7–10 cm above crop leaf height, and then 14 days later to prevent lateral shoots and late flowering plants from producing seed, cutting is repeated—this time at crop leaf height.

Instead of mechanical cutting, the "weed wiper" fitted to the front or rear of the tractor can be used. This consists of a boom containing a herbicide (at present glyphosate is recommended) which permeates through a nylon rope "wick" attached to the boom. As it goes through the crop the "wick" wipes the taller weed beet and bolters (taller, that is, than the beet plants) and so deposits the herbicide. Usually more than two treatments are necessary, starting before flowering at the beginning of July and then repeating it at 10–14 day intervals.

In addition to preventing the propagation of seed from bolters or weed beet which has already contaminated the crop, it is important that weed beet in crops *other* than sugar-beet should also be controlled. Weed beet found growing on old clamp and loading sites, headlands and waste land, must also be destroyed.

Finally, with a bad infestation in a field, it will be necessary for the grower to widen his rotation to 7–8 years, rather than the more usual 3–4-year interval.

Future development with the sugar-beet crop

The bed system. The developing work on weed control by overall spraying and the use of multi-row harvesters have opened up the way towards the bed growing of sugar-beet. It is a well-established practice with some field-scale vegetables and there is no reason why it should not be used for sugar-beet as well.

Although it is not yet on a commercial scale for sugar-beet, the bed system could offer quite a few advantages, one of the more important being that wheeling compaction is considerably less (at the most only 15% of the land surface) which should lead to better-shaped roots. Work on the crop should also be easier with the wheelings already marked out. And, where applicable, to reduce wind erosion, straw could easily be planted between the beds.

Transplanting. Transplanting of the beet plants, using soil blocks and mechanical transplanting equipment, is also being examined. It was, in fact, first tried out with sugar-beet at Nottingham more than 20 years ago. Trials with transplanted beet showed a 23 per cent increase in yield compared with conventionally-grown beet. Only about 28 m² of seedbed appear to be necessary to provide one hectare of transplants at a plant population of 83,000/ha.

Work is now being carried out by ADAS to examine the economic feasibility of such a practice.

Irrigation. Although it is a deep-rooted crop, if it is allowed to grow in uncompacted soil, the beet plant will respond to supplementary water in the majority of seasons. But in spite of this, less than 10% of the British sugar-beet crop is irrigated.

If, by irrigation, the plant can develop a reasonable canopy of leaves early on in its life it will be able to make more efficient use of sunlight for growth and sugar production (see page 101). Irrigation can also help to improve the root shape as well as preventing wilting in the day which, if it continues, will bring about leaf loss and thus low yield.

However, irrigation should not be necessary after August even if there is a water shortage because, by then, the deep root should easily be able to cope.

Harvesting. In theory, early November is the right time to harvest the beet crop. Although it is still slightly increasing in weight the sugar percentage is beginning to fall off.

But, of course, if all beet were delivered at this time there would be tremendous congestion at the factory. Therefore a permit delivery system is used, whereby each grower is given dated loading permits which operate from the end of September until about the end of January. It means that growers must be prepared to have their beet delivered to the factory at intervals throughout the period upon receiving prior notice from the factory.

The crop is now all mechanically harvested. Topping must be done accurately immediately above where the lowest leaves or buds on the crop have grown (see Fig. 40). Overtopping can result in a serious loss of yield, and under-topping means paying extra carriage for unwanted material. The modern complete harvester either tops, lifts and delivers the beet into a vehicle running alongside, or after topping

and lifting collects the beet in a self-unloading tank on the harvester. Most sugar-beet harvesters can be fitted with "top-savers" which leave the tops in clean condition for subsequent feeding. The tank-harvester (one- and two-row) has many advantages, not least of which is that harvesting can be operated on a one-man system. Multi-stage machines have now been re-introduced. They deal with either three, five or six rows. With the six-row, for example, the system consists of a six-windrowing topper in the first stage, followed by a six-row lifter in the second stage, with harvesting completed in the third stage by the roots being picked up from the six-row windrows. It is capable of harvesting at 0.5 ha per hour.

To minimize dirt tares, direct delivery from field to factory should be avoided, especially when the beet are harvested in wet conditions. If the beet can be clamped at least a week before delivery, and then reloaded using an elevator with a cleaning mechanism, dirt tares will be considerably reduced.

Beet stored on the farm should be done carefully, with the clamp adjacent to a hard road and accessible at both ends so that beet which has been clamped longest can be delivered first. Plenty of air circulation should be allowed (use a ridge-type clamp) for beet which is clamped early whilst the temperature is still high. However, as it gets colder this is less important and large square clamps, or heaps against a wall, can be built.

It is important to prevent the beet being damaged by frost and, depending on the district and incidence of frost, the clamp may have to be strawed.

MANGELS AND FODDER BEET

Both mangels and fodder beet are grown for feeding to cattle and sheep in the more southerly parts of the country where they are more certain croppers than swedes and turnips. The latter are preferable in the cooler and wetter regions.

About 6000 hectares are grown and there is at present a renewal of interest in these crops,

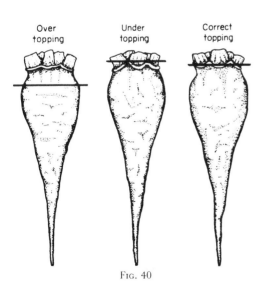

Over topping   Under topping   Correct topping

FIG. 40

principally because it is now possible to grow, harvest and feed them mechanically.

Mangels and fodder beet are palatable and highly digestible and, provided they are harvested before the risk of heavy frost, they store very satisfactorily.

Mangels

Yield per hectare. 100–125 tonnes—fresh crop; 12.5 tonnes dry matter.

Varieties.

Low dry matter group (10–12% DM)—English origin—e.g. *Red Intermediate, Prizewinner, Yellow Globe*.

Medium dry matter group (12–15% DM)—Continental origin, e.g. *Peramono*—a monogerm variety.

The low dry matter group tend to grow with less root in the ground which makes mechanical harvesting more dificult. The NIAB Leaflet No. 6, *Fodder Root Crops*, will give more detail on varieties.

Seed rate per hectare. 1.5–2 kg—rubbed and graded. 4 kg—pelleted, i.e. *Peramono*.

Sowing. April to early May precision seeded in rows 50–60 cm apart; unnecessarily wide rows reduce the yield; the ultimate aim should be a plant population of 65,000 plants/ha.

Fertilizers. 125 kg each of nitrogen, phosphate and potash. If salt has been applied at 370–400 kg the potash can be reduced to 60 kg.

Pre- and post-emergent herbicides can be used for keeping the crop growing in weed-free conditions.

The crop may have to be protected against the flea beetle and mangold fly as well as the aphid (see Table 53).

Other treatment during the growing period can follow the same pattern as for sugar-beet. An increasing number of crops are now drilled to a stand which will obviate any singling; crops not grown this way may need to be rough-singled.

Harvesting. November is the normal time to harvest the crop. Mangels are difficult to top and harvest by machine because they grow almost on top of the soil. Therefore they are simply pulled out and because they "bleed" very easily the tops (not the crowns) are either cut or twisted off. The roots are left untrimmed. Ideally, they should be left for a period in the field to "sweat out" in small heaps covered with leaves. Following this they can be carefully clamped. This should be built up as high as can be conveniently heaped (3 m) and then covered with a good layer of straw. This should be removed if the mangel is still in store the following spring when the temperature starts to rise.

Fodder beet

This has been bred from selections from sugar-beet and mangels. It has a higher dry matter content than the mangel averaging 15 tonnes/ha, and a yield of fresh crop of 85 tonnes/ha. More interest is being taken in fodder beet again following its decline in popularity in the middle 1950s prior to which it was chiefly fed to pigs. Now it is considered more as a crop for cattle.

Varieties.

Medium dry matter group (15–17% DM) e.g. *Kyros, Monoval*.

High dry matter group (17–20% DM) e.g. *Trestel, Monorosa*.

All varieties are monogerm.

These roots (particularly the high dry matter varieties) grow mainly below the ground and they closely resemble sugar-beet.

The production of fodder beet is similar to that of mangels, although 75,000 plants/ha is the target plant population.

Harvesting. October and November. The crop is harvested by machine and, as far as possible, the tops which are bigger than the mangels should be utilized to get the maximum production from the crop. The tops can be fed fresh to cattle or they may be ensiled and, in either case, it is important that they are kept as clean as possible.

Unlike the mangel, the fodder beet can be fed fresh to stock. It is, however, normally clamped in a similar way to mangels.

TURNIPS AND SWEDES

They are the most widely grown root crops in the northern and western regions of the country. Although their popularity has declined markedly in the last 20 years, new methods of growing and harvesting these crops is seeing a reversal of this trend. At present about 40,000 hectares are grown annually.

In appearance, the difference between the two crops is that turnips have hairy, grass-green leaves which arise direct from the bulb itself, and swedes have smooth ashy-green leaves which grow out from an extended stem or "neck". Nutritionally, swedes are more valuable than turnips as they have a higher dry matter content—swedes 9–12%, turnips 6–10%. The turnip has a shorter growing period and some varieties (notably from the Continent) are being increasingly used for catch-cropping (page 109).

Both are valuable for cattle and sheep and, depending on the variety grown, they can be used as table vegetables. A limited market is developing for contract growing for freezing, and for turnips particularly this is quite a specialist crop.

Main crops—yield per hectare—turnips, 60–80 tonnes; swedes, 70–80 tonnes.

Seed rate per hectare: 0.3–4.5 kg (lower amount with the precision drill which is now more often the case) in rows 17–35 cm for subsequent chemical weed control and 50–55 cm for inter-row cultivation. Turnips can also be broadcast at 4.5–7 kg.

Dual-purpose seed dressings, containing gamma-HCH as a protection against flea-beetle, and thiram for protection against soil-borne fungus, should be used.

Time of sowing: mid-April to end of June; the earlier the better for swedes. Any mildew which is more prone with earlier sowing, can be minimized by careful chemical treatment, e.g. tridemorph. In addition an increasing number of swede varieties are showing quite good resistance to mildew.

Climate and soil. The crops like a cool, moist climate without too much sunshine. Most soils (except heavy clays) are suitable.

Seed-bed. A fine, firm and moist seed-bed is necessary to get the plant quickly established. In very wet districts the crops can, with advantage, be sown on the ridge.

The plant foods required can be summarized as in Table 32.

TABLE 32

	N	P_2O_5	K_2O
	kg/ha	kg/ha	kg/ha
With FYM	50	50–75	50
With slurry	50	25–50	0
Fertilizers only	50–100	100–125	60–100

FYM, 25–40 tonnes/ha is applied in the autumn. This is especially important for improving the water-holding capacity of the lighter soils. Slurry can be used instead at about 35,000 litres/ha (undiluted).

The lower amount of nitrogen is used in wetter areas and more phosphate is needed on heavier soils.

Lime is most essential. Soil pH should be above 6.0 as Clubroot (finger and toe disease) can be prevalent under acid conditions. However, over-liming is equally serious as it can cause brown-heart (raan), see page 202.

The stale seed-bed technique using a contact herbicide to kill the weeds at the time of drilling, followed up, if necessary, by the use of a post-emergence herbicide such as propachlor applied when the crop has 3–4 true leaves, should keep it clean of weeds. However, inter-row cultivations are still carried out by many growers, and this will follow the same lines as with most root crops.

The low seed rate with precision drilling should mean a spacing of between 7 and 12 cm without any thinning. However, if it were considered necessary, this would be carried out at the 2–4 leaf stage at about 25 cm apart.

Harvesting turnips. The main crop is ready for lifting and feeding in October when the outer leaves begin to decay. They can now be harvested by machine and then stored in one main

Varieties

TABLE 33

Varieties	Feed value	Remarks
Turnips		
White-fleshed	Low	Heavy croppers; poor keepers; can be grown as a catch crop, as well as a main crop
Yellow-fleshed	High	Slowest to mature; keeping quality good; require fertile conditions for best results
Swedes (grouped according to skin colour)		
Light-purple	Generally low	Normally quickest to mature and heaviest yielder
Dark-purple	Medium to high	Not heavy yielding. Highly resistant to frost damage. Suited more to conditions in Scotland
Green	High	Later to mature and fairly hardy, many varieties are resistant to club-root disease
Bronze	Very mixed varieties as regards yield, time of maturity and hardiness	

For more details of varieties see the NIAB recommended list No. 6, *Fodder Root Crops.*

clamp (the leaves may be left on) in a similar way to mangels.

In *mild* districts yellow turnips can be left growing in the field and removed as required, or they may be stored in small, roughly covered heaps in the field.

Both white and yellow turnips, but particularly white turnips, are often grazed off in the field.

Harvesting swedes. In most districts swedes are lifted about November, before they are fully matured. It is therefore advisable to allow roots to ripen off in a clamp to minimize scouring when later feeding to stock. In fact, in *mild* districts they may be left growing to mature in the field. Special machines are now available for the complete harvesting of the crop. A 70-tonne/ha crop can be topped, lifted and elevated into a trailer at a rate of about one-tenth of a hectare (nearly one-third of an acre) per hour.

Swedes are vulnerable to rotting under poor storage conditions and when inadequately protected against the frost.

Swedes, like turnips, are very often grazed off in the field, but both crops can deteriorate under wet and frosty conditions.

KALE

Kale is grown for feeding to livestock, usually in the autumn and winter months. It can be fed either on the field, or, it can be cut (normally with a forage harvester) and carted off for feeding green or making into silage, and in these cases a heavier yielding crop is needed. It is mostly grown in the south and south-west of England, although it is not as popular as it was 10 years ago.

Yield. 35–75 tonnes/ha.

Time of sowing. End of March–mid-July. Early sowing gives heavier crops although later sowing may fit better into a cropping sequence.

Seed rate. 1-4 kg/ha *drilled* in 35–60 cm rows; 4.5–7 kg *broadcast.* The seed should be dressed with gamma-HCH as a protection against flea-beetle.

Types of kale

Marrow stem
  1 metre tall. Thick stems and large leaves. Produces heavy crops of good digestibility but not winter-hardy. Use before the new year.
Thousand-head
  About 1 metre tall. Much branched plant with numerous small leaves. Hardier, but less palatable and digestible than marrow stem.
Dwarf-thousand-head
  Useful for feeding in late winter. Shows a

considerable growth of new shoots late in the winter. Shorter stemmed but more palatable and digestible than thousand-head.

Hybrid kales

A number of triple cross hybrid varieties have been bred at the Plant Breeding Institute at Cambridge. They are not so tall as the traditional marrow-stem and thousand-head types which makes them more suitable for grazing. They are winter-hardy and also show higher D values than the older varieties. Maris Kestrel is at present the most widely grown of the hybrids, but the recently-introduced variety, Bittern, a cross between marrow stem kale and Brussels sprouts looks extremely promising with its very high yield and outstanding winter-hardiness.

The NIAB recommended list No. 2, *Green Fodder Crops*, will give advice on varieties.

Climate and soil. Kale is a very adaptable crop, although under very dry conditions it may be difficult to get it well established. For grazing it is preferable to grow it under drier, lighter soil conditions, or on well-drained soils.

Manuring. FYM—up to 50 tonnes/ha is applied in the autumn. It is especially important when a heavy-yielding crop is the aim. Slurry (undiluted) at 35,000 litres/ha can be used instead, and this is normally applied in the autumn.

Other plant foods required can be summarized as in Table 34.

TABLE 34

	N	P_2O_5	K_2O
	kg/ha	kg/ha	kg/ha
With FYM	125	63	63
With slurry	150	75	0
Fertilizers only	150	125	125

The fertilizer is usually applied during final seed-bed preparations. The nitrogen can be split and part top-dressed, but in most seasons it is unlikely to prove worthwhile.

Seed-bed. A fine, firm, and clean seed-bed is required. But the crop can be direct-drilled and this has the important advantages of conserving moisture at sowing time and leaving a much firmer surface for grazing and good annual weed control.

Treatment during the growing period. With drilled crops, steerage hoeing may have to be carried out a number of times from an early stage, but the herbicide desmetryne has eased the problem of keeping the crop weed-free (see page 176) (see GFO24).

If necessary, the crop should be dusted against flea-beetles.

The drilled crop is seldom singled. With a precision-drilled crop, the aim should be to produce a plant every 7.5–15 cm in the row. With an unthinned crop a higher proportion of leaf to stem is obtained which produces more succulent plants. Sometimes the crop is harrowed across to make it thinner.

Utilizing the crop. With cattle, grazing in the field using the electric fence avoids the laborious job of cutting and carting the crop. But light or well-drained soils are essential, otherwise both stock and soil suffer; cutting before fencing reduces waste.

To avoid wet and sticky conditions which normally get worse towards the end of the year, the tendency these days is to start feeding kale to dairy cows much earlier in the autumn, although in a mild season it will continue to grow on well into early winter.

If strip-grazing is not possible, the forage-harvester may be used to chop the crop coarsely and blow it into a trailer, or it can be cut with a drum mower and wilted before feeding.

STUBBLE TURNIPS

These quick-growing, white turnips which originated from the Netherlands derive their name from the practice of growing them on the stubble following an early-harvested cereal crop. However, they can be shown as early as April when they will provide a useful mid-summer feed, although they are not often used in this way.

As far as possible, stubble turnips should be sown by the end of July. In most seasons it will hardly be worth while to sow later than mid-August. Direct-drilling is worth considering; it helps to conserve moisture at what is usually a dry time of the year.

Yield. 50–60 tonnes/ha—fresh crop; 5 tonnes dry matter, produced 100 days after sowing by end of July. Up to 100 tonnes from crops drilled in Spring and early Summer.

Seed rate. 1–3 kg/ha drilled in rows 18–25 cm; 4–5 kg/ha broadcast. The seed should be dressed with gamma HCH as a protection against flea beetle.

Varieties. Early—*Civasto R, Debra, Marco.* Late—*Ponda, Taronda, Tigra, Appin.*

Notes.

(i) Some varieties when sown earlier in the year are susceptible to bolting.

(ii) *Appin* is a cross between *Tigra* and a Chinese Cabbage. Its growth is similar to stubble turnips, but is has a very high leaf to bulb ratio, although with good root anchorage. It has good disease and club root resistance.

(iii) The NIAB recommended list No. 6, *Fodder Root Crops,* will give more details of varieties.

Fertilizer. The emphasis with stubble turnips should be on nitrogen when up to 100 kg can be broadcast when sown. Phosphate and potash will not always be necessary although this will depend on soil indices.

Weed control. Should not be necessary.

Utilization. The crop is ready for grazing 3 months after drilling.

FORAGE RAPE

This is a very quick-growing palatable crop which is ready for grazing after 12 weeks.

Yield. 33–35 tonnes/ha; 4 tonnes dry matter. There are two main types:

(i) Giant with well-developed leaves and stems and an average height of 78 cm and a D value of about 69. Most varieties show reasonable resistance to mildew, but are susceptible to club root.

(ii) Dwarf—shorter, more prostrate habit of growth; average height 66 cm and a D value of about 70. However, generally not so palatable and not, in fact, so popular as the Giant type, mainly because of a lower yield of digestible organic matter. The mildew resistance is also not so high and only Nevin shows some resistance to club root.

Examples of varieties: *Winifred, Windel, Nevin.* The NIAB recommended list No. 2, *Green Fodder Crops,* will give more information on varieties.

Seed and sowing. The seed is either drilled at 2 kg/ha or 4.5 kg/ha broadcast from the end of April until mid-August, usually following in this last case an early harvested grain crop. Direct-drilling is often carried out. All seed should be dressed with gamma HCH as a protection against flea beetle.

Fertilizer. Up to 100 kg nitrogen can be used. Phosphate and potash are not always necessary, but this will depend on soil indices.

FODDER RADISH

This is grown for forage purposes, green manuring (see page 59) and game cover. It is very quick growing and is suited to most soil conditions. *Neris* and *Slobolt* are the most commonly-grown varieties. Fodder radish is normally sown in July and August for forage at 8 kg/ha drilled, or 13 kg broadcast. For green manuring it may be sown earlier with a heavier seed rate of 17 kg. Up to 75 kg nitrogen can be applied with the seed.

It should be grazed off before flowering and whilst it is still palatable, it being normally utilized within 8–12 weeks of sowing.

Fodder radish is very susceptible to frost but resistant to club root.

RAPHANOBRASSICA

This is a new forage plant, resulting from a cross between fodder radish and kale. It looks like fodder rape. Trials suggest that it can out-yield both forage turnips and rape. It also shows good disease resistance.

FORAGE PEAS

This appears to be an extremely flexible crop. It can be sown alone, following for example winter barley, or as part of an arable silage mixture. It is also useful in establishing leys—acting as a form of nurse crop.

Forage peas are ready for utilization within 12–14 weeks of sowing.

CATCH CROP (FORAGE CATCH CROPS)

This is the practice of taking a quick-growing crop between two major crops. With the dairy herd and sheep flock it can be very profitable provided:

(1) It does not interfere with the following main crop. This can happen when the main crop is planted late in badly prepared seed-bed.
(2) The catch crop is not grown when more attention should have been given to cleaning the land of weeds.
(3) It is grown cheaply, without expensive seed-bed preparation, and with limited use of fertilizers, except possibly nitrogen.
(4) The fodder catch crop fits the farming system. In rotations where brassicas and sugar beet are grown, the catch crop can propagate clubroot and beet eelworm, although there are varieties resistant to club root.

There is now an increasing interest in catch cropping, chiefly through the development of direct-drilling the autumn stubble with crops such as rape and the continental turnip.

Direct-drilling (page 41) has the advantage of conserving what moisture there is available in the soil at that time of the year. This helps to establish the catch crop that much more quickly.

Broadcasting gives a better cover and it should still be contemplated in the wetter areas.

Examples of catch-cropping with forage crops.

(1) Main crop—Winter barley.
Catch crop—for feeding until Christmas:
  (i) stubble turnips such as *Debra* at 3 kg/ha, with 75 kgN/ha;
  or
  (ii) *English Giant* forage rape at 6 kg/ha, with 75 kgN/ha. For feeding after Christmas, more frost-resistant crops such as the English yellow turnip at 5 kg/ha should be grown with 60 kgN/ha.
Main crop—Spring cereal or possibly kale.

(2) Main crop—Cereal undersown.
Catch crop—Italian ryegrass at 30–35 kg/ha with 60 kgN/ha.
This grass can provide:
  (a) stubble-grazing;
  (b) first bite, after having applied 75 kgN/ha (see Table 44);
  (c) possibly another grazing or silage before planting.
Main crop—Kale.

(3) Main crop—2nd early potatoes (harvested in July) or a ley (ploughed in July).
Catch crop—Rye (e.g. *Rheidol*) at 190 kg/ha with Italian ryegrass at 10 kg/ha and 50 kgN/ha to provide:
  (a) possibly an autumn graze (depending upon the season);
  (b) first bite, after having applied 80 kgN;
  (c) silage, after having applied 75 kgN, before planting.
Main crop—Kale.

(4) Main crop—Early potatoes or vining peas.
Catch crop:
  (i) Stubble-turnips, e.g. *Debra* at 2.5 kg/ha to allow at least a 10-week growing period: 75 kgN/ha applied at sowing.
  or
  (ii) Fodder radish, e.g. *Neris* at 8 kg/ha to allow at least an 8-week growing period: 75 kgN/ha applied at sowing. Either will provide grazing until end September before sowing.
Main crop—Wheat.

(5) Main crop—Winter wheat, undersown in autumn with
Catch crop—3–4 kg/ha Italian ryegrass.

This will produce and shed seed before harvest and this usually provides:

(a) first bite, after having applied 75 kgN;
(b) possibly another grazing, or silage cut after applying 75 kgN before planting.

Main crop—Roots or kale.

CABBAGES

Cabbages have a wide range of uses. They are a useful food for all classes of stock and, when suitable varieties are grown, high-value crops can be produced for human consumption. These can be fed to stock when the market price is too low.

Cabbages prefer moist, heavy soils, and seasons with plenty of rainfall, but apart from this the same growing conditions suit cabbage as kale. They respond to plenty of fertilizer, usually 200 kg/ha P_2O_5 (depending on soil index) and 225 of K_2O (depending on soil index and whether the crop is to be cut and removed from the field).

Varieties. There are a large number of varieties which can be roughly grouped as in Table 35.

A reliable variety should be chosen within these groups, and the NIAB recommended list will give details of varieties of flat poll cabbage. The seed is normally sown in specially prepared seed-beds. About 1 kg of seed gives sufficient plants for planting 1 hectare. 75 kg/ha of P_2O_5 should be broadcast before sowing. For planting out, a transplanter can be used. They are best planted on the square, i.e. approximately 60 cm between the plants and 60 cm between the rows. This will help in later cultivations. There is now an increasing tendency to space-drill the crop with the seed placed either three at a time—25 mm apart, or single-spaced 10 cm in the row. The plants are eventually thinned down to the required distances apart.

BRUSSELS SPROUTS

Brussels sprouts can be grown on a commercial scale in most parts of the country except South-west England and West Wales where the climate particularly favours ringspot—a fungus disease which at present cannot be controlled.

The ideal soil is a well-drained medium to heavy loam with a pH of at least 6.5.

The growing and marketing of the crop should now be considered in two different ways:

(1) for the fresh market or successive picking,
(2) for quick freezing or single harvesting.

1. The fresh market

Yield. 15–20 tonnes/ha.

Transplanting. The management is similar to that for cabbage growing. Plants are raised from seed in a seed-bed with a seed rate of 0.5–2.0 kg for every hectare to be eventually planted out. The actual rate will depend on the seed count (1000s/10 g), "field factor" (conditions which affect seedling emergence and growth) and eventual plant population required for the crop.

The seed should be treated with a combined gamma HCH/thiram or captan seed dressing against the flea beetle, cabbage stem weevil and soil-borne fungi.

The seed is sown:

(1) in August (early enough to withstand the winter) to be planted out in March and April, ready for picking late August and September;

TABLE 35

	Sown	Planted out	Ready for consumption	
Early	March	April, May	August–September	Can be used for stock and human consumption
	May	June, July	October onwards	
	August	Following April September	July March–June	
Late *or*	August	Following April	September	Chiefly used for stock feeding
Maincrop	August	October	July onwards	

or

(2) from mid-January to mid-March in batches to be planted out from April to June ready for picking October to March, according to variety and season.

The plants are normally, but not always, planted "on the square" at 60–90 cm apart, depending chiefly on the variety. The non-hybrids are better suited to a wider spacing and hence a lower plant population. This last will vary from about 12,000 to 45,000 per hectare.

Machine transplanting is very common now and, operated properly, it can ensure a very satisfactory stand. However, to ensure a full and uniform stand any gaps should be "dibbled in" by hand as soon as possible (within 10 days) after the main plant.

Cabbage root fly is a big problem for summer-planted crops, i.e. after the end of April. Granules applied at the time of planting or worked into the top centimetre of soil before space-drilling will give quite useful control of the larvae as they emerge from the eggs.

Space-drilling. This is often somewhat confusingly referred to as direct-drilling.

Brussels sprouts may be space-drilled in rows at anything from 45–90 cm apart (where high-yielding large sprouts are required, although the latter is becoming unusual), with a seed rate from 0.25–0.75 kg/ha, depending on seed count, germination percentage and required spacing. The seed is normally dropped in groups of three or four seeds at 25–35 mm apart, or alternatively one may be placed every 10 cm apart. In both cases, surplus plants are removed when large enough.

There are points for and against space-drilling compared with transplanting the nursery-reared plants.

At present it takes a total of about 37 hours per hectare to drill and single out the plants, as against 67 for raising and transplanting. But, as far as the farmer is concerned, he has none of the critical risks attendant in raising the plants from very costly seed. Also the sprout field is not occupied for so long with the crop. However, in spite of this, transplanting is now less popular.

Varieties. Medium to large sprouts—Early up to mid-October, Early F1 hybrids, e.g. *Topscore*, *Lancelot*. Mid-season mid-October to end December, Roofnerf selections from Holland and mid-season F1 hybrids, e.g. *Achilles*, *Lumet*. Late after December, F1 hybrids, e.g. *Ulysses*, *Wellington*.

Vegetable Growers leaflet No. 3 will give further details.

Place in rotation. Brussels sprouts are fairly flexible in this respect. Provided the land is clean and fertility is reasonably high, the crop will fit well into the cereal rotation. The earliest crops are cleared by the end of September and so it is possible to follow with winter wheat, although normally it is spring barley. Because of club root it is inadvisable to grow sprouts less than at least 1 year in 3.

Manuring. The plant foods required in the seed-bed are shown in Table 36.

TABLE 36

	N	P_2O_5	K_2O
	kg/ha	kg/ha	kg/ha
With FYM	150	75	105
With slurry	190	75	75
With fertilizer only	190	125	225

Nitrogen top-dressing will be required to make total N up to at least 300 kg/ha depending upon the appearance of the crop. The F1 hybrids generally require heavier top-dressings.

Treatment during the growing period. This will generally follow the same procedure as for the cabbage crop. Instead of inter-row hoeings and depending on the weeds present, trifluralin worked into the seed-bed and/or propachlor sprayed on surface will deal with most annual weed problems.

Control measures may be necessary against the cabbage root fly (gets inside the sprout as well), aphid and cabbage caterpillar (see Table 54).

Powdery mildew is at present the most serious disease likely to affect Brussels sprouts (see Table 54).

Stopping or cocking the plants. By removing the growing point (or terminal bud) from the plant or by destroying the bud with a sharp tap using a rubber hammer, sprout growth is stimulated and this will produce 5–10% higher yield, especially for the earlier varieties, as well as enabling harvesting to be carried out at an earlier date. Stopping takes place from early August to about the third week in October. Plants should be stopped about five weeks before the expected picking time, but there is no point in stopping after October.

Harvesting. Picking usually starts in early September and extends until March according to the variety and season. For maximum yield the plants are picked over four to eight times (at intervals of about 4 weeks) during the season.

EEC Standard for Fresh Sprouts came into operation in February 1974 for produce sold on the domestic market.

A summary of the *Standard for Crop Quality* shows:

Size: 20–40 mm diameter without other parts of the stem, i.e. clean cut.

Colour: dark green.

Texture: firm, dense without "wings" and smooth.

Shape: spherical and entire for the market.

Absence of defects: good bud resistance to powdery mildew, no frost damage, no internal browning, no soil, insects or excess moisture present.

Flavour: free from foreign smell or taste.

2. For freezing

Varieties producing medium-size sprouts are necessary for these crops grown on contract for the frozen food market.

Yield. 12–15 tonnes/ha (usually picked once only).

Varieties. F1 hybrids such as *Lancelot, Topscore,* for early single harvest; *Ladora, Ulysses,* for late single harvest.

F1 hybrids are particularly important for single once-over harvesting because of their evenness of maturity.

The growing of the crop is similar to that for the fresh market, although present experience indicates February nursery sowings (in cold frames) for transplanting in May, for harvesting up to the middle of October, and outdoor sowings in March for picking from mid-October till the end of the year.

Space-drilling is normally carried out in April and May.

For mid-season optimum yields, the plants should be grown on a 52.5 cm square, but up to a 67.5 cm square for the earlier August harvesting.

Harvesting of the single-pick crop involves de-leafing the stems immediately before they are cut by hand or machine for transport to the sprout-stripping machine. Mobile strippers are now being increasingly used in the field.

Occasionally the earliest maturing varieties are picked over by hand before the whole plant is cut for stripping.

(See MAFF, Reference Book 323, *Brussels Sprouts.*)

CARROTS

The common yellow carrots may be grown on contract for canning or quick-freezing but most of the crop is sold as a vegetable—either as fully grown clean roots or as bunches of young carrots. Unsaleable and surplus roots can be fed to stock (usually cattle); the white cattle carrots are not grown in this country now.

Yield of roots. Average, 30 tonnes/ha; good, 60 tonnes/ha.

Soils and climate. The climate in most parts of this country is suitable for carrots but the main growing areas are limited by soil conditions. Carrots can only be grown successfully on a farm scale on deep sandy loams and black fen soils, mainly because it is easy to lift and clean the roots. Stony and shallow soils produce badly shaped roots which may be unsaleable.

Seed-bed. Should be fine, firm, clean and level (as for sugar-beet). If there is a pan in the soil it must be broken by subsoiling.

Varieties. There are many good varieties which are grouped by the shape of root as *long, stump-*

rooted and *intermediate*. The latter are the most common type.

Sowing. The seed, which usually has the bristles machined off it, may be sown shallow 2–3 cm in:

(a) wide rows 45–50 cm—weeds controlled by inter-row cultivations;
(b) narrow rows 15–20 cm or in beds 1–1.5 metres wide—weeds controlled by chemicals.

The yield from (b) is very much greater due to the high plant population per ha (up to 370 plants/m^2). (See STL 137.)

Seed rate. Up to 5 kg/ha is sown where the crop is singled. This is not a common practice now when it is more usual to sow about 1–2 kg/ha with a precision drill and leave all plants to mature.

Time of sowing. Main crops—April, but varies from late February to May.

Manuring. FYM can be beneficial on sandy soils but it must be well rotted and ploughed in deeply in the autumn. The fertilizers used on average fertility soils without FYM should contain about 55 kg/ha N, 100 kg/ha P$_2$O$_5$ and 125 kg/ha K$_2$O; 375 kg/ha of salt may be applied several weeks before sowing and the potash reduced to 60 kg/ha.

Chemical weed control. This has greatly simplified carrot growing on many farms. Contact chemicals can be used to kill seedling weeds before the crop emerges (as in sugar-beet), or, most tractor vaporizing oils and some specially selected mineral oils can be used, when the carrots are in the seedling stage, to kill a wide range of weeds. Pre-emergent treatment with linuron, chlorbromuron, and prometryne gives very good weed control; these herbicides can also be used as post-emergence sprays. Metoxuron can also be used. (See STL 52.)

Harvesting. This usually starts in October and may continue well into the winter in areas where severe frost is not likely to cause damage. The crop is either lifted by hand or by various types of modified root harvesters.

Storage. Usually in small clamps with the soil in direct contact with the roots to help keep them fresh looking. Some are now stored indoors. For many markets it is now necessary to wash the carrots (usually by machines) before sale. They are often sold in net bags. (See HE 4 and STL 52, 73 and 107, AL 62.)

BULB ONIONS

Both bulb and salad onions are grown in this country, but the latter are mainly grown by market gardeners because of the large amount of casual labour required for harvesting (see MAFF Advisory leaflet No. 358, *Salad Onions*).

Bulb onions have increased in importance very rapidly recently because of the very high prices paid in some years, and because of precision drilling, improved chemical weed, pest and disease control, and better methods of harvesting and storage.

About half of our total requirements—350,000 tonnes—of bulb onions are home-grown and there is plenty of scope for increased production provided it is good quality.

The ripened onion consists of edible, swollen, leaf bases surrounded by scale leaves with withered tips. Bulb onions can now be supplied to our markets throughout the year because of modern developments in storage, and the production of autumn-sown crops, which in mild areas can be harvested from June onwards until the spring-sown crops are ready.

Yields. These are variable, usually 30–50 tonnes/ha.

Soils and climate. Bulb onions can be grown in many parts of this country, but do best in the eastern and south-eastern counties. They can be grown on a wide range of mineral and peat soils provided they are well drained, have a good available-water capacity, have a pH of 6.5 or over, and are friable enough to produce good seed-beds: shallow soils, thin chalk, dry sands, and sticky clays are not suitable. If growth is stopped by drought and a secondary growth follows after rain, the outer layers of the bulb may split open.

Rotation. Onions and crops such as field and broad beans, oats, and parsnips, which carry the same pests, should not normally be grown more often than 1 year in 5 on the same field because of the risk of eelworm, onion fly and white rot disease.

Seed-bed. A fine, firm, moist, clean, level and pan-free tilth is required.

Varieties. A wide range of varieties are available—mostly of the Rijnsburger type—for spring sowing; and Japanese, non-bolting types for autumn sowing (see NIAB list).

Time of sowing. As soon as possible after mid-February provided the seed-bed is in good condition; the seedlings can stand slight frosts. The formation of new green leaves and the start of bulb formation is associated with long days of 14–16 hours, i.e. late May and June, so it is important to get the crop established early. Over-wintering crops are usually sown in late August or early September.

Sowing. Onions should be sown with a precision drill in shallow rows (1–2 cm deep), which may be spaced in various ways, e.g. double rows at 50 cm centres.

Seed rate. Graded seed 4–6 kg/ha, or pelleted seed up to 28 kg/ha. The aim should be about 75 plants per square metre.

Fertilizers. For good to average conditions, 100 kg/ha of N and P_2O_5 and 200 kg/ha of K_2O.

Weed control. Good weed control over a long period is essential for good yields. This usually means using a pre-emergence herbicide such as *propachlor* ("Ramrod") or pyrazone/chlorbufam ("Alicep"), to be followed by post-emergent treatment with Alicep, Ramrod, *ioxynil* ("Totril"), or *methazole* ("Paxilon"), to control the common annual weeds. Couch and other perennial weeds should be destroyed in previous years. (See MAFF leaflet STL No. 73.)

Harvesting. Bulb onions are ready for harvesting when most of the tops have fallen over—usually in early September. If left later, yields might be slightly better, but the outer skins are likely to crack and so cause serious disease losses in store. Maleic hydracide, sprayed on the ripening crop, can prevent much sprouting in storage. The crop is usually harvested in two stages—first, the bulbs are lifted out of the ground by a horizontal tine or share and several rows put into one windrow. After drying for a few days they are elevated into a trailer. Direct harvesting produces better quality onions.

Storage. The bulbs can be stored in bulk stores up to 3 metres deep with ventilation ducts underneath, or in boxes or bins of various sizes. They should be dried as soon as possible by blowing with slightly heated air. Temperature and humidity during drying and storage are very important; onions remain dormant (i.e. no sprouting) at low (0°C) and high (27°C) temperatures; humidity can affect skin colour. (See MAFF leaflet STL No. 136, *Buildings for Onion Drying and Storage.*)

Onions are usually graded into various sizes before sale, e.g. over 60 mm, 45–60, 30–45 mm and picklers.

Diseases and pests. Neck rot (bulb rot)—a very serious problem—especially in some seasons; control benomyl ("Benlate") seed dressing and/or foliage sprays, and quick drying in store. *White rot*—soil-borne fungus causing rotting of roots and base of bulb; control by wide rotation and *calomel* seed dressing. *Onion fly*—maggots eat into developing bulb—especially in June and July; control with dieldrin seed dressing. (See MAFF leaflet AL Nos. 62, 85 and 163 also HE 1 and STL 52.)

VEGETABLE PRODUCTION ON FARMS

Farm-scale vegetable production is becoming increasingly important in many areas since the introduction of precision drills, safe and efficient herbicides and pesticides, mechanical harvesting and improved storage methods. This type of crop production is much more demanding of mental energy but can be very satisfying if well done. Profits can be higher than for ordinary farm crops, but costs and risks of failures are also high. It is very important to know what the market wants and to produce only what is wanted, when it is wanted, and in the quantities and quality required. If possible contracts should be arranged with buyers such as processors, pre-packers, supermarkets, chain stores and co-operatives. Limiting factors may be soil type, distance from markets, suitable labour and supervision, and if pre-packing on the farm a suitable building and a clean water supply may be necessary.

The main vegetable crops now being grown on a field scale are: broad, green and navy beans; vining and dried peas; Brussels sprouts; cabbages, carrots; cauliflowers; celery; onions; turnips and swedes.

It is outside the scope of this book to deal with these very specialized crops in detail, but some information on the more important ones is included. Further details can be obtained from *MAFF Bulletins* and leaflets, from ADAS and commercial advisory services, and from The National Vegetable Research Station.

SEED PRODUCTION

Under EEC regulations, in being or being introduced, all seeds sold to farmers must be officially controlled, certified, tested, sealed and labelled.

Detailed standards for acceptable varieties, purity and trueness to type, freedom from seed-borne pests and diseases, tolerances for contamination with weed seeds and rubbish, and viability, are laid down for each type of crop. (See NIAB Tech. Leaflet No. 1.)

Seed production from crops such as *cereals, beans* and *peas* is carried out in much the same way as growing ordinary grain crops but special care must be taken with regard to the previous cropping of the field, the grade of seed sown, avoidance of weed seed contamination (especially wild oats in cereals), diseases such as loose smut in cereals and ascochyta in beans, and

priority at harvest, drying temperatures, careful cleaning and storage to avoid damaging or contaminating the seed. Very good prices are now paid for seed crops which are produced to the required standards (see Booklet 2064).

Potatoes. Seed potato production has, traditionally, been mainly carried on in Scotland, Northern Ireland, and the hill areas of England and Wales, and this is how it will continue under EEC regulations. In these areas, the lower temperatures and stronger winds keep the greenfly (aphid) populations in check and so the severe virus diseases (leaf-roll and mosaics), which are spread from diseased to healthy plants by aphids, are less likely to occur. All seed crops must be certified during the growing season to ensure that they are true to type and variety, and are as free as possible from virus and other diseases. The certification grades of seed potatoes are:

(1) VTSC   Virus-tested stocks grown from stem cuttings.

(2) FS   Foundation seed produced from VTSC or FS stock; a number indicates its origin, e.g. FS2 means 2 years from VTSC.

(3) SS   Stock seed is a high grade of seed (not in Scotland).

(4) AA   First-quality commercial seed.

(5) CC   Healthy commercial seed for ware production.

Only grades (1)–(4) can be used for further seed production.

Over 30% of the total U.K. ware crop is produced from home-grown seed. Such growers should have a sample (100) of the tubers tested before planting to check on the amount of virus

TABLE 37   TOLERANCES FOR SEED POTATO CERTIFICATION AT FINAL INSPECTION IN THE FIELD

	VTSC %	FS %	SS %	AA %	CC %
Rogues, undesirable variations, wildings and bolters	0	0.05	0.05	0.1	0.5
Severe virus diseases	0	0.01	0.02	0.25	2.0
Mild mosaic	0	0.05	0.25	2.0	10.0
Blackleg	0	0.25	1.0	1.0	4.0

Note: 1.0% = 500 plants/ha approx.: 0.01% = 5 plants/ha.

present and be advised on its suitability for planting; this is a service provided by ADAS in England and Wales.

Full details of certification are obtainable from MAFF or Dept. of Agriculture (Scot.). Fields used for seed production must also be certified free of cyst eelworms. Strict grading standards must be observed when selling seed potatoes, and ideally, the seed containers should be inspected and sealed on the farm before despatch to other growers. Some growers now treat the seed with thiabendazole ("Storite") to help control of tuber rots.

The top-grade seed—VTSC—is produced by rooting stem cuttings of virus-free plants in sterilized compost in isolated glasshouses and then growing-on the resulting plants and their progeny in isolation from other stocks. This method ensures freedom from viruses and tuber-borne diseases such as gangrene, skin spot and blackleg. Great care must be taken when handling these stocks to prevent contamination from diseases on other stocks or equipment or from buildings. Using stem cuttings produces about 30 times as many tubers as would be obtained by planting tubers in the usual way. Micropropagation techniques are being used for increasing disease-free seed stocks—very much quicker than the VTSC method but also more expensive.

Ideally, seed potato crops should be planted with sprouted tubers, and at a high seed rate to produce a good yield of seed-size tubers (30–60 mm). The haulm should be destroyed chemically or otherwise when most of the tubers are seed size because this reduces the risk of spread of virus diseases by aphids, and spread of blight to tubers. It is an undesirable practice to let the crop grow to maturity so that additional income can be obtained from the large, ware size tubers.

Seed potatoes should not be grown in the same field more often than 1 year in 5 (1 in 7 would be better), because of the possible carry-over of groundkeepers from the previous crop. The field must be sampled and certified free of eelworms (nematodes).

Grasses. Grass seed production can be very profitable in favourable seasons and if done properly. It requires considerable skill and perseverance, and good yields are very dependent on good growing conditions, and dry weather during the critical harvesting period. The Italian ryegrasses can be harvested for one season only, but most of the perennial grasses will produce two or more seed crops. This type of ley is deep-rooting and is very good for building up humus and good soil structure as well as providing hay and some grazing (the latter is improved if some white clover is sown with the grass). It is very important that there are no grass weeds in the field—especially blackgrass, and other cultivated grass species with similar sized seed. About 200 metres isolation is necessary to avoid cross-pollination with grasses in other fields. The ryegrass and fescues are usually undersown in spring cereals or oil-seed rape, in narrow rows or broadcast, whereas cocksfoot and timothy are usually sown in wide rows (about 50 cm) and not undersown. A good ryegrass crop will usually "lodge" about a fortnight before harvest and this will reduce losses by wind. Harvesting is in July or early August (timothy in late August or September), and the crop may be combined direct or from windrows—timing is critical. The seed must be carefully dried and cleaned. (See also *MAFF Bulletin* No. 204, NIAB literature, and Booklet 2048 (*harvesting*) and Booklet 2046 (*drying seed*.)

Red clover. This is a good arable break crop and grows best on well-drained, high pH (6.5) soils with high phosphate and potash levels, but moderately low nitrogen (high N favours foliage growth). The crops must be isolated from other red clovers by at least 50 metres. Seed yield is very dependent on pollination by bumblebees (main pollinators) and so small areas (up to 5 ha), near the bees' nests in waste ground, are preferred. The seed is best drilled in 15–18-cm rows under a stiff-strawed cereal crop at 8–10 kg/ha (11–13 for tetraploids) and may require a phosphate and potash fertilizer after harvest. Broad-leaved weeds can be controlled with MCPB, 2,4,DB and benazolin in the establishment year but not in the harvest year. Carbetamide or paraquat should be used for blackgrass in winter and topping will kill some weeds in spring.

Single-cut and late-flowering red clovers produce best seed yields from the first cut and so are closed up for seed early in the year—unless the soil is very fertile—when a grazing or cutting in April might be helpful. The first cut of broad red clover is best taken for hay or silage, but if this is delayed after the third week of May pollination is poorer and harvesting may be too late (October) and difficult.

Red clover should be ready for harvesting in late August or early September, when the heads turn brown and the seed (then purple) is easily rubbed out. It usually ripens unevenly so it may be necessary to desiccate the crop with diquat, sulphuric acid, or dinoseb, and combine direct 2–3 days later. Alternatively, cut with a mower and pick up the swath with a combine 7–10 days later. The combine must be carefully set to avoid damaging the seed and, if necessary, the seed should be carefully dried to 12% moisture.

White clover (only 10% home-grown). This is sometimes sown and harvested for seed with a late ryegrass but yields are low. It is best grown alone in a clean, level field (stones well rolled into the surface) with high phosphate and potash levels, moderate nitrogen, and 50 metres isolation from other white clovers. It is susceptible to drought so moisture-retentive soil or irrigation is required. White clover is sometimes undersown but it is better to drill alone in 15–22-cm rows at about 2 kg/ha. Weed control is as for red clover, but more liberty can be taken with spraying in the seed year—spraying delays flowering. It should be cut or grazed (but there is a bloat risk) up to the beginning of May (when the flower stalks start to elongate). It is in full flower by mid-June and it is worthwhile having one hive of honey bees per hectare for pollination—brought in at end of May. It is ready for harvest when 80–90% of the heads are brown (seed easily rubbed out and yellow, orange and brown)—usually in August, but in a dry season it may be possible to harvest two crops—in July and at the end of September. The crop is mown very close to the ground and windrowed and picked up with the combine about a week later (set carefully and check for escape holes). Bad weather at harvest time can be disastrous (dry on a grass drier!).

If kept free of weeds, the crop may be seeded for several years.

Sugar-beet. Seed crops are grown under contract to the British Sugar-Beet Seed Producers Association who sell the seed to the British Sugar Corporation. There are two important aspects: (1) growing the crop and (2) processing the seed, e.g. rubbing and grading, blending mixtures of hybrids, pelleting, etc.

The seed crops are mainly concentrated in Lincs., Cambs. and Essex, but some are also grown in parts of Glos., Oxford and Northants. (all reasonably dry areas for harvesting). Seed crops must be isolated (in zoned areas) from other sugar-beet seed crops, mangolds, fodder and garden beet (1000 m). Sugar-beet is a biennial plant, so seed is produced in the year following sowing. There are two main methods of production:

(a) Transplanted crops—stecklings (young plants) are grown in seed-beds in summer and transplanted in autumn or the following spring (about 37,000/ha). This method is now mainly used for elite or special seed production because it returns about six times as much seed as method (b).

(b) *In-situ*—the seed is sown at 8–11 kg/ha to produce about 370,000 plants/ha, 10 times (a)—and these are left to grow on to seed. These crops may be drilled under a cover crop of barley, one row beet, two rows barley, one row beet, etc. (this is a useful way of checking downy mildew and virus yellow in the eastern counties), or it may be sown in July after a fallow or another crop, e.g. winter barley, arable silage, peas, early potatoes, etc. The sowing can be done with a special drill, and sometimes mixed varieties, by the contract firm. Isolated areas, such as the Cotswolds, usually produce healthy seed because of the absence of ordinary beet crops.

Fertilizer requirements are similar to ordinary beet, but soil type is not so important because the roots are not harvested.

The seed is harvested in late August or early September. It is cut, left in windrows on a high stubble (30 cm) and combined 7–10 days later.

Yield. Range 2–5 tonnes/ha. Mangolds and other beet are grown in a similar way.

Kales, swedes, turnips, rapes, etc., are also biennial plants which produce seed in the year following sowing. Swedes and turnips sown in August and September are usually hardy enough to over-winter. Special precautions have to be taken to avoid cross-fertilization. They are harvested in late summer—usually combined direct or from a windrow.

Detailed information on the production of seed from these and many other crops are given in NIAB seed production leaflets.

SUGGESTIONS FOR CLASSWORK

1. If possible, visit a flour mill, a maltings, an oat-processing mill, a sugar-beet factory, a seed-cleaning plant, a potato grading store and a sprouting house.
2. Examine good and bad samples of all the cereal grains.
3. Visit growing crops of cereals, potatoes, roots, peas and beans, etc.
4. When visiting farms see and discuss the various methods used for handling and storing cereals and potatoes.
5. Make notes on the fertilizers used for the crops you visit.
6. Learn how to recognize the different cereals in the leafy stage.
7. Examine potato tubers stored in sprouting boxes and note the controlling effect of light on size of sprouts; see the colour differences between sprouts of different varieties.
8. Note that the haulms of different potato varieties vary in many ways—this is the only sure method of recognizing potato varieties.
9. Visit growing crops of peas and discuss how they are grown and harvested and the problems involved (e.g. damage by pigeons).
10. Examine the different types of sugar-beet seed.
11. Visit sugar-beet crops at singling and harvesting times and discuss the problems involved.

FURTHER READING

Culpin, *Farm Machinery*, C.L.S.
Robertson, *Mechanical Vegetable Production*, Farming Press.
Eddowes, *Crop Production in Europe*, Oxford Univ. Press.
Cox, *The Potato*, Collingridge.
Potatoes, MAFF Bulletin No. 94.
Kent, *Technology of Cereals*, Pergamon Press.
Harris (editor), *The Potato Crop*, Chapman and Hall.
Dr. D. G. Hessayon, *The Cereal Disease Expert*, P.B.I. publications.
Bishop and Maunder, *Potato Mechanisation and Storage*, Farming Press.
Various publications of the British Sugar Corporation.

GRASSLAND

GRASS is the most important crop in this country. it occupies about two-thirds of the total area of crops and grass grown in the United Kingdom, although it is only about one-third in the drier, sunny, eastern areas. Moreover, it is well to remember that grass is a crop, not something which just grows in a field!

CLASSIFICATION OF GRASSLAND

In the British Isles the total area of grassland, including rough grazings, is just over 11 million hectares. It can broadly be divided into two groups.

1. Uncultivated grasslands

This represents about 44% of the total area of grassland. It consists of:

(i) *Rough mountain and hill grazings.* The plants making up this type of grassland are not of great value. They consist mainly of fescues, bents, nardus and molinia grasses, as well as cotton grass, heather and gorse. Only sheep and beef cattle at very low stocking rates are possible. In some areas where the soil is extremely acid, reafforestation is being successfully carried out.

(ii) *Lowland heaths.* Sheep's fescue is very often the dominant grass on the prevalent acid soils. These heaths are found in the South and East England and some of them have now been reclaimed.

(iii) *Downs of South England.* Apart from herbs, grasses such as sheep's fescue and erect brome are found on these chalk and limestone soils.

(iv) *Fen area in East and South-east England.* Unreclaimed areas are mostly poorly drained and are dominated by water-loving plants such as molinia, rushes and sedges.

2. Cultivated grasslands

This represents 56% of the total area of grassland. It consists of:

(i) *Permanent grassland.* This is grass over five years' old and it represents 41% of the total area of grassland. As a group, permanent pastures cover a wide range, the quality depending chiefly on the amount of perennial ryegrass in the sward. A first grade permanent pasture contains more than 30% perennial ryegrass with a small amount of agrostis; second-grade pasture has between 20 and 29% ryegrass and more agrostis, and third-grade pasture tends to contain very much more agrostis species with less than 20% ryegrass in the sward. Very poor permanent pastures will be dominated by agrostis, and on poorly-drained soils most of the sward will be made up of agrostis, rushes and sedges. Many of these permanent pastures can be improved (page 147).

(ii) *Leys.* These are temporary swards which have been sown to grass and clover for a limited period up to five years. Leys represent about 15% of the total area of grassland. In most cases, depending on management, they will produce more than permanent pasture due to the more productive plants which make up the sward. But generally this sward will not stand up so well to treading as the permanent pasture.

PLANTS MAKING UP THE LEY

These can be grouped as follows:

Grasses.
Legumes: (a) clovers,
          (b) lucerne and sainfoin.
Herbs.

The grasses and clovers can be divided into *varieties* (cultivars) and further divided into *strains.* There are often important growth differences between strains, for example, the hay strains are earlier to start growth in the spring, earlier to flower, have a taller habit of growth, more flowering stems, and are less leafy than the grazing strains of the same variety. Some varieties also have dual-purpose strains which are suitable for either grazing or cutting.

Apart from these differences in strain, varieties can also be classified as:

Commercial strains. These are early to start growth in the spring; they flower early, and have the typical upright habit of growth of the hay strain. Most of them do not live long, but they may be useful in short leys.

Pedigree strains. These are strains which have been carefully selected and bred, and which will do well in most parts of the country. The S strains from Aberystwyth, and the Continental strains are good examples. Most of them are not so early in the spring as the commercial strains, but they are leafier and they live longer, and will normally grow better in the autumn.

It should be noted that the term "strain" is not used to the same extent now. Although technically incorrect, "variety" is used instead.

GRASS IDENTIFICATION

Before it is possible to recognize plants in a grass field (see Table 38), it is necessary to know something about the parts which make up the plant.

Vegetative (leafy) parts

Stem.

(a) The *flowering stem* or *culm.* This grows erect and produces the flower. Most stems of annual grasses are culms.
(b) The *vegetative stem.* This does not produce a flower, and has not such an erect habit of growth as the culm. Perennial grasses have both flowering and vegetative stems.

Leaves. They are arranged on two alternate rows on the stem, and are attached to the stem at a node. Each leaf consists of two parts (see Fig. 41).

(a) The *sheath* which is attached to the stem.
(b) The *blade* which diverges from the stem.

The leaf sheath encloses the *buds* and *younger leaves.* Its edges may be *joined* (*entire*) or they may *overlap* each other (*split*) (see Fig. 42). If the leaves are *rolled* in the leaf sheath, the *shoots* will be *round* (see Fig. 43) but when *folded* the *shoots* will be *flattened* (see Fig. 44).

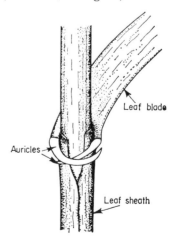

FIG. 41. Parts of the grass leaf.

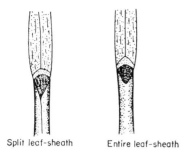

Split leaf-sheath   Entire leaf-sheath

FIG. 42. Parts of the grass leaf.

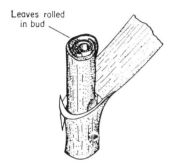

Leaves rolled in bud

FIG. 43. Parts of the grass leaf.

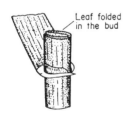

Leaf folded in the bud

FIG. 44. Parts of the grass leaf.

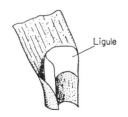

Ligule

FIG. 45. Parts of the grass leaf.

At the junction between the leaf blade and leaf sheath is the *ligule*. This is an outgrowth from the inner lining of the sheath (see Fig. 45).

The *auricles* may also be seen on some grasses where the blade joins the sheath. They are a pair of clawlike outgrowths (see Fig. 41).

In some species, the leaf blade will show distinct veins when held against the light.

According to the variety, the underside of the leaf blade may be shiny or dull.

Other features of the leaves are more variable, and are not very reliable; they can vary with age.

Inflorescence—the flower head of the grass

The inflorescence consists of a number of branches called *spikelets* which carry the flowers. There are two types of inflorescence:

(1) The *spike*—the spikelets are attached to the main stem without a stalk (see Fig. 46).
(2) The *panicle*—the spikelets are attached to the main stem with a stalk (see Fig. 47). In some grasses the spikelets are attached to the main stem with very short stalks to form a dense type of inflorescence termed *spike-like* (see Fig. 48).

A spike inflorescence

FIG. 46. Grass inflorescence.

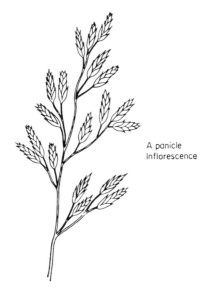

A panicle
Inflorescence

FIG. 47. Grass inflorescence.

The spikelet is normally made up of an *axis*, bearing at its base the *upper* and *lower glumes* (see Fig. 49). Most grasses have two glumes.

Above the glumes, and arranged in the same way, are the *outer* and *inner pales*. In some species these pales may carry *awns* which are usually extensions from the pales (see Fig. 50).

Within the pales is the *flower*.

The flower consists of three parts (see Fig. 51).

(1) The male organs—*three stamens*.
(2) The female organ—the rounded *ovary* from which arise the feathery *stigmas*.
(3) A *pair of lodicules*—at the base of the ovary. They are indirectly concerned with the fertilization process, which is basically the same in all species of plants (see page 11).

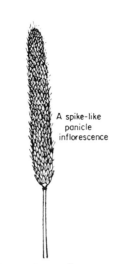

A spike-like panicle inflorescence

FIG. 48. Grass inflorescence.

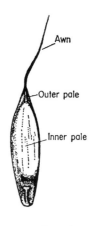

Awn

Outer pale

Inner pale

FIG. 50. The pales.

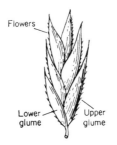

Flowers

Lower glume

Upper glume

FIG. 49. The spikelet.

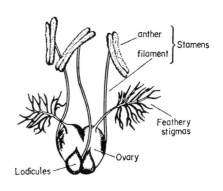

anther
filament
} Stamens

Feathery stigmas

Ovary

Lodicules

FIG. 51. The flower of the grass.

TABLE 38   HOW TO RECOGNIZE THE IMPORTANT GRASSES

	Short duration ryegrasses	Perennial ryegrass	Meadow fescue	Cocksfoot	Timothy
Leaf sheath	Definitely split Pink at base Rolled in shoot	Split or entire Pink at base Folded in shoot	Split Pink at base Rolled in shoot	Entire at first, later split Folded in shoot	Split Pale at base Rolled in shoot
Blade	Broad Margin smooth Dark green	Narrow Margin smooth Dark green	Narrow Margin rough Lighter green	Broad Margin rough Light green	Broad Margin smooth Light green
Lower side	Shiny	Shiny	Shiny	Dull	Dull
Ligule	Blunt	Short and blunt	Small, blunt, greenish-white	Long and transparent	Prominent and membranous
Auricles	Medium size and spreading	Small, clasping the stem	Small, narrow and spreading	Absent	Absent
General	Veins indistinct when held to light. Not hairy	Not hairy	Veins appear as white lines when held to light. Not hairy	Not hairy	Base of shoots may be swollen. Not hairy

IDENTIFICATION OF THE LEGUMES
See Table 39

Leaves. With the exception of the first leaves (which may be simple) all leaves are compound. In some species the midrib is extended slightly to form a *mucronate* tip. Other features on the leaf may be *serrated* margins, presence or absence of marks, colour and hairyness (see Figs. 52 and 53).

The leaves are arranged alternately on the stem, and they can consist of the *stalk* which bears two or more leaflets according to the species (see Fig. 54).

Stipules. These are attached to the base of the leaf stalk. They vary in shape and colour (see Fig. 53).

Flower. The flowers are brightly coloured, and being arranged on a central axis form an indefinite type of inflorescence (see page 12).

TABLE 39   HOW TO RECOGNIZE THE IMPORTANT LEGUMES

	Leaves, etc.	Stipule	General	Species
Mucronate tip	Centre leaflet with prominent stalk Leaflets serrated at tip	Broad, serrated and sharply pointed	May be hairy	**Lucerne**
	6–12 pairs leaflets, plus a terminal one	Thin, finely pointed	Stems 30–60 cm high. Slightly hairy	**Sainfoin**
No mucronate tip	Trifoliate, dark green with white half-moon markings on upper surface	Membranous with greenish purple veins Pointed	Hairy	**Red clover**
	Trifoliate, serrated edge with or without markings on upper surface	Small and pointed	Not hairy	**White clover**

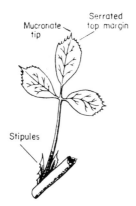

FIG. 52. Parts of the legume.

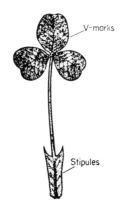

FIG. 53. Parts of the legume.

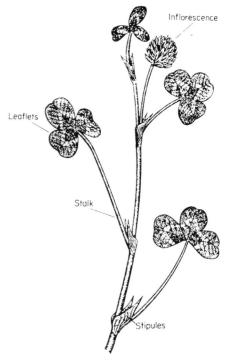

FIG. 54. Parts of the legume.

SOME TERMS USED IN GRASSLAND

Before discussing grasses and grassland management it is important to understand what is meant by the following terms:

Seeding year. The year in which the seed mixture is sown.

First harvest year. The first year after the seeding year, and thus the second and third, etc., harvest years.

Undersowing. Sowing the seed mixture with another crop (a cover crop). It is usually a cereal crop.

Direct sowing or seeding. Sowing on bare ground without a cover crop.

Direct re-seeding strictly means sowing without a cover crop, and putting the field straight back to grass, the previous crop having been grass. Very often it is used in the same way as direct sowing.

GRASSES

Over 150 different varieties of grasses can be found growing in this country, but only a few are of any importance to the farmer:

(i) The short-duration ryegrasses—this class consists of varieties which fall into three main groups: Westerwolds ryegrass, Italian ryegrass, and Hybrid ryegrass.

(ii) Perennial ryegrass.
(iii) Cocksfoot.
(iv) Timothy.
(v)   Meadow fescue.

Short-duration ryegrasses (see Fig. 55)

Westerwolds. This is an annual and the quickest growing of all grasses. A good crop can often be obtained within 6–8 weeks of sowing. It should not be undersown (see page 126), and is best direct sown (see page 126) in the spring and summer as it is not at all winter-hardy. Dutch-bred varieties are used.

Italian ryegrass. This is short-lived (most varieties persist for 18–24 months); very quick to establish. Sown in the spring, Italian ryegrass

can produce good growth in its seeding year and an early graze the following year, but for optimum production in its harvest year (particularly the spring) it is best sown in summer or early autumn. It does well under most conditions, but responds best to fertile soils and plenty of nitrogen. Like all ryegrasses, its winter hardiness is improved when surplus growth is removed in the autumn. Although stemmy it is palatable with a high digestibility (see page 133).

R.V.P. and *Trident* are leafy varieties. *Sabalan*, a tetraploid (see page 128) from the Welsh Plant Breeding Station at Aberystwyth, is of special value because of its resistance to mildew and ryegrass mosaic virus.

Hybrid ryegrass. Some strains in this group are similar to the more persistent Italian ryegrass strains and others are similar to perennial ryegrass. A feature of hybrid ryegrass is the relatively high D value of a second conservation cut.

Augusta and *Sabel* are both tetraploids (see page 128) from Aberystwyth.

Perennial ryegrass (see Fig. 56)

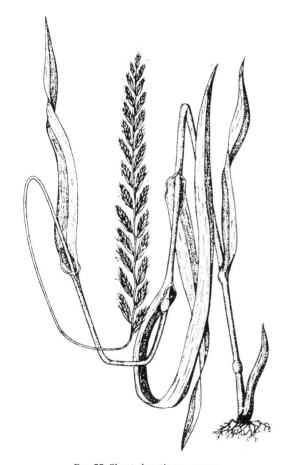

FIG. 55. Short-duration ryegrass.

FIG. 56. Perennial ryegrass.

This forms the basis of the majority of long leys. The most important grass found in good permanent pastures. It is quick to establish, and yields well in the spring, early summer and autumn. It does best under fertile conditions, and responds well to nitrogen.

Perennial ryegrass is highly digestible.

Very early and early varieties: *Barvestra* (tet.), *Gremie, Cropper, Monta*.

Intermediate varieties: *Tove* (tet.), *Talbot*.

Late and very late varieties: *Meltra, Wendy, Melle, Angela*.

Tetraploid ryegrasses

The chromosome numbers in the cell nuclei have been doubled. The seeds are larger, and this will normally mean a slightly higher seed rate. They generally produce a bigger plant than the diploid (ordinary grass), but in yield of dry matter there is no great difference.

Most of the recommended tetraploids are from the Netherlands (although Aberystwyth has produced some short duration ryegrass tetraploids), and they do appear to have a better resistance to frost than the British diploids, but this is probably due to the country of origin.

Tetraploids are characteristically deep green, and are slightly more palatable and digestible than diploids. But because of their higher moisture content and larger and thicker-walled cell structure, tetraploids are perhaps not quite so suitable for conservation in the wetter parts of the country, particularly with high nitrogen use.

Sown alone they are not so persistent, as they tend to produce rather too open a sward.

Cocksfoot (see Fig. 57)

This is quick to establish and is fairly early in the spring. It is a high-yielding grass, but unless it is heavily stocked, most of the existing varieties will soon become coarse and unpalatable. Cocksfoot is one of the most deep rooting of all grasses, and is therefore an excellent drought-resister. It does not need really

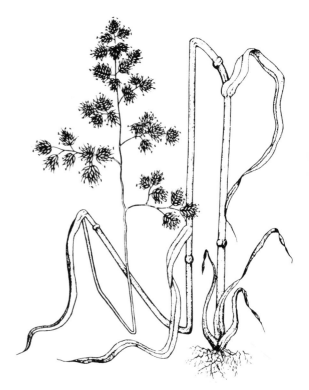

FIG. 57. Cocksfoot.

fertile conditions. It should be considered as a special-purpose grass on the drier lighter soils in areas of low rainfall. The existing varieties are not as digestible as the other important grasses, but new varieties are promised with D values as high as the ryegrasses.

Early varieties: *Sylvan, Modoc*.

Late varieties: *S26, Prairial*.

Timothy (see Fig. 58)

This is fairly quick to establish. It is not particularly early in the spring. It is less productive than the other commonly used grasses, but is very palatable, although its digestibility is not as good as the ryegrasses. It is winter-hardy, and does well under a wide range of conditions except on very light, dry soils.

Early varieties: *S352, Scots, Kampe 11*.

Intermediate varieties: *Motim, Pecora*.

Late varieties: *S48, Intense, Olympia*.

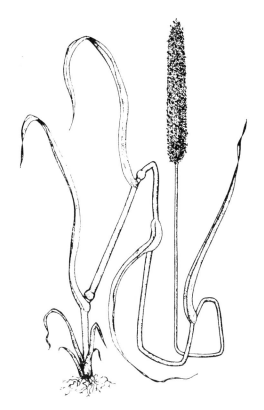

FIG. 58. Timothy.

FIG. 59. Meadow fescue.

Meadow fescue (see Fig. 59)

Meadow fescue is not used very much now. This is because the once famous Timothy/ Meadow fescue mixture is no longer considered to be so valuable as it is not so productive compared with the ryegrass sward.

Meadow fescue is slow to establish, but once established it is fairly early in the spring and it has a high digestibility.

Early varieties: *Salfat, S215*.

Late variety: *Bundy*.

OTHER GRASSES

Tall Fescue is a perennial. Once it is well established, strains such as *S170* and *Festal* are very useful for early grass in the spring, but production after this is not very high. It is very hardy and it can be grazed in the winter.

The *Small Fescue* grasses, such as Fine-leaved fescue, Red fescue, Creeping Red fescue and Sheep's fescue, are all useful under hill and marginal land conditions, and in some situations they will produce more than perennial ryegrass. They have no practical value under farm lowland conditions. They are typical "bottom" grasses, and they produce a close and well-knit sward. For this reason they are often included in seeds mixtures for lawns and playing fields.

Rough-stalked Meadow Grass is indigenous to the majority of soils, but it prefers more moist conditions. In the later years of a long ley, it can make a very useful contribution to the total production of the sward after the white clover has been killed out by heavy nitrogenous manuring. It is never included in a seeds mixture now.

Smooth-stalked Meadow Grass (Kentucky Blue Grass) spreads by rhizomes, and can withstand quite dry conditions.

Crested Dog's Tail is not very palatable because of its wiry inflorescence. Now it is only used in seeds mixtures for lawns and playing fields.

New grasses

The next generation of grasses developed by the plant breeders could well see more emphasis on persistency, rather than just on improved yield performance. New varieties are needed to stand up to the more intensive grazing systems common now, compared with 25 years ago. This intensification of grassland farming brought about by much heavier nitrogen use and increased stocking has meant that varieties which were persistent for 8–10 years are now dying out much sooner. It is now recognized that long leys need plants better able to withstand this hard management as well as leafier varieties with a more prostrate habit of growth to form a good bottom to the sward. High nitrogen applications are also altering the balance of protein and carbohydrate in the plant. A plant with a higher protein content is resulting in greater susceptibility to winter kill. These new varieties will generally have a higher carbohydrate content. Furthermore it seems that these carbohydrates could be more effectively used by selecting ryegrass material with a lower rate of respiration. This should lead to an increased dry matter yield, particularly during the summer when temperatures are higher.

A susceptibility to winter kill has always been a problem with ryegrass varieties. Eventually, however, this could be overcome by breeding from material collected in Switzerland. These ryegrasses appear to have a higher degree of winter hardiness combined with a characteristic of early spring growth.

Apart from the Italian/perennial ryegrass hybrids, which have been developed at Aberystwyth and which are now in commercial use, hybrids between other grasses are also showing promise. An Italian ryegrass/meadow fescue cross should mean the combination of the winter hardiness of the fescue, and the rapid growth and earliness of the ryegrass. Perennial ryegrass is also being used with meadow fescue which would obviously show more persistence than the Italian ryegrass cross.

The possibility of a hybrid between Italian ryegrass and tall fescue with the potential of a persistent conservation plant is again being examined. Poor seed production has been a problem in the past. New plant breeding techniques could alter the picture.

Cocksfoot varieties which are easier to manage under grazing conditions and which, at the same time, have higher digestibilities, are still awaited. Material from the Iberian Peninsula is showing promise in both these respects.

Only recently has it been appreciated that grass diseases and pest attack on swards can have a considerable adverse effect on herbage production. Probably up to 30% production is lost every year.

Resistance to *rynchosporium, crown rust* and *mildew* of the short duration ryegrasses and *crown rust* of perennial ryegrass has now developed with an increasing number of varieties. Varietal resistance to these and other diseases will offer the best and cheapest form of control.

Short duration ryegrass is more severely affected by ryegrass mosaic virus (see Table 53) than perennial ryegrass varieties. Here again varietal resistance or tolerance would seem to be the answer.

Frit fly (see Table 53) is an important pest of grass plants. The plant breeder is now searching for resistant ryegrass material to cope with this pest.

See NIAB leaflet No. 16 for further details of grasses.

WEED GRASSES

The majority of grasses naturally growing in this country are of little value. But some of them are extremely persistent and are able to grow under very poor conditions where the more valuable grasses would not thrive. However, their production is always low, they are usually unpalatable, and under most conditions they can be considered as weeds.

Well-known examples of weed grasses are:

The bents. They are very unproductive and unpalatable and are found in poor permanent pastures.

Brome grasses. They are found in arable fields and also in short leys. There are many species, and none are of any value.

Yorkshire fog. This is extremely unpalatable except when very young. It is prevalent under acid conditions, although it can be found growing almost anywhere. It is sometimes used in reclaiming hill pastures.

The weed grasses—couch, slender foxtail (blackgrass), wild oat and barren brome—are all associated with intensive cereal growing.

CLOVERS

Clovers are essential plants for the longer ley and permanent pasture. In the short ley they are used for conservation. Apart from their ability to fix nitrogen from the atmosphere, clovers are also useful in that, especially in the longer ley, they act as "bottom plants". With their creeping habit of growth they knit the sward together and help to keep out weeds. Although the majority of clovers are palatable with a high feeding value and digestibility (particularly white clover), they are not so productive as the grasses, and they must not be allowed to dominate the sward at the expense of the grasses. The clovers of agricultural importance are the red and white clovers.

Red clovers (see Fig. 60)

These are short lived. They are included in short leys and sometimes in long leys to improve bulk in the early years. Although used more for conservation, some of the more persistent improved strains are useful for grazing. There are two main groups.

FIG. 60. Red clover.

Early red clover (broad red clover: double cut cowgrass). The tetraploid varieties such as *Redhead* and *Trifomo* are higher yielding and generally more persistent than the diploids. They usually show good resistance to sclerotinia (clover rot, see Table 54), but with one exception *Norseman* (only partially resistant to sclerotinia) they are susceptible to stem eelworm (see Table 53).

Late red clover (late-flowering red clover). These varieties are later to start growth in the spring than the early red clovers. They are also more persistent and, under good conditions, they can last up to 4 years. *Grasslands Pawera* is a new tetraploid from New Zealand. It is a high yielder in the first harvest year, and it also shows good resistance to stem eelworm.

White clovers (see Fig. 61)

These should be regarded as the foundation of the grazing ley. They are generally not so pro-

FIG. 61. White clover.

ductive as the red clovers, but are more persistent. There are three types, classified according to leaf size.

Medium large-leaved white clover. These varieties are becoming more popular because of their ability to compete better with the large amounts of nitrogen now used on grazing swards. *Aran* is the most tolerant of this group of clovers to nitrogen fertilizer, but it is not very resistant to sclerotinia. *Sabeda, Olwen, Kersey* and *Blanca* are other varieties in this group.

Medium small-leaved white clovers. These are extremely useful in leys of up to 4 years and are very quick to establish. *S100* is the best-known variety, although it is susceptible to clover rot. The Aberystwyth variety *Donna* is a better all-round variety.

Small-leaved white clovers. These are the most essential of all plants in the long ley. They are rather slow to establish but can become dominant. *Kent wild white* clover is a good local variety and *S184* is another very productive and creeping type. *Promitro* is a Dutch variety with reasonably good resistance to sclerotinia.

OTHER LEGUMES

Lucerne (see Fig. 62)

This is a very deep-rooting legume and it is therefore useful on dry soils, although it can be grown successfully under a wide range of soil conditions provided drainage is good. For better establishment lucerne should be grown with a companion grass such as perennial ryegrass, Timothy or meadow fescue. Although it is very productive lucerne is not very palatable and it is the least digestible of all the important legumes. It is probably best utilized by drying through the grass drier, although excellent barn dried lucerne hay can be made without much loss of leaf. There are many varieties of lucerne, and in this country they can be grouped into early and mid-season types. Where possible the early type should be grown. At present *Europe* is the heaviest yielding variety, but where verticillium wilt is suspected a tolerant variety such as *Sabilt* or *Sverre* (also reasonably resistant to stem eelworm) should be used.

FIG. 62. Lucerne.

Sainfoin (See Fig. 63)

As regards its economic use, sainfoin is very similar to lucerne although it is not so productive. It has not been very popular in the past, due to difficulties in establishment (it does not compete very well with weeds), but sown with grasses such as cocksfoot, this problem has been overcome. For forage purposes especially it should be seriously considered for calcareous soils. Sainfoin hay is highly valued for horses, but it is difficult to make without losing leaf. There are two main types of sainfoin:

English giant. This is heavy yielding, but short-lived.

Common sainfoin. This lasts for several years, but is not so heavy yielding.

See NIAB leaflet No. 4 for further details of herbage legumes.

Fig. 63. Sainfoin.

HERBS

These are deep-rooting plants which are generally beneficial to pastures. But to be of any value, they should be palatable, in no way harmful to stock, and they should not compete with other species in the sward. They have a high mineral content which may benefit the grazing animal.

Yarrow, chicory, rib grass and burnet (see Figs. 64–67) are the most useful of the many herbs which exist. They can be included in seeds mixture for a grazing type of long ley. They are not cheap, however, and as one or more of these herbs will usually get into the sward on its own accord, there does not seem much point in buying them in the first place.

HERBAGE DIGESTIBILITY

One of the main characteristics which determine the feeding value of herbage plants is its digestibility, i.e. the amount of the plant which is actually digested by the animal. Until recently, the digestibility of foods was only determined by animal feeding experiments. But, with the development of the laboratory *in vitro* technique which simulates rumen digestion, it is now possible for the digestibility of any species of herbage plant to be assessed very much more easily without using animals.

The digestibility of herbage plants is now expressed as the D value *which is the percentage of digestible organic matter in the dry matter.*

High digestibility values are desirable as it means that the animal is able to obtain the greatest amount of nutrients from the herbage being fed. It is now accepted that the digestibility of herbage plants is a major factor affecting intake.

D value will vary according to species, variety and management, notably stage of growth for grazing or cutting. And once the maximum D value has been reached, unless the plant is defoliated it falls quite quickly depending on the species or variety.

With *grasses*, for conditions approximating to *grazing*, the D value for perennial ryegrass is about 70%, i.e. 70 D indicates 70 kg of digestible energy-producing organic matter per 100 kg of dry feed. Cocksfoot has a D value of approximately 66%. Short-duration ryegrass and

FIG. 64. Yarrow.

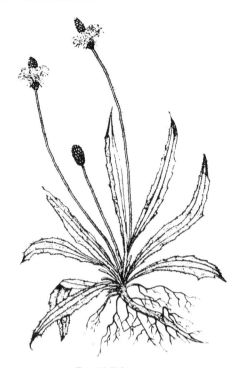

FIG. 66. Rib grass.

FIG. 65. Chicory.

FIG. 67. Burnet.

meadow fescue have D values just below 70% with Timothy at about 68%. For *silage*, higher yields will result when the crop is cut at a later stage of growth. In farm practice and on average the D value is 7% lower in all the varieties mentioned. However, with greater emphasis being placed on quality silage, advice is now for a D value of 67% as far as is possible (see page 133).

With legumes, white clover has a higher D value (about 75%) than the grasses, and it does maintain its maximum value over a longer period. For these reasons alone it should be seriously considered in grazing leys of 3 or more years' duration even with high nitrogen use. Sainfoin has a D value of about 65%; red clover and lucerne are approximately 62%. Lucerne over the season will show a higher yield of digestible dry matter than the other legumes, especially if when made into hay, the loss of leaf is kept to a minimum.

SEEDS MIXTURES

Many farmers depend on reliable seed firms to supply them with standard seed mixtures, whilst others prefer to plan their own mixtures which the merchant will then make up for them.

The following points must be considered when deciding upon a seeds mixture.

The purpose of the ley

Varieties and strains of herbage plants have different growth characteristics, and because of this the type of stock using the ley will influence the choice of seed. This is shown in Table 40.

Soil and climatic conditions

The majority of herbage plants will grow under a wide range of conditions. However, on heavy, wet soils there is no point in growing early grasses or planning for foggage grazing. Like the majority of crops, grasses and, particularly, legumes will not thrive where there is a lack of lime in the soil.

The length of the ley

Perennial ryegrass, cocksfoot, Timothy and meadow fescue, being persistent, are suitable for long leys, whilst the less persistent and

TABLE 40

Purpose	Varieties which should be used
For early grazing, i.e. early bite	Mainly Italian ryegrass and hybrid ryegrass
For optimum production throughout the season	Ryegrass (supported by either/or non-ryegrass, permanent pasture)
For winter grazing cattle—foggage (provided conditions allow)	Non-ryegrass
For the grazing block	Herbage plants which produce a closely knit sward
For the cutting block	Herbage plants which produce a relatively tall habit of growth

Note: The heading dates (number of days after 1st April) will generally coincide if the above points are borne in mind when deciding on a seeds mixture. As far as possible, varieties should correspond because, with only a few exceptions, the digestibility of a plant starts to decline when the ear emerges (see page 133). A grass crop cannot be so valuable if its various plant components head at different times. It is also important that one variety or strain of a variety does not dominate the sward at the expense of all other varieties.

quick-growing early types, of which Italian ryegrass is an outstanding example, will make up the short ley.

Cost of the mixture

This will vary from season to season depending upon the previous year's seed harvest, the strain of plant used and the seed rate. It is false economy to buy unsuitable strains just because they are cheap. It is equally unwise to be persuaded to buy expensive and often unproved seed. Generally speaking, it will be found that varieties required for the short-term ley are cheaper than those for the longer ley.

Tables 41–43 give examples of different seed mixtures.

The Ministry publication, Booklet 2041, gives more details of seed mixtures.

MAKING A NEW LEY

A case can be made for sowing the seeds either in the spring or late summer/early autumn period.

Spring sowing

If direct sowing without a cover crop, the maiden seeds can give valuable production in the summer. Establishment can be enhanced by stock being able to graze the developing sward within a few weeks of sowing. This is not possible to the same extent with autumn sowing. A limiting factor with spring sowing may be moisture, and in the drier districts the seeds should be sown in March if possible. The plant should thus establish itself sufficiently well to withstand a probable dry period in late spring. Of course, undersowing in corn (which is only possible in the spring) does mean that the fullest possible use is being made of the field, although with the slower-growing grasses, establishment is usually not so good.

Late summer/early autumn sowing

There is usually some rain at this time; heavy dews have started again, and the soil is warm. But with clovers in the seed mixture earlier sowing may have to be carried out so that the plants have developed a good tap root system before the onset of frosts. Whether earlier sowing is possible depends on when the preceding crop in the field is harvested.

Undersowing or direct sowing

There are points for and against the practice, but *direct sowing* is now becoming quite widespread. It is certainly preferable for early bite following sowing the previous late summer/early autumn period. It is also the case following spring sowing where conditions for establishment are not so good and when extra grass is required in the summer.

Green crops as cover crops are excellent, e.g.

(1) Rape sown at 4–7 kg/ha with the seeds mixture, and grazed off in 6–10 weeks.
(2) oats or barley sown at 60 kg/ha with the seed mixture.
(3) Oats as an arable silage crop cut in June.

SOWING THE CROP

Direct sowing

Reference has already been made to the cultivations necessary for preparing the right sort of seed-bed for the seeds. But it must be re-emphasized that grass and clover seeds are small, and therefore they must be sown shallow, and that therefore a fine and firm seed-bed is necessary.

With ample moisture the seed can be broadcast. This should be on a ribbed-rolled surface, so that the seeds tend to fall into the small furrows made by the roller. Most fertilizer distributors can be used for broadcasting.

TABLE 41   1-YEAR LEYS (Amounts in kg/ha)

	Westerwolds	I.R.G.	Early red clover	Late red clover	Total	
A	25–30				25–30	Not suitable for undersowing; extremely quick to establish; will produce two good cuts or grazings, especially if the first one is taken before the seeds are set
B		20–40 *R.V.P.*			20–40	The lower seed rate is used in the wetter parts of the country. Stemmy but palatable. Chiefly used for grazing
C		15 *R.V.P.* 15 *Augusta*			30	Becoming quite popular, for general use, although chiefly for grazing
D		16 *Augusta*	11 tetraploid		27	When undersown will give autumn grazing; in the following year two good crops, usually, but not always, for conservation
E		8 *R.V.P.* 8 *Sabbalan*	4 tetraploid	3 *Britta*	23	Similar to mixture D—but will give better aftermath grazing because of the inclusion of the late red clover

In the examples given ryegrass is the only variety of grass used. The emphasis must be on quick establishment with good production from an early stage, and with an ability to respond well to nitrogen. As persistency is not important, the short-duration ryegrasses will amply fulfil these requirements, provided that management is correct.

TABLE 42   2–3-YEAR LEYS (Amounts in kg/ha)

	Short-duration ryegrass (not Westerwolds)	Perennial ryegrass	Cocksfoot	Early red clover	Late red clover	Medium small-leaved white clover	Total	
A	4 *R.V.P.*	12 S 24		2 tetra-ploid	3 *Britta*	1 S 100	22	This mixture should give early grazing, an extremely useful hay or silage crop and good aftermath grazing, but plenty of nitrogen will be needed for maximum production
B		12 *Talbot* 12 S 23					24	A mixture which should quite easily last 3 years. This is an easy ley to manage, and can be extremely productive when liberally fed with nitrogen. Very good for set stocking
C	10 *R.V.P.* 10 *Augusta*	10 S 24 or 10 *Talbot*					30	A very productive ley. With no clover to depress, up to 400 kg/ha of nitrogen can be economically applied in the season
D		4 S 24	14 *Sylvan*		3 *Britta*	1 S 100	22	For light poor soil types only. A mixture more suitable for general purpose use

TABLE 43   LONG LEYS (Amounts in kg/ha)

	Perennial ryegrass	Cocksfoot	Timothy	Meadow fescue	Late red clover	Medium small-leaved clover	Small-leaved white clover	Total	
A	20 *Melle*					2 S 100	0.5 S 184	22.5	For general purpose use. A very productive and hard-wearing ley. Varieties similar to *Melle* could be used
B	11 *Talbot* 7 *Wendy*		4 S 48			2 *Donna*		24	An earlier mixture than A. Responds to intensive management
C			8 S 48	10 S 215		2 S 100	0.5 S 184	20.5	For general purpose use. This ley will tie in well with the previous ryegrass mixture, and by managing the two leys together it should be possible to get fairly even production throughout the season. However, it is not so productive as the ryegrass ley
D	14 *Talbot*	7 S 26	4 S 48		3 *Britta*	1 S 100	0.5 S 184	29.5	This is the well known Cockle Park type general purpose mixture. It is meant to give even production throughout the whole season, but eventually, depending upon the soil type and management, it will tend to become either ryegrass or cocksfoot dominant

It will be noted that wild white clover is included in these mixtures. After about 3 years it should have established itself sufficiently well to fill the bottom of the ley and give a nice well-knit sward. Only a small amount is needed, otherwise it could dominate the whole sward at the expense of the more productive grasses.

In the drier areas, and on lighter soils, drilling is safer. The seed is then in much closer contact with the soil. The 10-cm coulter spacing of the ordinary grass drill should give a satisfactory cover of seeds, but if using the corn drill with 18-cm spacing, the seeds should be cross-drilled. After either broadcasting or drilling, except on the wetter seed-beds (when the seeds harrow will be used), rolling will complete the whole operation. Where necessary, 300–350 kg/ha of a compound fertilizer, e.g. (15:15:21) can be broadcast and worked in during the final seed-bed preparations, or, depending on the drill used, it can be applied with the seed.

Direct drilling

An increasing amount of grassland is now reseeded by direct drilling. Although it is no cheaper than conventional reseeding, the practice does allow a much quicker turnround from the old grass to the new sward. It also reduces poaching and, most important, it enables reseeding in situations where it would not be possible to plough or even do minimum cultivations, e.g. water-meadows.

Glyphosate is increasingly associated with the technique, although paraquat, dalapon and aminotriazole can be used.

Undersowing

Any of the *autumn-sown* cereals may be used as a cover crop but they compete more with the seeds than spring sown cereals. Harrowing of the ground will be necessary, and then the seed should preferably be drilled across the corn drills followed by the light harrows. Alternatively, the seed can be broadcast and harrowed in, but this is not so satisfactory.

With a *spring-sown* cover crop the cereal is sown first and it can be immediately followed by the seed mixture drilled or broadcast. This is desirable with slow establishing mixtures but with vigorous species, e.g. Italian ryegrass and red clover, it may be preferable to broadcast the seed after the cereal is established.

MANURING OF GRASSLAND

Grass, like all crops, needs plant food for its establishment, maintenance and production. *Nitrogen* is essential for maximum production from grass. Provided other essential plant foods are available to the plant nitrogen is the nutrient most responsible for the grass crop's production.

Ryegrass and good permanent pasture swards will normally receive more nitrogen than non-ryegrass and other permanent pastures. With a straight ryegrass sward (i.e. with no clover) the response to nitrogen is almost linear up to about 400 kg, but clover (especially the larger-leaved types) complicates the response to some extent, although undoubtedly the total production of dry matter from a given amount of nitrogen will be higher from a grass/clover mixture than from a grass only sward.

The amount of nitrogen used each year will also depend on the type and intensity of management (see Table 44).

Phosphate and *potash* will both help in the establishment of the sward, and on most soils they are important in helping to maintain the general vigour and well-being of the crop. There is no doubt that the response to nitrogen by the grass crop is far greater if there is an adequate supply of phosphate and potash present.

Phosphate can be applied at any time of the year, although it is not generally advised during the winter non-growing period. How much to apply will depend to some extent on the soil phosphorus index, but it can be the same, irrespective of whether the crop is cut or grazed. (See Tables 44, 45.)

Potash should not be applied in the spring. The uptake of magnesium by the plant is always slow at that time of the year, and as potash tends to act as a buffer against magnesium a spring application will accentuate a shortage of magnesium in the plant. This could well lead to hypomagnesaemia or grass staggers in the animal.

For grazed swards little potash is required although this depends on the potash status of the soil. The grazing animal itself recirculates potash back to the soil, and at normal stocking rates this can amount to more than 125 kg/ha in the year. If the soil's annual natural contribution (60–250 kg/ha, depending on soil type) is added, it will be seen that it is not always necessary to apply potash. Luxury uptake of potash (the plant containing potash surplus to its needs) should be avoided. It is wasteful, and it does increase the risk of hypomagnesaemia.

The cut crop needs more potash than the grazed sward. A vigorous growing crop will take up about 360 kg/ha in the year. Depending upon soil type, up to 300 kg/ha might be needed to replace that removed. For efficient utilization, it should be applied at intervals through the season *after* the crop has been taken.

Lime. The grass, or grass/clover sward, like any other crop cannot thrive in acid conditions. The legumes, particularly red clover and lucerne, are very sensitive to a low pH. Maintenance dressings of up to 4 tonnes/ha calcium carbonate are very often necessary for the crop, although this depends on soil type and whether the crop is cut or grazed.

Farmyard manure (FYM). The composition of FYM has already been discussed. If FYM cannot be used for an arable crop then it is best applied to those swards which are going to be taken for hay or silage. Cutting will help to even out the uneven effects of growth following

FYM application, due to the variable plant food content of the manure.

It should be applied at 25–30 tonnes/ha and, from the point of view of better utilization of the nitrogen, ideally in the spring, although this is hardly practicable.

Slurry. The plant food content of slurry, and the problems of applying it to the grazing sward, have been discussed on page 56. On the all-grass farm unseparated slurry must obviously be used on the cutting block. Used in this way (rather than on a "sacrifice" field), it is best applied at rates of up to 60,000 litres/ha at intervals throughout the summer.

THE MANAGEMENT OF A YOUNG LEY

A direct-sown long ley in its seeding year— sown in the spring or summer

The sward should be grazed 6–10 weeks after sowing, depending upon weather conditions which naturally affect growth. This early grazing helps to consolidate the developing sward, and encourages the plant to tiller out. It should not be too hard, but equally it is important not to undergraze. The sward should then be rested, and followed by periodic grazings throughout the season.

It is unwise to cut the ley in its seeding year. Plants which are allowed to grow too tall before being cropped never develop very strongly. The essential tillering is not encouraged to the same extent, and the sward is left "very open" into which weeds may soon gain a foothold.

If grazing is not possible, the plants should be topped before they grow too tall.

A direct-sown ley—sown in the late summer

This should either be grazed or topped in the autumn.

An undersown ley

This should either be grazed or topped in the autumn. If the seeds look "thin" after the corn has been harvested 40 kg/ha of nitrogen will help to stimulate growth.

MANAGEMENT OF THE ESTABLISHED LEY AND PERMANENT PASTURE

The ryegrasses now make up more than 90% of the grass species used in leys.

It would no longer be true to say that there is such a bad "summer gap" period with perennial ryegrass because, with nitrogen fertilizer treatment, the later varieties of perennial ryegrass will now grow quite well in July and August, although this will depend on the season. Direct sown ryegrass leys seeded down in May/early June will also help to avoid a fall in production later on in the Summer, and so, too, will good permanent pasture if properly managed. This all contrasts with previous years when, in order to obtain a fairly uniform supply of grass throughout the season, non-ryegrass swards were managed alongside ryegrass mixtures. But, as perennial ryegrass under most management conditions is the most productive grass, it is obviously sensible to use it wherever possible, although the value of good permanent pasture should not be forgotten. However, new and more productive non-ryegrass varieties and the eventual introduction of hybrids (page 130) could change the picture in the future. Of course, there are still some farmers who will use a Timothy/meadow fescue mixture as an insurance against poor production of the ryegrass. And non-ryegrass swards are useful for winter grazing or foggage.

Table 44 shows ways of managing grassland. Naturally there will have to be modifications according to the season, growth and type of stock available. The management chiefly refers to the dairy cow, although it can apply to an intensive beef-grazing system.

1. Management of long leys

TABLE 44   PERENNIAL RYEGRASS—GRAZING BLOCK

FEBRUARY MARCH	Apply 80–100 kg nitrogen for first graze following early bite on a short-term ley. If have to use PRG for early bite, nitrogen applied earlier (see pages 142, 143).
APRIL	First graze taken.
MAY JUNE JULY AUGUST	Apply nitrogen at rate of 2.5 kg/ha for each day's interval between grazing, e.g. a 28-day paddock system requires 70 kg nitrogen. When set stocking, the whole field can either be top-dressed with nitrogen every 4 weeks or each quarter of the field receives nitrogen in successive weeks. The August application of nitrogen can be reduced by about 20% to help maintain or improve the sward's clover content. Late nitrogen also increases the protein content of the grass which makes it more vulnerable to winter kill.
SEPTEMBER	If there is too much clover in the sward the grass should be allowed to grow up a little more before taking the final graze.
OCTOBER	It is important to see that eventually the sward is grazed down tight (sheep are ideal, and in any case autumn-calving cows should be off grass by this time) to avoid winter kill. Apply 0–40 kg phosphate and 0–40 kg potash annually; 1–1.25 tonne/ha of K slag (or its equivalent as ground mineral phosphate) can be used every 4–5 years.
NOVEMBER DECEMBER JANUARY	Lime if required

Notes

(i) Under dry conditions such frequent nitrogen applications are unnecessary in the March/August period, but at least 250 kg should have been applied by early June. This will help to overcome the effects of a soil moisture deficit as a soil with a high nitrogen content increases the efficiency with which the soil water is used.

(ii) After a few years of intensive management, using high nitrogen applications, the soil fertility will have increased because of heavier stocking and a consequent increased amount of re-circulated nitrogen as well as soil organic nitrogen. This should mean that less nitrogen can be used in the year.

(iii) Except with irrigation or possibly set-stocking, nitrogen, at an annual rate of more than 250 kg, will have an adverse effect on the white clover in the sward. When a more competitive clover plant is bred it will certainly make a useful contribution to the grass crop's nitrogen requirements.

2. Management of non-ryegrass leys and permanent pasture

On non-ryegrass and permanent pasture, management is generally not as intensive as the ryegrass sward. Total nitrogen use could well be above 300 kg, although obviously there is no hard and fast rule on this one. Early bite would not be expected but, as far as possible, these swards should be managed so that they can produce a worthwhile crop when the ryegrass may have fallen back in production. With nitrogen applied from the end of March—up to 160 kg in two equal splits—for silage made 8–10 weeks later, or 80–90 kg nitrogen applied again in late March for a hay cut from mid-June onwards, and with 80 kg nitrogen following the silage or hay, the sward should be able to provide a useful crop again for conservation or grazing. The swards are also very useful for autumn grazing, and

nitrogen applied late August–early September at 80 kg would help for foggage or winter grass, provided ground conditions are suitable.

If these swards are mainly grazed, only about 40 kg each of phosphate and potash are used each year (usually in the autumn) or as 1–1.5 tonne/ha of K slag or equivalent every 4–5 years. Do not graze stock after slag application until it is well washed-in. If conservation cuts are taken, potash is particularly necessary at anything from 50–120 kg, depending on soil type. It should be top-dressed immediately the crop has been taken.

As with the ryegrass swards, attention should be given to maintaining a reasonably correct balance between the clovers (20–25% plants making up the sward) and the grasses. Too much clover in the sward should be avoided, although too little will lead to an unproductive ley.

Lime can be applied in the autumn if it is required.

3. Management of short leys

TABLE 45   MANAGEMENT OF SHORT LEYS

	Seeding year	First year	Second year	
1-year ley	After the cover crop has been removed, 30–40 kg/ha N can be applied to give useful grazing in September and October Apply 50 kg/ha of phosphate and potash after this grazing	If early bite is required, 80–90 kg/ha N, applied in February Following early bite, 75 kg N can be applied either for conservation or grazing followed by a further 75 kg N for a second cut or graze		Lime should not be necessary. If possible, the arable break in the rotation should receive the lime Rolling will probably be necessary in early spring as the leys are usually cut, and it is wise to press the stones down firmly
2-year ley	As for the 1-year ley	As for the 1-year ley, with the addition of 40 kg/ha phosphate and up to 120 kg/ha potash in the autumn (depending on soil k index)	As for the previous year, except that the autumn application of phosphate and potash will not be needed	

GRAZING EARLY IN THE SEASON

The early graze in the spring is usually (but not always) associated with the short ley. As far as possible an "early" field should be chosen for this first graze. A field which is sheltered, and with a south-facing slope (but not susceptible to frost) and which has a light, free-draining soil to warm up early in the spring is ideal. This, of course, is not always possible, but whatever field is chosen, it is really only possible to get early grazing from a ryegrass sward. 80–90 kg nitrogen is usually applied for early bite, but the timing of this first nitrogen is now considered important. Obviously, if it goes on too early serious losses can occur in a wet spring through leaching as the herbage plants are not in a condition to take up the nitrogen. On the other hand, too late an application will mean that potential growing time has been lost.

The T Sum method from the Netherlands is now receiving attention in this country. This relates to the sum of the mean daily air temperature (mean of the maximum and minimum) from 1st January each year. Minus temperatures are not subtracted; they are disregarded. When the total reaches 200°C then it is suggested that the nitrogen should go on—at least, that is the advice for Dutch farmers.

The work done in this country so far indicates that generally (but not always) T200 is reached before the nitrogen would normally be applied. However, it is not too far off the normally accepted time for conditions over here. More trial work is being carried out, including whether T Sums other than 200°C are more appropriate for different regions in the United Kingdom.

An alternative method is based on the start of the growing season which, for many crops, is still accepted at 6°C soil temperature, usually at a 15-cm depth. So that the nitrogen is readily available for the plant roots when 6°C is reached, the advice is that it should go on at 5°C which, under average conditions, is about 10 days before the 6°C temperature. Meteorological records in South-west England, excluding Devon and Cornwall, show that the first nitrogen application could be on about 7th March, whilst in Northern England it would be 15–20 days later. Adjustments have to be made for altitude, with the start of the growing season delayed by

5 days for each 100 m above the average height for the area.

Forage rye can also be used for early spring grazing. 10 kg Italian ryegrass can be added to the 150 kg rye for a more palatable crop, although the ryegrass is largely wasted if it is sown after mid-September.

The crop is top-dressed with 75 kg nitrogen in February, and following the first graze an additional 75 kg nitrogen is applied which preferably should be taken for silage, as the cattle, having been on grass, find the second graze of rye rather unpalatable.

Normally, the field will then be prepared for kale.

GRAZING BY STOCK

All stock do not graze in the same way. Some are much better grazers than others. The most efficient are store cattle, followed by dairy cows and fattening cattle, sheep and young cattle. Horses are notoriously bad grazers!

Although mixed stocking is ideal, in practice it is seldom possible. But what should be done is to see that the most profitable stock get the best. This usually means the dairy cow or fattening beast, and at certain times of the year, the sheep flock.

GRAZING SYSTEMS FOR THE DAIRY HERD

No one grazing system will suit all conditions of dairy herd management. But whichever is chosen it is important that the cow in milk has priority over the grass. She should never be regarded as the conditioner for the pasture. It is really preferable to waste grass rather than to expect the cow to work hard to get it (but see Leader/Follower system, page 145). Grazing down from an optimum height 20–30 cm to 6–10 cm is quite sufficient. In fact, under a paddock system, herbage intake and thus yield is depressed below 8 cm, and below 6 cm with set-stocking.

Although any grazing system needs to be flexible, nevertheless it is possible to work out, at least in theory, the requirements of any size herd over the season, using an appropriate amount of nitrogen.

Assuming that the cow requires about 12 kg dry matter from grass (total daily DM requirement is 16 kg) each grazing day, 1200 kg is needed for a 100-cow herd—20 hectares (0.2 ha/cow) would have to be used to provide about 240,000 kg dry matter to meet the grazing requirements of a 200-day season. Under average conditions this would be met by about 380 kg nitrogen/ha. This does not allow for wastage, but any extra production of grass required could, in theory, be provided by soil and re-circulated nitrogen.

Basically, there is a choice of three grazing systems which can be used for the dairy herd.

1. Strip grazing (Fig. 68)

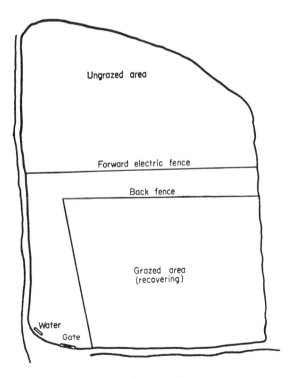

FIG. 68. Strip grazing.

Although now not so popular, strip-grazing is particularly suitable for the smaller dairy herd— suggested less than sixty cows. The system allows the animal access to a limited area of fresh crop either twice daily, daily, or for longer intervals. Wherever possible, the back fence should be used, whereby the area once grazed is almost immediately fenced off. This is to protect the recovering sward from constantly being nibbled over. Without the back fence the recovery rate is very much slower, although few strip-grazing systems use it these days. Properly managed, it is the most efficient of any grazing system for the dairy herd. However, it is labour-consuming, and it also requires a daily decision as to how much grass is needed for the cows. Until experienced (and even then it is not easy) there is a tendency to allow too little to satisfy the herd's requirements. In addition, wet soil conditions can lead to serious poaching along the line of the fence.

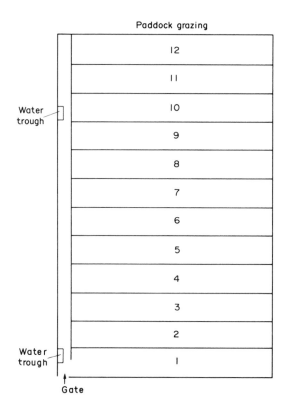

FIG. 69. Paddock grazing.

2. *Paddock grazing* (Fig. 69)

The principle of the paddock system is rotational grazing alternating with rest periods. The grazing area is divided into equal sized paddocks which will normally occupy the fields near the milking unit where access is easy.

The 21-day system was the first of the modern paddock grazing systems which have been evolved in the last 30 years. Twenty-one 1-day paddocks are used because it is calculated that, by allowing each cow 0.2 hectare per season for grazing and applying at least 50 kg nitrogen after each grazing, average growing conditions will bring the grass back to the right stage for grazing in 3 weeks.

The paddock size will obviously depend on the number of cows in the herd, but as a rule of thumb and on the basis of daily dry-matter intake, 100 cows will need about 0.8 hectare (50 cows/acre) per day of fresh grass. Thus, on this basis, as an example, 20 hectares divided into 21 paddocks will be needed for a 100-cow herd.

There can well be a loss of nitrogen efficiency if the grass is cropped before the full response from the plant food has been allowed to take place, and this is tending to prompt a move away from the 21-day grazing cycle, particularly in drier areas where a 28-day cycle is considered to be more appropriate.

The Morrey 1-day, 1-sward system has thirty-six paddocks with a cycle of 32 days, and the Wye College 28-day system of grazing comprises four plots, each of which is divided into seven sub-plots, a fresh one being grazed daily, but the previously grazed sub-plots are not back-fenced off.

Similar length grazing cycles are also being achieved with a fewer number (up to twelve) of larger sized paddocks which are grazed for 2–3 days. In the early part of the season especially it may be necessary to strip graze individual paddocks, but because fewer paddocks are required the main fencing can, with advantage, be more expensive and reliable. It is also more flexible than using 1-day paddocks.

At the peak of spring growth it may be possible to alternate round less paddocks, in which case any surplus grass should, if possible, be cut for silage or even hay. (Extra potash—average 85 kg—should be applied *after* the crop has been taken.) Naturally, adjustments will have to be made when the crop is not so productive, with other grass brought into the system.

Usually this will come from the cutting block following the silage or hay cut.

As far as possible, any uneaten herbage should be removed by topping after two or three cycles of the paddocks. Tighter grazing (above the rates already suggested is not the answer with dairy cows unless on the Leader/Follower System). However, it is important to prevent an accumulation of uneaten herbage, as otherwise it will cause a rapid deterioration of the sward as well as leading to aerial tillering.

The *Leader/Follower system* can be superimposed on the 1-day paddock system.

It is best used for calves and heifers (page 146), but with dairy cows it consists basically of dividing the herd up into four groups of cows (and so it is really only applicable for the larger herds) which will use four paddocks at any one time. The freshly calved cows are placed in the first paddock, followed next day (when the first group of cows has moved on) by the mid-lactation group, and then the stale milkers, with the dry cows grazing the first paddock on the fourth day. The stocking rate is as before—0.2 hectare per cow.

With this system, the third and fourth groups are being used as scavengers, and good utilization of the sward is being achieved, but obviously management is more complicated.

3. Set stocking

Set stocking, as understood from the past, was associated with extensive grazing, a corre-

sponding low rate of stocking and little or no nitrogen.

Intensive set stocking uses the same area of grass as for paddock grazing. The only difference is that there are no internal fences.

To help prevent the grass getting out of control, it has been found necessary to start grazing in the spring a little earlier than is usual with a controlled system. A stocking rate of 0.2 ha is advised at the start, but as the season proceeds and aftermath grass from the cutting area becomes available, so the grazing can become less intensive. It is also preferable to have persistent and prostrate perennial ryegrass varieties as the main constituents of the sward.

Nitrogen fertilizer is applied every 4 weeks over the whole grazing area or to a quarter of the area each week. Because the grass is kept fairly short all the time, even with 370 kg nitrogen in the season, the clover seems to stay in the sward that much better compared with controlled grazing.

Provided there is sufficient grass available for the cows, any grazing system is as efficient as any other from the point of view of animal performance. But whatever method of grazing, the yardstick must be the production and condition of the stock coupled, as far as possible, with the condition of the sward, although too much emphasis should not be placed on this second point. Given a reasonable rest period a good sward will soon recover although, as already mentioned, it may be necessary to top over the uneven effects of grazing, but this is not often the case with set stocking. Recent trials are now showing that set stocking under intensive conditions does not compare so favourably with controlled grazing systems, particularly from mid-summer onwards. Under less intensive conditions of two or less cows/ha there is little to choose between controlled grazing and set stocking. Once the correct allocation of grass has been made, management should be easier with a set-stocked system.

Irrespective of the grazing system, it is preferable if the grazing block is kept separate from the conservation area (also about 0.2 ha per cow). This is known as the two-sward system. Management is easier and different types of swards can be grown to suit the different methods of utilization.

It is seldom easy to fit the followers in behind the dairy cow, particularly on a paddock system. In fact, whatever method of grazing, management is easier if the followers are kept separate from the herd grazing area. They are usually set-stocked, although paddocks can be used. The stocking rate with followers must obviously depend on the class of stock being grazed (lower with in-calf heifers compared with bulling heifers) and the time of the year of grazing (more in the spring). It is in the range of 12 to 5 head/ha.

The *Leader/Follower system* has now been developed for calves and heifers on grass, although a similar principle was used in New Zealand many years ago. In this system the calves rotationally graze paddocks in front of older heifers which are in their second grazing season. Compared with conventional grazing, the calves have been shown to produce higher growth rates. This is because there is less of a build up of stomach worms in the sward, and the intake of grass is greater. The growth rate of the heifers may be slightly less. Stocking rates can be up to four calves and four heifers/ha.

Because of the greater control of stomach worms, the system is suitable for permanent pasture. Clean land is not required; the same field can be used for a number of years.

Zero grazing or the feeding of fresh cut grass to stock is an old practice. It used to be called green soiling, then being associated with a much wider range of crops than is the case these days. Grass or grass/clover mixtures are now mainly used, although the future could well see other crops being grown again.

It is not easy to see a revival of interest in the technique. However, when maximum stocking rate has been achieved by orthodox grazing, zero grazing, in most situations, can increase stocking intensity by 5–10%. But it is a twice-a-day all-the-season operation involving expensive machinery which of necessity must be extremely reliable. In addition there is the problem of dealing with the slurry from stock housed all the year round. Overall extra costs incurred could amount to about £20.00 per cow per year.

Grazing systems for other stock

Beef cattle

The *traditional system* of finishing off stores on first grade fattening pastures such as used to be found in the Midlands is now not often seen. Where it is practised the cattle are turned into the field (which may have received up to 100 kg nitrogen) at the beginning of the season and are kept there all the time. For finishing, a stocking rate of $2\frac{1}{2}$ bullocks/ha is used, and they should be able to put on a liveweight increase of 100 kg in the season May until July. The field then takes a second group of cattle, although they can never be finished off without supplementary feeding.

The *semi-intensive system* has, until recently, been associated with paddock grazing, but this has largely been replaced by set stocking.

Usually the field is, to start with, divided into one-third for grazing and two-thirds for silage. The grazing area is normally grazed continuously until July and then the stock are moved on to the silage area to graze the aftermath. The hitherto grazed area is then put up either for silage or hay, and when this has been cut the stock, at what will then be at the end of the season, graze the whole field. Nitrogen should be used at 50–60 kg on the grazing area every four weeks. For silage 120–140 kg nitrogen should be sufficient, but for hay not more than 50 kg should be used, particularly in the wetter parts of the country. Depending upon the soil indices, however, phosphate and potash will normally be applied after the conservation cut at about 50 and 70 kg respectively.

Stocking rates at about 11 to the hectare will be higher at the start of the season, but the increase in weight and appetite of the stock coincides with a fall in production from the grass as the season continues. This should mean a reduction in stocking rate to about 6/ha and then 4/ha which will fit in well with the new grazing area made available in July and September. Live-weight gains of about 0.7 kg per day should be obtained.

Sheep

For quite a large part of the year the grass sheep flock should be regarded as a scavenger flock. There is no reason to treat it otherwise.

It is impossible to give any hard and fast rules on the rates of stocking. The grass will obviously not be at its best, as other stock would have already been over it. The sheep are simply clearing up. Therefore the condition of the stock must be watched and stocking rates adjusted accordingly. But do not overgraze the pasture with sheep.

Better grass is needed for ewes being flushed, and for ewes and lambs for fat-lamb production. For all practical purposes *set stocking* is now the only method used for the ewes and their lambs during the fattening period. Rotational and forward creep grazing are no longer considered necessary now that anthelmintics are so widely used. Worm infestation is not now a serious risk, but pasture should, as far as possible, be clean (see MAFF Booklet 2324). Set stocking is the least complicated of any form of fat lamb production off grass, and under good growing conditions a stocking rate of at least 20 ewes and their lambs per hectare should be possible.

75–125 kg nitrogen can be applied at the start of the season, but subsequent growth will dictate if any more should be applied later on in the year. Not always will it be necessary.

The *"Follow N" System* is, in a sense, a compromise between rotational grazing and set stocking. Nitrogen is applied to each quarter of the field in successive weeks, but the flock will tend to keep off the most recently top-dressed quarter and concentrate on that part of the field which had its nitrogen three weeks earlier. In this way the field is grazed in successive quarters. It has the benefits of a rotational grazing system, but with less stress and no internal fencing.

GRASSLAND RENOVATION AND RENEWAL

Poor permanent pasture can be recognized by a mat of old decaying vegetation at the base of the sward. Although a certain amount is inevitable in a permanent pasture, a heavy mat is not required as it will adversely affect the performance of the sward.

Before any programme of renovation is attempted it is important to understand why the grass has got into its present state. It is usually the result of indifferent management over a long period, such as bad grazing or continually cutting for hay with insufficient fertilizer, although it can be due to bad drainage as well as lack of lime. Any of these reasons can lead to the inferior, indigenous grasses choking out the better plants.

There are a number of ways in which a poor pasture can be made more productive. At one time, apart from replacing it by ploughing and direct reseeding, the only way that a sward could be improved was by a slow process involving drastic harrowing to pull out the old mat, followed by lime and fertilizers, particularly phosphate, to encourage the better grasses and clovers. The use of stock to bite down the new growth and encourage tillering would bring about a gradual improvement of the sward, albeit over a number of years.

Broad-leaved weed killers can quicken up the rate of improvement and oversowing, especially of small-leaved white clover (up to 6 kg/ha) and ryegrass, is often incorporated into the programme. But, even when much of the mat has been destroyed, oversowing is not always successful, mainly because usually there is insufficient soil cover for the seeds. Alternative methods of renovation can, therefore, be considered.

Dalapon at a low rate—3 kg in 350 litres/ha—can be used as a selective weedkiller in a grassland environment, provided the perennial ryegrass is evenly distributed in the sward—at least one plant per 30 cm^2. Nitrogen at 50 kg should be broadcast immediately after the dalapon with grazing normally 3–4 weeks later. The clover is not affected by the dalapon, although there may be a temporary check to the ryegrass. As a means of pasture improvement, results have been variable, but the Weed Research Organization suggests that, by using dalapon in the summer every 2 to 3 years the ryegrass can be maintained quite satisfactorily and cheaply in a sward at the expense of weedgrasses and broad-leaved weeds.

Ethofumesate, which was originally marketed for weed control in the sugar-beet crop, can also be used to improve a poor grass sward with little adverse effect on the better grasses such as ryegrass. It is particularly useful against blackgrass, barleygrass, brome, annual meadowgrass and rough-stalked meadowgrass, but not couch or watergrass. It also controls some broad-leaved weeds, notably chickweed and cleavers, but it will kill out any clover present.

Ethofumesate is best applied at 10 litres/ha in the late autumn to swards where there is at least one perennial ryegrass plant to 30 cm^2. It is slow acting and it can be at least 8 weeks before any visible effect is seen. The cost is also expensive at about £60/ha.

Partial sward replacement using the Weed Research Organzation's *one-pass* technique at about £50/ha should be considerably cheaper than conventional ploughing-up and reseeding. In one operation it involves spraying a 10-cm wide band of herbicide—paraquat, glyphosate or dalapon; turning aside a strip of turf about 2.5 cm wide and deep in the middle of the sprayed band, and sowing into the trench—seed, fertilizer and slug pellets. Row widths (distance between the band spraying) can, of course, vary. The Gibbs slot-seeder is an example of a machine that can be used.

The sward should be closely defoliated before sowing at any time from April to September. Grazing can start up again a few days after sowing.

Apart from actual renovation the technique can also be used to introduce clover into a hitherto all-grass sward, and the production from a permanent pasture could be improved in the spring by sowing short-duration ryegrass into the sward the previous August.

Ploughing and reseeding is still the established method of renewing a sward. It does mean a clean reseed, albeit a rather slow establishing one. But, in quite a few situations it is a difficult, if not impossible, operation and this is often the main reason why so many poor permanent pastures have remained as poor permanent pastures. However, in quite a few instances, these can now be remedied by methods other than ploughing.

Minimum cultivations are usually carried out with the rotary cultivator and in conjunction with herbicides.

Dalapon, aminotriazole, paraquat and glyphosate can be used. Both paraquat and glyphosate have the advantage of leaving no toxic residue in the soil, although sufficient time (7 days) should be allowed for the chemical to move through the leaf and kill the grass. Aminotriazole has a soil toxicity of 5–6 weeks and dalapon 7–8 weeks.

As dalapon and paraquat do not control broad-leaved weeds very satisfactorily, it is recommended that, if necessary and where these herbicides are being used, MCPA or 2,4-D is applied at least 6 weeks before the grass weed killers are used.

Any of these herbicides should be sprayed onto 10–12 cm of actively growing herbage with application rates in the range as given on the leaflets.

After about 2 weeks, a shallow rotovation (normally two passes) will complete the destruction of the old sward. The field is then rolled, after which drilling can take place.

This method of grassland renewal is usually a summer operation. But any of the herbicides (particularly aminotriazole and dalapon) can be applied in the autumn to be followed by rotovation and seeding in the spring.

Direct-drilling (see also page 138) is another way of renewing an old sward. The technique is now usually associated with either paraquat or glyphosate, both of which have the very real advantage of rapid inactivation in the soil which means that there can be a relatively quick turn-round from the old pasture to the new sward. It also produces a firmer sward compared with the plough or rotary cultivator;

and there is, of course, less loss of moisture compared with other methods. However, direct-drilling will mean more regeneration of grasses.

It is usually carried out in the summer with an application of glyphosate or paraquat on, as far as possible, an even regrowth of grass. After 14 days the desiccated herbage can be burnt off, followed by fertilizer prior to drilling the seeds, plus slug pellets. Thorough rolling should complete the operation. It ought to be possible to graze the new herbage within six to eight weeks of sowing.

GRASS CONSERVATION

Grass is by far the most commonly used crop for conserving either as hay, silage or dried grass. In terms of dry matter the approximate percentage of the total amount conserved is:

Hay	48–50%
Silage	50%
Dried grass	1%

Any form of green crop conservation will lead to a loss of digestible nutrients, and Table 46 shows the percentage loss in terms of metabolizable energy and digestible crude protein.

TABLE 46

	Percentage loss	
	ME	DCP
Hay	45	41
Silage	23	24
Dried crop	5	16

Green crop drying is not generally applicable to the average farm system. It should be considered separately, see page 162.

Both hay and silage-making techniques have improved in the last 10 years. This is particularly the case with silage and, in the same period, there has been a marked swing towards this method of conservation at the expense of hay. Silage is now playing a much more important role in the diet of ruminants because of the ever-increasing cost of concentrate feed.

HAY

Too much poor-quality hay is made in this country. Apart from the wet weather, poor-quality herbage and inefficient methods of making are mainly responsible. Although crops such as lucerne, sainfoin and cereal/pulse mixtures can be used for hay, the cheapest and generally the most satisfactory crop is the grass/clover mixture.

The NIAB leaflet *Grass for Conservation* discusses in detail the stage for cutting grass consistent with digestibility and yield. In theory optimum digestibility is suggested at 63D (page 133), but in practice with haymaking, the D value is likely to be nearer 60. In the majority of grass species this will be at the ear-emergence stage.

The critical period for hay occurs when the crop is partly dried in the field, and therefore there is a *golden rule for haymaking*. No more hay should be ready for picking up in one day than can be dealt with by the equipment and staff. If the cutting outstrips the drying and collection, the hazards of weather damage are greatly increased. If the baler can only deal with, say, 8 hectares in the day, then cutting should be in near 8-hectare lots.

The object in haymaking should be to dry the crop as rapidly as possible without too much exposure to sun and the least possible movement after the crop is partially dry. Consequently, there are only two methods of making hay worth considering:

(a) The quick haymaking method whereby the crop is baled in the quickest possible time consistent with its safety.

(b) Barn drying of hay.

Quick haymaking

With this method, there can be two or three stages in the making of hay.

(i) Fresh crop 75–80% moisture content. Curing in the field.

(ii) Approximately 25% m.c. Drying in the bale in the field.

(iii) Approximately 20–23% m.c.
Drying in the stack.

Hay—safe for storage 18% m.c.

(i) *Curing in the field.* The crop should be cut when it is dry and when the weather appears to be fine. The local meteorological office may be able to give a weather forecast for 2–3 days ahead. If possible the headlands of the field should be cut earlier for silage. Hay on the headland usually takes longer to dry out than the rest of the field. So that the crop should be cut quickly when the forecast is right, high speed uninterrupted mowing is imperative. The ordinary reciprocating knife mower will not give the best performance with heavy laid crops, and although double-knife mowers will achieve fast cutting under almost any conditions, maintenance is more difficult.

The flail cutter has been developed from the forage harvester. It will certainly increase the rate of drying in the field, but unless special care is taken, dry-matter loss through shattered leaf can be very high.

The drum or disc mower has a high performance under all conditions. It works fast, and has the additional important advantage of causing little loss of leaf, and although there is little curing effect, the swath is left in good condition for subsequent treatment.

Mower conditioners can set the pattern for quick curing without a heavy loss of leaf in the field. The principle is that, after the crop has been cut, it is immediately treated with a conditioner in tandem. There are many types now available such as a disc drum cutter with a rotor fitted with swinging flails or tedder-like tines, or a drum rotary mower with spirally-ribbed steel crushing rollers at the rear.

The quick haymaking method should work on the principle that, as soon as the crop is cut, it is mowed and thereafter, as soon as the broken swath is drier on top than underneath, it is mowed again. Under most conditions this could involve a total of two passes through the crop in the day (excluding the cutting). The tedder is the ideal implement to use on the first day before the leaf dries out to any extent. The next morning,

when the dew is off the ground, it should again be moved as on the previous day, but as it gets drier (and/or depending on the amount of leaf present) more gentle handling is necessary, using only the turner. On the third day, when the dew is gone, the crop should be turned, perhaps twice, and then it may possibly be ready for baling. But this does depend on the weather, the size and type of crop, and it may be *at least* another day before the hay is fit to bale.

Only general principles have been discussed for field curing. The importance of preserving the leaf cannot be over-emphasized. In the past there has perhaps been a tendency to stress the green colour of the hay as the only index of well-cured hay. Sometimes this has been achieved at the expense of the leaf.

Although it is convenient to talk in terms of moisture content in connection with stages of drying, there is at present no reliable moisture meter to test the moisture content of the crop in the field. It is a question of experience in deciding when the hay is fit to bale. As a guide, Table 47 from the National Institute of Agricultural Engineering shows the condition of the hay crop at varying stages of moisture content.

Another way of assessing moisture content is to select and weigh at intervals random lengths of the swath in the field. When the swath weighs 25% of its fresh weight it is fit for baling in the field. When it weighs 30% it can be treated with an additive (see page 154). When it weighs 35–40% it can be baled for barn drying (see pages 152, 153).

(ii) *Drying in the bale.* It is a matter of opinion as to whether the bales should be left in the field after baling to continue drying. Certainly mechanical handling of bales has generally resulted in fewer of them being left out in the field as a recognized stage of the drying process. However, it should be understood that the immediate carrying of the bales must mean baling at a lower moisture content—say 23%.

Round bales are virtually weatherproof, and they can be left to dry slowly in the field for some weeks although, obviously, aftermath recovery

TABLE 47   ASSESSMENT OF MOISTURE CONTENT OF HAY

Moisture content (%)	Condition
50–50	Little surface moisture—leaves flaccid, juices easily extruded from stems or from leaves if pressed hard
40–50	No surface moisture—parts of leaves becoming brittle. Juice easily extracted from stems if twisted in a small bundle
30–40	Leaves begin to rustle and do not give up moisture unless rubbed hard. Moisture easily extruded from stems using thumb-nail or pen-knife, or with more difficulty by twisting in the hands
25–30	Hay rustles—a bundle twisted in the hands will snap with difficulty, but should extrude no surface moisture. Thick stems extrude moisture if scraped with thumb-nail
20–25	Hay rustles readily—a bundle will snap easily if twisted—leaves may shatter—a few juicy stems
15–20	Swath-made hay fractures easily—snaps easily when twisted—juice difficult to extrude

of the grass will be very much retarded under the bales.

(iii) *Drying in the stack.* There are various ways in which the bales can be picked up from the field prior to stacking.

The stack is usually built under a Dutch barn. With any type of stack a good level bottom is necessary, one formed of substantial rough timbers is ideal to keep the first layer of bales well clear of the ground.

The stack walls must be built carefully and firmly. Within the walls, the bales should be stacked leaving air spaces between them. They should not be squeezed in. There are two reasons for this:

(a) The moisture content at stacking will generally be about 22%. It will eventually drop to about 16%. This moisture must be allowed to escape.

(b) As the bales are not very dense, they will, in the lower layers of the stack, tend to be squeezed out by the weight of the bales above. Therefore, cavities between the bales will reduce the tendency of the walls to "belly out".

A good layer of straw should be put on top of the stack to soak up the escaping moisture. This will help to prevent the top layers of bales from going mouldy.

The big bale

The most obvious advantage of the big bale is its facility for speeding up baling and carting from the field. It means that more grass can be cut and cured at any one time: i.e. more hay can be made in fewer days.

There are two types of big bale—the round and the rectangular bale, and with both present experience suggests that even more care is needed to see that the crop is uniformly cured in the field, and then formed into a swarth of rectangular cross-section to match the width of the baler intake.

For field-cured hay, rather than barn-dried, the hay should be dried down to 25–30%, and then baled to about 34.5 bars (500 lb/in^2) pressure—the higher the moisture content the more compact should be the bale. There is some evidence to suggest that the round bale requires a lower moisture content on baling. Actual baling is also slower with the round bale, but it is more weatherproof, and it can safely be left out in the field, although this should never be for too long. Because big bales weigh anything from 600–900 kg their handling must obviously be completely mechanized. "Grippers" fitted to the front end loader on the tractor are suitable for the rectangular bale. The round bale can be handled in a similar way, and it can also be lifted with a single spike driven into one end of the bale.

The barn drying of hay

This is a process whereby partially-dried herbage is dried sufficiently for storage by blowing air through it. When making hay in the field, it is in the final stage of curing—the reduction of moisture from about 35% to 25%—that demands all the skill and attention of the farmer. It is at this critical period that the major losses of dry matter take place through too severe handling, and damage by the weather. If the hay can be carried at an earlier stage for curing to be completed in the barn, it should be much leafier and more nutritious.

Basically, *storage drying* is now the only system of drying, and there are three ways in which it can be carried out.

1. In walled barns

With this system of drying, the bales (at 35–45% moisture content) are placed on a flat evenly ventilated floor through which unheated air is forced by one or more axial fans. Drying takes place in a building with airtight sides so that the air will flow through the holes (Fig. 70a).

The drying platform should be strongly constructed of welded mesh supported on timbers at a height of about 60 cm above ground level.

Drying commences as soon as the first layer of bales has been completed and initial loading continues until three or four layers have been built. It is generally recommended that all bales are stacked on edge as tightly as possible, and successive layers cross-bonded to prevent a continuous vertical seam between the bales. It is important to ensure that air passes through and not between the bales.

Normally, after 2 to 3 days, a further three to four layers can be loaded, and the process is repeated until the barn is full. At least ten layers of bales are built in this way, but it is necessary to ensure that the top layer is at least 30 cm below the eaves to permit the damp air to escape.

Drying will normally take 10 days after the completion of building, i.e. probably up to 3 weeks from the start of building.

2. In open barns—the Dutch method (the radial-drying system) (Fig. 70b)

Bales are loaded round a central "bung", and as stacking proceeds, the bung is raised, leaving a central vertical chimney or air duct through which air is blown by an axial fan normally mounted on top of the bung. Where the roof

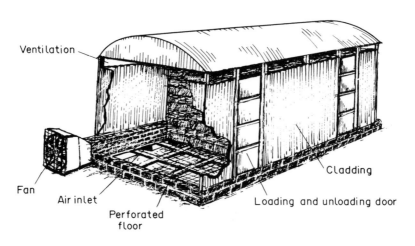

Fig. 70a.

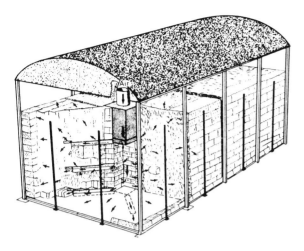

FIG. 70b. Drying in the barn (Dutch system).

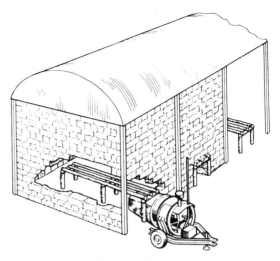

FIG. 71. A form of storage drying.

structure may not support the bung/fan unit, the air can be blown up from an underground air duct. The air then flows readily outwards through the bales, with blowing commencing when at the most three layers of bales have been stacked. Branch ducts in the diagonals of the stack ensure a more even flow of air to all the bales.

Continuous blowing is normally necessary for up to 12 days. This is followed by intermittent blowing to give a total overall drying period of about 21 days.

3. In open barns—with a centre duct

In this system—an above- or below-ground centre duct runs the length of the barn. The bales are stacked in the form of a tunnel over the duct through which unheated air is blown. Drying commences when three layers have been laid across the top of the duct, after which loading will depend on the rate of field curing, bearing in mind that the moisture content of baling should not exceed 40%. Drying will take up to 10 days to complete after the final loading, with again an overall drying period of about 21 days.

There are various modifications of these barn-drying systems. For instance, rather than dry along the full length of the barn, the hay can also be dried in bays as in Fig. 71.

Whatever method is employed for drying, the importance of even wilting in the field cannot be over-emphasized. This is to ensure that subsequent drying of the bales will be as uniform as possible. Under normal conditions, 1–1½ days' curing in the field prior to baling will be necessary for the hay to be reduced to a moisture content of between 35–45%.

Blowing should be continuous with any form of barn drying. Especially with the mobile moisture extractor unit, there may be a temptation to dry for a few days, and then to move on to another stack for a time before coming back to the first one. In the intervening period, the partially-dried hay will have started to heat up and valuable digestible nutrients will be lost.

The drying of big bales (see Fig. 72)

FIG. 72. Bale arch for tunnel drying.

The moisture content for drying big bales is in the range 25–35%.

At present, tunnel drying is recommended. In total eleven bales are used to make up one arch of a double thickness of bales as shown in Fig. 72. A complete tunnel will usually consist of four or five arches. The tunnel duct should be 1.2 metres wide at the fan end, but tapered to the far end which is best plugged with conventional bales. This will help the fan air to go out through the bales, rather than quickly flowing down the tunnel. Some form of supporting grid is necessary to prevent the tunnel collapsing as the bales settle during drying. Under average summer conditions continuous blowing will be necessary for anything from 14–16 days depending on the moisture content of the crop when baled.

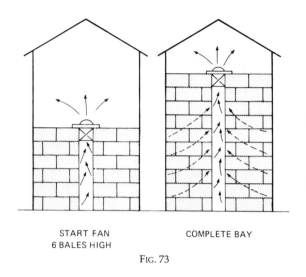

START FAN   COMPLETE BAY
6 BALES HIGH

FIG. 73

Hay ventilation

Barn drying is, of course, a form of ventilation in that air is forced through the bales. It is quite different from an old principle which was used with haystacks, where a hole was put up through the centre of the stack, and this drew air through the hay in the form of convection currents. The idea has now been revised, but this time with a hole built up in the middle of a stack of bales. In addition, a fan is positioned at the top of this vertical plenum chamber to suck out the warm air from the middle of the stack. This creates a vacuum which, in turn, draws the moisture carrying convection currents through the bales and out through the top of the stack as in Fig. 73.

Where there is insufficient headroom to build the stack in the normal way, the fan unit can instead be laid on its side at the entrance of a tunnel built into the stack.

This ventilation, as described, would normally be necessary for up to 6 weeks.

The main idea of this form of ventilation is to enable the farmer to bale and immediately carry his hay at about 30% moisture content (approximately a day earlier than would normally be advised for complete field curing) for curing to be completed in the barn.

Hay additives

Hay additives are a possible compromise between quick haymaking and barn drying in that their use permits the hay to be baled at a higher moisture content than would be possible with quick haymaking, although lower than if for barn drying. Compared with quick haymaking, leafier, better-quality hay should result simply because it is brought in more quickly from the hazards of the weather.

The chemical used as the additive is based on propionic acid. It is normally applied as a fine spray on the hay entering the baler at an application rate, depending on the formulation, which varies between 4 kg and 14 kg/tonne hay baled. Properly applied, the additive will inhibit bacterial fermentation, and will also suppress mould growth during hay storage. These micro-organisms will consume valuable nutrients, and so reduce the feeding value of the hay. At moisture contents over 25% and temperature in excess of 40°C they can also cause farmer's lung disease and mycotic abortion in cattle.

One of the problems of hay additives has been the difficulty of even application, but once this has been overcome their use could be a valuable additional aid to haymaking. However, they are no substitute for sound haymaking techniques.

Tripoding of hay

Although good-quality hay can be made by curing the crop on wooden frameworks in the field, it is a very slow and laborious method. It is generally only seen now on small stock farms in the wetter western parts of the British Isles.

SILAGE

Silage is produced by conserving green crops in a succulent state. The actual conservation process is known as ensilage and the container in which the material is placed is a silo.

Crops for silage

Grass is the ideal crop for silage, provided it is cut at the right stage, and the fermentation is satisfactory.

The effect of digestibility of the cut crop has now been quantified by calculating that, with the dairy cow, for every unit (%) increase in D value, there is an average increase in milk yield of 0.23 kg per cow per day. Taking an improvement from 63 to 67 D, the increase in yield would therefore be about 1 kg per day.

Lucerne. For ensilage it is preferable to grow lucerne with a companion grass crop, but even then, in spite of wilting and using acid additives, it is not easy to make really well-fermented silage. The reason is that the alkaline minerals contained in the plant:

  (i) tend to neutralize the desirable acidity required which will increase the risk of a clostridial butyric fermentation (page 163) which, in turn, will normally mean poorer intake;
  (ii) will make the silage rather bitter which again will lead to a reduced intake by the animal.

Lucerne for silage is normally cut at the pre-flowering stage.

Red clover as a crop is not usually taken on its own for silage. However, early red clover can be used, with the seed taken from the second growth, usually in September or October.

Compared with lucerne, a more palatable silage should be made from early red clover but, again, wilting is necessary as well as an additive.

Arable crops. A typical mixture is 140 kg oats and 50 kg vetches. It is more often sown in the spring using the appropriate varieties, although an autumn-sown crop should yield up to 30 tonnes/ha, 10 tonnes more than when spring-sown.

These crops are harvested at the end of June to mid-July (depending on time of sowing) in the milky stage of the oat with the straw still green.

Arable silage is preferably fed to beef cattle, rather than the dairy cow. It is not very "milky".

Whole-crops are usually associated with high dry-matter silage for the tower silo. They have a high dry matter (40–45%) which do not require wilting (page 158). Because of its high sugar content it also ferments well. However, the digestibility and crude protein levels are lower than good grass silage—D value 60, and digestible crude protein 5% in the dry matter.

Whole-crop silage should be distinguished from arable silage because, apart from its higher dry-matter content, much of the nutrient value is within the grain.

Barley makes the best whole crop. It is preferred to oats as, although the latter is slightly higher yielding, it contains a higher proportion of straw, and this will reduce its overall feeding value. Whole-crop barley should be cut at the mealy ripe stage of the grain.

Forage maize makes a very palatable silage but, because of its inherently low protein content, it is better suited for a beef system. The protein content can be increased, albeit rather expensively, with the addition of non-protein-nitrogen compounds, usually when the crop is being ensiled.

For clamp silage, maize should be ensiled when the dry-matter content is between 25% and 35% and, at this stage with the D value about 68, the grain is cheesey and doughy and is beginning to dent (dimple) at the top. This normally means harvesting from the end of

September to the middle of October, depending on the season.

A moderate frost before the crop is harvested will not seriously affect the yield or nutritive value. In fact, it can help to increase the dry matter of the crop simply by destroying the plant cell and increasing the rate of evaporation of cell moisture.

Precision-chop harvesters must be used when the crop is ensiled. In order to get really good consolidation the aim should be a chop between 10–22 mm. Additives are unnecessary for fermentation, but they can be used (i.e. based on propionic acid) to slow down aerobic deterioration when the clamp is opened for feeding. A narrow feeding face will also help to reduce aerobic deterioration.

For tower silos, 35–45% dry-matter contents are preferred and this will mean later harvesting. In these cases, the chop should be less than 10 mm. The yield of maize silage varies between 25 and 60 tonnes/ha.

By-product silages, e.g. that produced from sugar-beet tops and and pea-haulms, can be of high quality, provided the material is ensiled clean.

FIG. 74. Dutch barn silo.

of various materials—reinforced concrete probably being the most popular now that railway sleepers are so expensive. It also helps to make more of an airtight seal at the sides of the silo. Safety rails above the walls should certainly be considered.

On a gross capital cost per tonne of silage stored, these silos are even more expensive than the tower silo, but, in addition, they can be used for storage of hay as well as possibly part of the building housing stock.

SILOS

The size of silos vary; the density of settled silage is in the range of 600–960 kg/m^3 according to the degree of compaction.

Choice of silo

Permanent roofed and walled clamp silo—the Dutch barn silo (Fig. 74)

This is generally accepted as the most satisfactory type of silo. The roofing allows protection from the rain whilst filling, during storage and when the silage is being used, but plastic sheeting is still necessary directly covering the silage. The roof should be not less than 5.5 metres from floor to eaves to allow adequate tractor movement when filling. The walls can be constructed

Unroofed walled clamp silo

This is the cheapest of the commonly used silos. The silage should be carefully covered on the top and shoulders to keep the rain out during storage. With careful making the side wastage can be kept to a minimum.

Safety rails above the walls are advised.

Unroofed and unwalled clamp silos (Figs. 75–76)

These silos without a permanent cover are not commonly made these days. Generally the losses are high and, within reason, the larger the silo the better as this should mean a smaller proportion of wastage. Heavy gauge plastic sheeting drawn right over the top and down the sides will help to reduce the losses by preventing convection currents going through the clamp, causing excessive oxidation.

FIG. 75. Run-over clamp.

FIG. 76. Wedge clamp.

FIG. 77. Pit silo.

FIG. 78. Airtight tower silo.

Trench or pit silo (Fig. 77)

This is not seen so often now, having in the past been associated with the buckrake system of harvesting.

Pit silos are cheaply constructed, and they may perhaps be considered for outwintered stock away from the buildings, in low rainfall areas and on well-drained soils.

Tower silo (Fig. 78)

This is normally made of galvanized steel and is glass-lined or treated with a protective paint to make it airtight and acid resistant. A domed metal top completes the seal.

Tower silos are either top or bottom unloaded. The latter, whilst being more expensive, does mean that the silo is more airtight when it is in use (when being emptied) and so there is less risk of aerobic deterioration. It has a plastic breathing bag and pressure-relief valve inside at the top to compensate for the difference in pressure between the outside and inside of the silo.

The top unloader has the advantage in that the feed auger is less prone to blockages, and if there is a breakdown it is possible to unload it by hand. Additionally, aerobic deterioration of the silage is virtually minimal if no less than 5 cm thickness is removed each day.

Conventional towers of wood, concrete or galvanized steel (but without protective covering to make them more airtight) are still in use. The silage is removed from the top.

Drainage of silos

Unless the silo is well drained the bottom layers of silage will soon putrify. A simple drain is all that is necessary to get the effluent away. The simplest method is to have the silo floor sloping from one end to the other, with a cross drain at the lower end, although there are more expensive modifications.

It is essential to prevent any effluent running into a watercourse.

HARVESTING THE CROP FOR SILAGE

Cutting the crop direct into a trailer without wilting and using either a single or double chop forage harvester is the simplest and cheapest system, but these days the crop is usually wilted in the field prior to its being ensiled (see this page) and, in this respect, the mower conditioner (see this page) is ideal for initial cutting.

Most crops are now picked up using either the precision or full chop (10–75 mm theoretical chop length) or the fine chop (10–150 mm theoretical, but less accurate, chop length) forage harvester. Although they are more expensive than the previously widely-used double-chop harvester, their use should mean better silage because the shorter crop results in the fermentable juices being more quickly released, producing a quicker fermentation.

However, there is some indication that the intake and performance of precision-chopped silage (as opposed to a longer chop) may be affected when fed with a large amount of concentrates to cattle.

The forage harvester is used with high-sided trailers which collect the grass as it is blown from the "spout" of the harvester. Normally the trailer will dump its load by the silo which, in the case of the clamp, is then filled or built with the small buckrake. With towers, the crop is thrown into a dump box, and then blown into the silo.

SILAGE MAKING (ENSILAGE)

The fermentation process

In silage making this is essentially a matter of the breakdown of the carbohydrate and protein.

The plant cells in the crop are not immediately killed as soon as it is cut. Respiration will continue for some time by the cells taking in oxygen from the air in the ensiled crop, and carbon dioxide is given off, and then, as the cells lose their rigidity, so the carbohydrate starts to oxidize and the protein begins to break down.

This respiration and breakdown of the cells brings about a rise in temperature. The more air there is present, the more the respiration and breakdown.

At the same time, the bacteria which are always present on the green crop act on the carbohydrate to produce organic acids. Two main acids can be produced by their respective bacteria, lactic acid which is highly desirable, and butyric acid which is very undesirable.

Thus the stage is reached when the green crop is "pickled" in the acid; this, in fact, is silage, and what type of acid dominates depends upon the type of fermentation. This can be controlled to a large extent by the farmer himself.

As a result of different types of fermentation, three main types of silage can be produced as in Table 50.

Cold silage

The development of heat brought about by respiration should, as far as is possible, be kept to a minimum. Provided the crop is wilted and dry, and/or additives are used to provide acid or sugar for the lactic bacteria, good silage can be made at temperatures of below 16°C with little risk of butyric bacteria predominating. At this low temperature, there is less breakdown of carbohydrate compared with the old hot fermentation technique, and so a more valuable silage is produced.

Wilting will:

(a) *Create more favourable conditions for the lactobacilli.* By wilting, water is removed which results in a higher concentration of sugars in the crop when it is ensiled. Ideally, it should contain at least 2$\frac{1}{2}$% soluble carbohydrate to produce a desirable lactic fermentation (below 2$\frac{1}{2}$%, an additive is suggested to help get the right fermentation).

There is no simple field test for determining sugar levels, but most of the ADAS regions offer a telephone service to farmers which will give a daily indication of the general pattern of changes of sugar levels in the grass crop, and the need or otherwise for an additive.

(b) *Reduce silage effluent.* Silage effluent is one of the most polluting agents that is produced in agriculture. It is 150 times stronger than human sewerage.

TABLE 48

Percentage dry matter of crop at ensiling	Amount of effluent per tonne silage
10–15	360–450 litres
16–20	90–230 litres
over 25	virtually nil

Table 48 indicates that the drier the crop, the less the effluent.

Grass at 10–15% dry matter ensiled in a 300-tonne clamp will produce about 135,000 litres (30,000 gallons) of effluent, and most of this will be discharged during the first 10–14 days after ensiling.

Through the various Prevention of Pollution Acts, prosecution is liable if effluent is allowed to run into a water course. Also, it should *not* be added to the slurry store if under cover because of the danger of releasing hydrogen sulphide which can cause human fatalities. It is probably best if, after a 1:1 dilution with water, it is put straight back on to the grass (as long as there is no risk of pollution); 10,000 litres effluent contains between 10 and 35 kg nitrogen.

(c) *Enable*, on average, *up to one-third more crop to be carried in from the field in the trailer*, compared with the heavier, unwilted crop.

The disadvantages of wilting are:

(a) When wilted to more than 28% dry matter for subsequent ensiling in a clamp, feed trials are now indicating that there is a reduction in dry matter intake and subsequent animal performance.

(b) It will bring about a greater loss of digestible nutrients as in Table 49, because respiration has been allowed to continue that much longer.

(c) It is more difficult to pack down a wilted crop in the silo, although this is largely overcome by using a precision-chop harvester.

TABLE 49

Moisture content of the crop when ensiled (%)	Loss of digestible nutrients in the field (%)
70–75	10
60–70	12–14
under 60	15–16

The advantages of wilting outweigh the disadvantages. The best advice when making clamp silage is to carry out wilting so long as it does not hold up actual silage making.

The rate of wilting will depend on the prevailing weather conditions, as well as the treatment that the crop receives. The moisture content can be reduced by up to 10% (from an initial 75–80%) in the first 24 hours under *good* conditions where the crop receives no treatment after cutting. But if a mower conditioner is used there can be a 15% reduction in 6 hours.

Additives. There are two main groups of additives:

1. *Inhibitors.* These reduce clostridial activity and, at the same time, they can cut down on the amount of fermentation.

There are at present two types:

(a) The *formic acid* formulae works on the principle that, provided the correct rate of application is used, it will bring about an almost immediate increased acidity for the lactobacilli (which prefer a more acid medium) at the expense of the clostridial butyric bacteria. To start with, this additive is encouraging fermentation. However, because of more acid production, there are more lactic bacteria and these will result in a quicker death of the plant cell which will, therefore, mean reduced fermentation and a lower loss of digestible nutrients compared with untreated silage. It is used at 2.25–4.5 litres per tonne green crop—the heavier rate (which is unusual) with a wet, leafy crop. It is very corrosive.

(b) Used at 4.5 litres per tonne green crop, the *formalin/sulphuric acid* or *formalin/formic acid* additive could be described as a partial sterilant. It cuts down on the rate of fermentation and thus the breakdown of both carbohydrate and protein by inhibiting, not only the butyric bacteria but also, to a certain extent, the lactobacilli. But there will still be sufficient acid to preserve the silage in a stable condition. The analysis of silage treated in this way will tend to show a pH near 5 (which is normally indicative of butyric fermentation) because there has been less acid production. If this additive is used at 2.25 litres per tonne green crop, the fermentation will tend to follow that of the formic acid principle and the pH, like that with formic acid, will be in the region 3.8–4.0. These additives are not as corrosive as formic acid.

Note: (i) The calcium formate/sodium nitrite mixture can also be classified as an inhibitor. It is in powdered form, being applied at an average 2 kg per tonne green crop. It is non-corrosive.

(ii) Most other acid additives are based on formic acid; some include propionic acid.

2. *Stimulants.* (a) Molasses has been used for many years, and it acts by encouraging lactic acid production from the sugar it contains. The recommended rate is 9–12 litres/tonne. However, it is not easy to use, although it can be applied from a tank mounted on the forage harvester, or diluted with warm water and sprayed on to the crop, either before or after cutting, by means of a low-volume sprayer with large jets.

It is non-corrosive, but is not the most effective additive, and it has largely been superseded by the acid inhibitors.

(b) An innoculant additive containing live lactobacilli is now under study. Early trials suggest that the fermentation produced following its use is neither better nor worse than that from using other additives, but it is not yet fully proven.

It is available as a dry powder and is considerably more pleasant to handle than the acid additive. There is obviously no corrosive effect.

Sealed silage (Waltham or Dorset wedge system)

To produce silage with less waste it is essential to kill the respiring plant cell as quickly as possible. Thus air should be prevented from entering the green crop in the silo. The oxygen originally present is used up, and with no further air, and adequate acid production, the plant cell will die.

The walls of the silo should be made as airtight as possible to prevent air getting into the ensiled crop. With a sleeper wall, a plastic sheet attached to the inside is quite satisfactory.

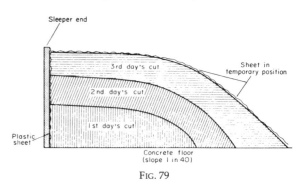

FIG. 79

The silo is filled in the form of a wedge as in Fig. 79. The first trailer loads are tipped up against the end wall and subsequent loads are built up with a buckrake to form a wedge. At the end of the day the wedge is covered with plastic sheeting (500 gauge) attached to the end wall of the silo. The next day, after the sheet has been rolled back more crop is added to the wedge exending its length until the silo is filled. Care is taken to keep the completed part of the wedge covered all the time and the uncompleted section covered after each day's loading. After the silo has been filled the cover should be well weighed down; old car tyres are ideal. To reduce side wastage particular attention must be paid to pressing the crop down at the sides. The tractor and buckrake will normally give sufficient consolidation although extra may be necessary at the start of each day's work. With this cold fermentation method, provided respiration is kept to a minimum, the temperature should not rise above 16°C.

The wedge-shaped principle can also be used for a clamp silo without walls. In this case a plastic sheet is used to cover the wedge completely after each day's building, and at the completion of the silo.

Big bale silage

The main advantage of the system is that the capital cost requirements in machinery and storage can be quite low. This is particularly the case if a big baler is already being used on the farm for hay and/or straw. No expensive building is required. The bales can, if necessary, be stored in the field although existing clamp silos can be used.

The baled grass is either stored in individual polythene bags or as unbagged bales stacked in a clamp. The former is more labour intensive but probably produces slightly better silage because, once sealed, there is less risk of air contamination.

The cut grass should be wilted to between 25 and 40% dry matter. It is necessary to reduce the effluent, particularly when ensiling in bags.

It is generally recommended that the width of the swath for baling should be the same as the baler pick-up. It is very important that an even rectangular swath is prepared to produce a neat compact drum-shaped bale of even density. Cone or barrel-shaped bales with lower densities at one or both ends can mean aerobic losses with over-heating. Subsequent stacking is also more difficult.

Because of the higher dry matter involved there does not appear to be the same necessity for the normal acid additive as an aid to fermentation. There is some evidence to suggest that a propionic-based mould inhibitor will help to prevent aerobic deterioration when the bales are fed. More trials will clarify this point.

The whole silage operation should be geared to the number of bales to make and move to the clamp for stacking in the day. They deteriorate fairly quickly in the field once made, especially at dry matters above 35%. An hourly output of

25–30 bales, weighing on average 500 kg, ought to be possible.

500-gauge black polythene bags should be used if ensiling in bags. At least two people are needed to slide the bag over the bale which is lifted up by a modified fore-end loader. An existing clamp silo is ideal but, in any case, the site should be carefully chosen and free from stones or other objects likely to pierce the plastic bags. To minimize vermin, it is preferable to site away from places which may harbour rats and mice. Bait around the stack, however, may be necessary.

The bags are best stacked in pyramid style as in Fig. 80a. So that the air is properly removed the bales in the first layer should not be tied (using soft polypropylene twine) until a layer of bales has been placed on them. Polypropylene netting, weighted down, is best used as a cover to minimize wind damage.

Unbagged bales can be ensiled as in Fig. 80b. They should be stacked as lightly as possible on a 1000-gauge plastic sheet and then, when two layers high, covered with a 500-gauge sheet. After allowing the stack to settle, a further top sheet, which should be kept as tight as possible, completes the cover.

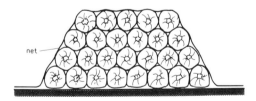

FIG. 80a. Bagged silage bales.

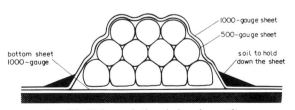

FIG. 80b. Unbagged silage bales after settling.

High dry-matter silage

High dry-matter silage of between 35% and 45% dry matter is normally made in a tower as illustrated in Fig. 78, although on a large scale it is possible to make it in a sealed clamp.

Silage made in a tower follows the same principle as that of sealed clamp silage. In order to justify the high capital cost involved of tower, field and feeding equipment, the silage produced should have a high nutritive value capable of not only maintaining the animal, but also of making a significant contribution to its production ration. This means that prior to ensiling, the crop must be very well wilted, and loss of dry matter in the field can be as much as 15–20% (see Table 49), particularly when wet weather holds up field operations. This is one reason why "whole cereal" crops are useful for towers. (See page 155.)

Maize is also a useful crop for high dry-matter silage, particularly when fed to beef cattle.

The silage from tower silos, whilst very palatable and showing little visible waste, is sometimes disappointing, especially as a milk-producing feed. But feeding is usually a very simple operation, and this fact alone could make the tower system more important in the future.

Silage analysis

A physical evaluation of the silage as indicated in Table 50 will help to determine its value. In addition, the following will give a guide as to dry-matter content when a sample is squeezed in the hand.

Wet silage—moisture squeezed out easily; dry-matter content less than 20%.

Medium dry silage—a little moisture expressed when squeezed firmly; dry-matter content 20–25%.

Dry silage—moisture not squeezed out; dry-matter content more than 25%.

But any silage should be properly analysed in the laboratory. A chemical analysis is very reliable, but one of the problems is to make sure that the sample analysed is truly representative of the silage to be fed.

GREEN CROP DRYING

This is more commonly referred to as grass drying. It is the most efficient method of conservation because, compared with silage and haymaking, there is far less loss of digestible materials (see Table 46).

Apart from grass, and lucerne as whole crop cereals, maize, beans and forage rye will probably be used to an increasing extent in the future.

Two types of driers can be used:

(1) *The low-temperature* (160°C) *conveyor type.* Basically this consists of a horizontal conveyor through which the wet material is conveyed on a moving bed. Heated air is forced up through the bed to drive off the moisture. Output varies according to the initial moisture content of the crop (usually between 75% and 80%). With a crop previously field wilted (although carotene will be lost) it can dry up to 0.7 tonne of dried material per hour, although the average is nearer 300 kg. Two men are needed to operate the drier, but only one extra would be required with two conveyor driers run side by side. Until fairly recently, this drier has been the main type used, but because of its low output it is being replaced by high-temperature driers.

(2) *The high-temperature* (up to 1100°C) *pneumatic rotary drum drier.* Drum driers are either single pass or triple pass, these terms describing the arrangement of the internal structure of the drum. The principle with this type of drier is that the wet crop, usually chopped into short lengths, is introduced into a stream of hot air. As the moisture is removed so the material gets lighter; it rises up and is blown away by another stream of air. An average output of about 3 tonnes of dried material per hour can be expected when the crop is dried from 80% to 10% moisture content. But there are now much bigger capacity driers in operation with outputs as high as 10 tonnes per hour. For operation two men are required, thus the labour charge per tonne of dried crop is considerably less than the conveyor drier. But these driers are extremely expensive with total capital investments from £100,000 to £500,000.

Small on-farm machines are being developed

TABLE 50   TYPES OF SILAGE

	Sample	Feeding value	Reasons	Prevention
Silage overheated	Colour: brown to black Smell: burnt sugar Texture: dryish	Although palatable, nutritionally value is poor. Carbohydrates have been burnt up and protein digestibility considerably impaired by the high temperature	Temperature remains at 49°C or more, due to an appreciable amount of air present in the silo. This happens with stemmy and/or over-wilted material	Do not let the crop get too mature before ensiling. When necessary fill the silo quickly and keep the air out
Butyric acid	Colour: drab, olive green Smell: unpleasant and rancid Texture: slimy, soft tissues easily rubbed from fibres Taste: not sharp, pH 5.0 or over	Reasonably palatable and nutritionally quite good, but this depends on the stage of butyric acid fermentation. With very butyric silage, palatability will be poor and much of the protein will have been broken down by the spoiling bacteria. In extreme cases the silage may become toxic, especially to younger stock	The butyric acid bacteria are allowed to dominate, conditions being unfavourable for the growth of the desirable bacteria, i.e. when young, leafy and unwilted crops with a high moisture content are put into the silo, and also when soil-contaminated crops (butyric acid bacteria most commonly occur in the soil) are ensiled. The growth of the lactic acid bacteria is slow under these conditions, and therefore they do not produce sufficient acid to prevent the butyric acid bacteria from maintaining and increasing their presence	Create unfavourable conditions for the butyric acid bacteria, i.e. encourage the lactic acid bacteria. Ensile dry, wilted crops, and, if necessary (when less than 2½% soluble carbohydrate —the dry matter) use an additive
Lactic acid	Colour: bright light green to yellow-green Smell: sharp and vinegary Texture: firm soft tissue not easily rubbed from fibres Taste, sharply acid, pH 4.2 or less, unless treated with formalin/acid additive at 4.5 litres/tonne green crop, when pH will be near 5	Good, and palatability should be excellent	The lactic acid bacteria have dominated the ensiling process. They have grown rapidly to produce sufficient acid to keep out the spoiling bacteria. Dry conditions have favoured the lactic acid bacteria and if the ensiled crop has been young and leafy, an additive has been added to help the desirable bacteria	Do not prevent, encourage!

and some of these are mobile units. Their maximum cost could be up to about £50,000.

At present grass makes up 75% and lucerne 25% of the tonnage dried in this country. To justify the high capital requirements for the operating plant and the high cost of producing dried grass it is essential that the crop should be cut at the correct stage of growth. If the dried crop is to be fed to ruminants rather than to poultry, the digestibility of the crop will assume a new significance. Hence it is important to see that as far as possible the crop is cut at its maximum digestibility. This should mean growing different species and varieties of species of herbage plants to give a sequence of maximum digestibility throughout the drying season from April to November. Plant breeders are now trying to produce varieties of herbage plants to give this ideal situation.

For the grass crop nitrogen usage up to 625 kg/ha per year (with the first 125 kg/ha going on in February) will normally be required. This in turn will necessitate potash application of up to 250 kg/ha depending on the soil type. Lucerne should not need any nitrogen but very often may require a total of 375 kg/ha of potash throughout the year, depending on soil type.

Because of the high capital cost involved with green crop drying the future must lie with the development of the large unit normally on a co-operative basis. To run the drier efficiently, and to ensure that the right sort of crop is being dried, it is essential that the manager of a plant must have complete control not only of the drier but also the crops to be dried.

THE DEWATERING OF GREEN CROPS

The dewatering of green crops was initiated chiefly with the idea of reducing the cost of artificial drying. However, as a result of this work, considerable interest has been generated on the protein extracted from the fractionation process.

Basically, the green crop is first shredded, then pressed to produce wet fibre and juice. The wet fibre still contains sufficient protein for it to be dried and graded as a dried crop. The protein in the juice is coagulated by steam, separated from the residual liquor (a brown juice), and then dried to produce leaf protein concentrate (LPC). The brown juice contains quite significant quantities of nitrogen and potassium, which can be returned to the land. Feeding trials are in progress to assess the performance of LCP incorporated in rations for ruminants and non-ruminants.

A 45% extraction of LPC is at present considered about right for the dried crop to contain sufficient protein for marketing. The 45% protein also competes with soya bean meal.

It is calculated that, if 1000 tonnes of green crop (20% dry matter) is processed as described, it will produce about 180 tonnes dried crop, 430 tonnes brown juice, and 25 tonnes LPC (45% protein).

Dewatering and LPC production is only now beginning to be carried out commercially.

SUGGESTIONS FOR CLASSWORK

(1) Identify the grasses and clovers of agricultural importance, both in their vegetative and flowering stage.
(2) Examine a seeds mixture, and identify the different types of seeds.
(3) Compare the periods of production of a ryegrass and non-ryegrass sward.
(4) Study and compare different techniques of grazing dairy cows and beef cattle.
(5) Study and compare different methods of fattening lambs off grass.
(6) Visit a barn hay-drying plant, and compare barn-dried hay with that made in the field.
(7) Examine different samples of silage, and make notes on the type of fermentation, dry-matter content and amount of wastage present in the silo.
(8) Visit a grass-drying plant.

FURTHER READING

Mcg. Cooper and Morris, *Grass Farming*, Farming Press.

Raymond, Shepperson and Waltham, *Forage Conservation and Feeding*, Farming Press.

Silage, MAFF Bulletin No. 37, HMSO.

ADAS, Booklet No. 9, *Silage*, MAFF.

William Davies, *The Grass Crop*.

Spedding and Diekmahns, *Grasses and Legumes in British Agriculture*, Publ. Commonwealth Agric. Bureau.

N. J. Nash, *Crop Conservation and Storage*, Pergamon Press.

Holmes (Ed.), *Grass. Its Production and Utilization*, Blackwell.

6

WEEDS

Weeds are plants which are growing where they are not wanted.

HARMFUL EFFECTS OF WEEDS

(1) Weeds reduce yields by shading and smothering crops.
(2) Weeds compete with crops for plant nutrients and water.
(3) Weeds can spoil the quality of a crop and so lower its value, e.g. wild onion bulbils in wheat; ryegrass in a meadow fescue seed crop.
(4) Weeds can act as host plants for various pests and diseases of crop plants, e.g. charlock is a host for flea-beetles and club-root which attack brassica crops: fat hen and knotgrass are hosts for virus yellows and root eelworms of sugar-beet. Couch grasses are hosts for take-all and eyespot of cereals.
(5) Weeds such as bindweed, cleavers and thistles can hinder cereal harvesting and increase the cost of drying the grain.
(6) Weeds such as thistles, buttercups, docks, ragwort, etc., can reduce the grazing area and feeding value of pastures. Some grassland weeds may taint milk when eaten by cows, e.g. buttercups, wild onion.
(7) Weeds such as ragwort, horsetails, nightshade, foxgloves and hemlock are poisonous and if eaten by stock are likely to cause unthriftiness or death. Fortunately, stock normally do not eat poisonous weeds (see MAFF Ref. book 161—*British Poisonous Plants*).

SPREAD OF WEEDS

Weeds become established in various ways such as:

(1) *From seeds:*

   (a) sown with crop seeds—this is most likely where a farmer uses his own seed and it is not properly cleaned;
   (b) shed in previous years; some weed seeds can remain dormant in the soil for up to 60 years;
   (c) carried onto the field by birds and animals, or by the wind;
   (d) in farmyard manure, e.g. docks and fat hen.

(2) *Vegetatively* from pieces of:

   (a) *rhizomes* (underground stems), e.g. couch, black bent, stinging nettles, corn mint, ground elder, bracken, coltsfoot;
   (b) *stolons* (mainly surface runners), e.g. creeping bent (watergrass), creeping buttercup, yarrow;
   (c) deep *creeping roots*, e.g. hoary cress, creeping thistle, field bindweed, perennial sowthistle;
   (d) *tap roots*, e.g. docks, dandelions;
   (e) *bulbs* and *bulbils*, e.g. wild onion;
   (f) *bulbous shoot bases*, e.g. onion couch (false oat grass).

These pieces of weeds are usually carried about by cultivation implements.

ASSESSING WEED PROBLEMS IN A FIELD

The seriousness of weed problems can be judged by:

(a) the amount and type of weed growth in crops—especially where no control measures have been taken, and
(b) between crops—particularly by the amount of rhizomes and root growth of weeds such as "couch" grasses, field bindweed, coltsfoot and others, which can be readily seen in cultivated areas.

It is possible to identify species and calculate numbers of viable weed seeds in a soil by taking samples, separating out the seeds and testing their germination capacity (dormancy creates problems here). This is very time-consuming work and only justified for research and survey purposes; it could not be regarded as a service which could be offered to farmers like the chemical analysis of a soil.

When decisions have to be taken on the use of herbicides, problems arise regarding:

(a) The possible weeds which are likely to appear—for example in a root crop; a record of the weeds present in the field in recent years can be a very good guide when selecting a suitable soil-acting residual herbicide.
(b) The identification of weed seedlings growing in a crop—young grass seedlings are particularly difficult to identify properly.
(c) The numbers of weeds present, e.g. per m^2 or per hectare.

There are no reliable guidelines which can be used to forecast accurately the possible harmful effects of a given population of weeds in a crop. The effects on yield also depend on the type of weed and its aggressiveness and on the density and vigour of the crop. Other harmful effects may be:

— trouble at harvest with climbing weeds such as cleavers and bindweed and combine sieve blockage with wild oat seeds,

— crop quality can be affected by weed seeds and/or bulbils of wild onion,
— the future; if allowed to grow and seed, a small population of weeds can create a large problem, e.g. the first wild oats, blackgrass, sterile brome or mayweeds to appear.

Where there are more than 100 weed seedlings per m^2 the damaging effects on cereal crops can be very serious and one or more carefully timed herbicidal treatments may be necessary: between 10 and 100/m^2, spraying costs are usually justified by the possible yield loss and contamination of grain problems; lower populations, e.g. 1–10/m^2, are much less damaging—especially in good vigorous crops but spraying could be worthwhile if the crop was thin and backward or aggressive weeds such as cleavers were present; at less than one per m^2 there is very little competition and spraying could usually only be justified to prevent seeding and future problems, and for appearance sake, e.g. wild oats; roguing can be justified when the numbers present are less than 500/ha.

The cost of a spray treatment—which will vary with the type of weeds involved—must also be taken into consideration when deciding on whether or not to spray. Advisors with wide experience of treatments and results over many years are best able to give sound advice.

In grassland, it is often difficult to decide on which plants should be regarded as weeds, but thistles, rushes, bracken, tussock grass and poisonous weeds such a ragwort, horsetails and hemlock should be destroyed.

CONTROL OF WEEDS

In recent years, the introduction of chemical *herbicides* or *weedicides* has greatly simplified the problem of controlling many weeds. Most of these chemicals can act in a *selective* manner by killing weeds growing in arable crops and grassland. The control of weeds with herbicides is now becoming an established and necessary practice on most farms. Nevertheless, it is worth remembering that other good husbandry methods can still play an important part in controlling weeds.

Methods used to control weeds are:

(1) *Cultivations* (see pages 45, 46).
(2) *Cutting*, e.g. bracken, rushes, ragwort, thistles. This weakens the plants and prevents seeding. The results are often disappointing.
(3) *Drainage*. This is a very important method of controlling weeds which can thrive in waterlogged soils. Lowering the water-table by good drainage will help to control weeds such as rushes, sedges and creeping buttercup.
(4) *Rotations*. By growing leys and various arable crops there is an opportunity of tackling weeds in many ways and at various times of the year; this method has become less important since herbicides were introduced.
(5) *Maintenance of good fertility*. Arable crops and good grass require a high level of fertility, i.e. the soil must be adequately supplied with lime, nitrogen, phosphates, potash and humus. Under these conditions crops can compete strongly with most weeds.
(6) *Chemical control*. It is outside the scope of this book to deal in detail with this very involved subject. However, the following is a summary of the main chemicals and methods which are used.

Most of the chemicals used have a *selective* effect, i.e. they are substances which stunt or kill weeds and have little or no harmful effects on the crop in which the weeds are growing. A severe check of weed growth is usually sufficient to prevent seeding and to allow the crop to grow away strongly.

Most of the common weeds found in cereals can now be controlled by selective herbicides. It is hoped, eventually, to have chemicals to control all weeds in all crops.

Herbicides are usually sold under a wide range of proprietary names which can be very confusing—especially if the common name of the active material is not stated. Throughout this book, the common name of the chemical is used when referring to herbicides and, occasionally, where there is only one proprietary product, the trade name is also given. A list of proprietary names of approved products is published annually by HMSO. The selectivity of a herbicide depends on:

(a) The *chemical* itself and its *formulation*, e.g. whether it is in water-soluble, emulsion or granular form; also, whether wetters, stickers or spreaders have been added.
(b) The amount of the *active ingredient* applied and the quantity of carrier (water, oil or solid). Most herbicides are applied in water solution.
(c) The *stage of growth* of the crop and the weeds. In general, weeds are easier to kill in the young stages of growth. However, treatment may have to be delayed until the crop is far enough advanced to be resistant to damage.
(d) *Weather* conditions. The action of some chemicals is reduced by cold air temperatures and rain after spraying. (See also Appendix 4.)

The chemicals now commonly used as herbicides can be grouped as follows:

(1) *Contact herbicides*. These will kill most plant tissue by a contact action with little or no movement through the plant; shoots of perennials may be killed but regrowth from the underground parts usually occurs. Some examples of contact herbicides are: phenmedipham, dinoseb, pentanochlor, sulphuric acid, bentazone, ioxynil, bromoxynil. Others, such as diquat and paraquat, are sometimes called contact herbicides but they are only effective on green plant material and involving complex chemical changes. Contact herbicides have very little or no residual action in the soil.
(2) *Soil-acting residual herbicides*. These chemicals act through the roots of the plant after being applied to the soil surface or worked

into the soil (volatile types); some of them are also absorbed by foliage, e.g. linuron. Most of this group act by interfering with photosynthesis. Some other examples are: atrazine, simazine, ametryne, prometryne, chloridazon, lenacil, monolinuron, trifluralin, propachlor, carbetamide, propyzamide, pendimethalin, tri-allate, isoproturon, chlortoluron.

(3) *Growth regulator ("hormone") herbicides.* These are a special group of translocated chemicals which are similar to substances produced naturally by plants and which can regulate or control the growth of some plants; susceptible plants usually produce distorted growth before dying. They are mainly used for controlling weeds in cereals and grassland. The more important ones are: MCPA, 2,4-D, mecoprop, dichlorprop, dicamba and benazolin.

(4) *Growth inhibitors.* These limit or stop the growth of susceptible plants, e.g. propham, barban, TCA, dalapon.

Translocated herbicides are those which can move through the plant before acting on one or more of the growth processes.

Herbicides may be classified by their chemical groups, or sometimes as a group which can be used to deal with some particular problem, e.g. wild oat herbicides.

WEED CONTROL IN CEREALS

Cereals used to be regarded as the dirty crops in the rotation until the introduction of MCPA in 1945. This, and the very similar 2,4-D, easily killed all the troublesome and aggressive broad-leaved weeds at that time—especially yellow charlock, poppy and fat hen, but it was necessary to use these chemicals for many years to destroy seedlings developing each year from the large numbers of dormant seeds in the soil. After a time these weeds more or less disappeared from many fields, but others, which were resistant to MCPA, were able to grow and set seed without competition from the more aggressive weeds such as yellow charlock. Weeds such as chickweed and cleavers then became troublesome and in 1957 two herbicides—CMPP (now known as mecoprop) and TBA/MCPA mixture—were introduced and these proved effective against the chickweed and cleavers as well as the weeds which were controlled by MCPA. Later, another resistant group has developed—the polygonums (redshank, black bindweed and knotgrass)—and to deal with these 2,4-DP (now known as dichlorprop) and dicamba/MCPA mixture were introduced in 1961; knotgrass is still proving a difficult one to deal with. Now, another group is developing—the mayweeds—and a new range of chemicals such as mixtures including ioxynil, bromoxynil and bentazone are being used.

In all these cases the chemicals used usually justify their use for several reasons such as:

higher crop yields when weed competition is removed;

easier harvesting because there is little or no green weed material in the crop to delay drying in showery weather, also, seed heads of weeds such as poppies, mayweeds and thistles are not allowed to develop and cause drying and cleaning problems.

However, although MCPA is a cheap chemical, some of the newer herbicides (called "broad-spectrum" herbicides), which kill most of the weeds occurring now-a-days, are much more expensive and in some cases the benefits may not justify the cost. Also, greater care is sometimes required when applying them if crop damage is to be avoided.

It is not possible to use herbicides such as MCPA, mecoprop, dichlorprop, TBA and dicamba mixtures, ioxynil and bromoxynil mixtures on cereal crops undersown with clovers or other legumes. However, it was discovered that the butyric chemicals MCPB or 2,4-DB did not damage clovers but killed many of the weeds controlled by MCPA with the exception of yellow charlock, runch and hempnettle. The addition of benazolin gives a wide spectrum kill of the most troublesome weeds in undersown crops—especially chickweed; dinoseb and bentazon may also be used (see page 170).

Perennial broad-leaved weeds are not so easily killed as the annuals—especially thistles, field bindweed, wild onion and docks—because the foliage usually develops after the normal spraying time for the annuals. Field bindweed is often very troublesome—causing lodging and great difficulty in combining because of the mass of green growth. However, all perennial broad-leaved and grass weeds can be effectively controlled by spraying "Roundup" on the ripening crop 1–3 weeks before harvest—provided the weeds are green and actively growing; stubble treatment is less effective .

Annual weeds are much easier to kill as seedlings. However, the safest time for spraying cereals is between the five-leaf and jointing stages of growth (see Fig. 81) when some weed seedlings are becoming well established. There are exceptions to this, e.g. oats can be sprayed with MCPA after the first leaf stage; also mecoprop, dichlorprop, and ioxynil, bromoxynil and bentazon mixtures can be used at the 3–4 leaf stage or earlier.

The following common weeds are easily and cheaply controlled by MCPA (or 2,4-D):

annual nettle	pennycress
buttercups	plantains
charlock	poppy
docks (seedlings)	rush (common, soft)
fat hen	shepherd's purse
hempnettle (not 2,4-D)	thistles
mustard	wild radish (runch)
orache	

Table 51 summarizes the safest or cheapest herbicides (but not the only ones) which could be used on weeds not controlled by MCPA or 2,4-D. For crops undersown with clover, a small amount of MCPA may be included in the spray to deal with weeds not controlled by MCPB or 2,4-DB, for example, charlock.

The Wild Oat is now a very serious weed problem on many farms, and whatever control measures are used it is likely to persist for a long time because of dormant seeds in the soil. This dormancy problem is made worse if stubble cultivations are carried out after harvest, because most of the shed seeds, if left on the soil surface, are destroyed or disappear in various ways. Hand roguing should be done when wild oats first appear in a field, but later, as the numbers increase, herbicides will have to be used in wheat and barley crops to avoid serious yield losses. The main herbicides which are giving good result are:

Tri-allate ("Avadex BW")—this is a relatively cheap, volatile chemical which has to be worked into the soil before planting any variety of wheat or barley in autumn or spring (not satisfactory on stony or cloddy soils); it can also be applied as tiny granules after sowing; surviving wild oats grow normally and so are easy to see for roguing.

Barban ("Carbyne")—this is also a relatively cheap chemical which can be used on most varieties of wheat and barley when the wild oats are at the 1–2½ leaf stage; surviving wild oats are

TABLE 51

MCPA controlled weeds plus	Straight cereals	Undersown with clover
(1) Chickweed, cleavers, fumitory	mecoprop	benazolin plus MCPA, 2,4-DB or MCPB
(2) Black bindweed, redshank, spurrey + (1)	dichlorprop, dicamba mixtures	2,4-DB (not spurrey), bentazon/MCPB
(3) Mayweed, knotgrass, gromwell + (1) and (2)	mixtures containing ioxynil, bromoxynil, dicamba or 3,6-dichloropicolinic acid	bentazon/MCPB
(4) Forget-me-not, pansy, parsley piet + (1), (2) and (3)	dicamba/benazolin/dichlorprop	—
(5) Corn marigold	ioxynil/bromoxynil mixtures, bentazon/dichlorprop	bentazon/MCPB
(6) Speedwell, dead-nettle	ioxynil mixtures	dinoseb

Note: Dinoseb is very poisonous; protective clothing must be worn. For fuller details on weed control in cereals see MAFF booklet 2253.

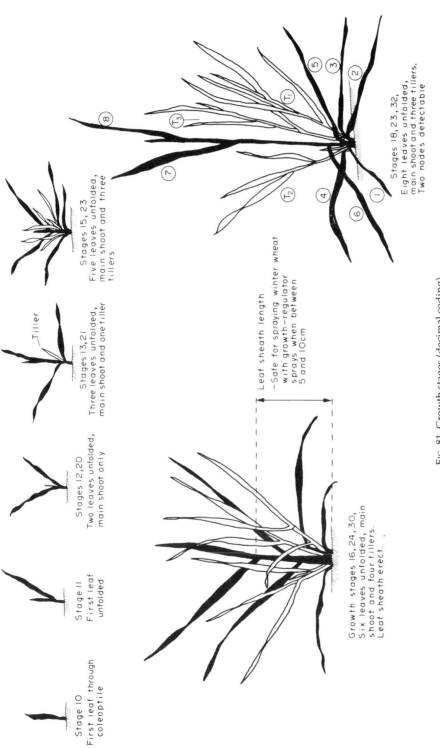

Stage 10
First leaf through coleoptile

Stage 11
First leaf unfolded

Stages 12, 20
Two leaves unfolded, main shoot only

Stages 13, 21
Three leaves unfolded, main shoot and one tiller

Stages 15, 23
Five leaves unfolded, main shoot and three tillers

Stages 18, 23, 32,
Eight leaves unfolded, main shoot and three tillers.
Two nodes detectable

Growth stages 16, 24, 30,
Six leaves unfolded, main shoot and four tillers.
Leaf sheath erect.

Leaf sheath length
—Safe for spraying winter wheat with growth-regulator sprays when between 5 and 10cm

Fig. 81. Growth stages (decimal coding).
See also Appendix 10.

FIG. 82a. Soft brome grass (*Bromus mollis*).
(Reproduced by permission, Fisons)

FIG. 82c. Couch grass (*Agropyron repens*).
(Reproduced by permission, Fisons)

FIG. 82b. Common wild oat (*Avena fatua*).
(Reproduced by permission, Fisons)

FIG. 82d. Black bent grass (*Agrostis gigantea*).
(Reproduced by permission, Fisons)

FIG. 82e. Creeping bent grass (*Agrostis stolonifera*).
(Reproduced by permission, Fisons)

FIG. 82g. Sterile brome grass (*Bromus sterilis*).
(Reproduced by permission, Fisons)

FIG. 82f. Onion couch (*Arrhenatherum elatius* var. *bulbosum*).
(Reproduced by permission, Fisons)

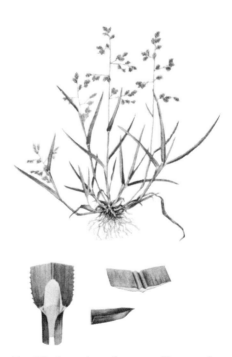

FIG. 82h. Annual meadow grass (*Poa annua*).
(Reproduced by permission, Fisons)

FIG. 82i. Blackgrass (*Alopecurus myosuroides*).
(Reproduced by permission, Fisons)

severely stunted. A different formulation ("B 25") is recommended for spring barley.

Flamprop-isopropyl ("Commando")—an expensive chemical for use on actively growing wild oats in the spring in all the common varieties of wheat and barley (including undersown crops) from leaf sheath erect stage to the 3rd node in barley and 4th node in wheat.

Flamprop-methyl ("Lancer")—an expensive chemical for use in spring on all varieties of winter and spring wheat (including undersown crops) up to 3rd node stage.

Difenzoquat ("Avenge")—a fairly expensive chemical for use in all varieties of barley and most varieties of wheat—mainly for spring use, but half dose may be used November to February; apply for wild oats from 2-leaves stage to end of tillering.

Dichlorop-methyl ("Hoegrass")—for use on all varieties of wheat and barley; dose rate (and cost) varies with the time of application—may be used at low rate up to 3 expanded leaves (w.o.)

and before end of February and double rates for older plants and later applications also gives good control of ryegrass seedlings.

All the above "wild oats" herbicides, except "Avenge", also provide a useful suppression of blackgrass; "Avenge" gives some control of mildew. Some of the "blackgrass" herbicides—chlortoluron, isoproturon and metoxuron also give a useful control of wild oats (see AL 452, *Wild Oats*).

Blackgrass has been a very serious problem on many farms where the soils are heavy and wet in winter and where autumn-sown crops are often grown. It is also spreading to lighter land farms where autumn cereals predominate. Blackgrass produces masses of viable seeds which mostly germinate in the autumn, but enough can germinate in spring in badly infested fields to seriously reduce yields. The following herbicides are recommended for controlling blackgrass (also some other seedling grasses and broad-leaved weeds) in wheat and barley:

Isoproturon ("Arelon", "Hytane", "Tolkan")—can be used on all varieties of winter wheat and winter barley, either pre- or post-emergence of the crop (not before mid-October because of rapid breakdown in the warm soil); well established blackgrass can be controlled in the spring but autumn treatment is better.

Chlortoluron ("Dicurane")—similar to isoproturon but likely to damage some varieties.

Metoxuron ("Dosaflo")—post-emergence only and likely to damage some varieties.

Metoxuron/simazine ("Fylene")—similar to metoxuron.

Methabenzthiazuron ("Tribunil")—mainly for pre-emergence use (up to the end of November) in winter crops of wheat, barley and oats; half-dose rate for broad-leaved weed control only.

Pendimethalin ("Stomp")—for pre-emergence use in all winter varieties of wheat, barley and rye; half-dose for annual meadow grass and broad-leaved weeds.

Terbutryne ("Prebane")—pre-emergence in all winter varieties of wheat and barley (not barley on light soils); half-dose for broad-leaved weeds (see also Leaflet 522—*Blackgrass*).

Sterile (barren) brome—this annual grass has become a serious weed on many farms where continuous winter cereals are being grown—the seeds germinate very easily in the autumn. Metoxuron gives reasonable control but this is improved if the metoxuron, isoproturon or "Fylene" follows tri-allate (see leaflet 777—*Barren Brome*).

A good straw burn can destroy large numbers of wild oats, blackgrass, sterile brome and other seeds lying on the soil surface.

The increase in early-sown winter barley crops has led to the use of many broad-leaved herbicides in the autumn, e.g. mecoprop for cleavers and chickweed; ioxynil/bromoxynil, cyanazine and others for broader spectrum weed control, or to supplement the control given by "blackgrass" herbicides where these are used.

Many perennial weeds have been controlled with varying success in the autumn by cultivations (page 45) or with herbicides such as aminotriazole and glyphosate; couch with dalapon or TCA, creeping bent with paraquat. The approved use of the very low toxicity, non-residual herbicide glyphosate ("Roundup") to cereal crops 1–3 weeks before harvest (grain moisture less than 30%) has greatly simplified the control of all perennial weeds which have green leaves and are actively growing at the time of spraying, and sprayer booms are set to give good weed coverage; tramlines, high-clearance wheels and a sheet under the tractor minimizes crop damage: winter barley offers a better chance to kill early senescing weeds such as "onion" couch; straw burning afterwards is easy.

CHEMICAL WEED CONTROL IN POTATOES, ROOTS AND KALE

Potatoes

Weed control by cultivations may be impossible in wet seasons and in dry seasons there can be considerable loss of valuable moisture and damage to crop roots; cultivations can also produce clods in some soil conditions. Consequently, in over 75% of potato crops weeds are now controlled by herbicides applied before and/or after planting.

Pre-planting. All perennials are best controlled by "Roundup" pre-harvest in the previous cereal crop or by autumn treatments. "Eptam" will control couch, wild oats and some broad-leaved weeds when worked into the soil (very volatile) a short time before planting; keep the ridges shallow for a few weeks after planting.

Pre-emergence of crop. Paraquat ± diquat will kill emerged seedling weeds: dalapon may control emerged grasses. The following are some examples of herbicides which kill seedling weeds by contact and/or residual soil action: *linuron* (or *monolinuron*) alone or in mixtures with *paraquat, cyanazine* or *trietazine; prometryne; metribuzin* ("Sencorex")—the latter can also be used post-emergence of maincrop (not *M. Piper*) to give a longer period of weed control (in a dry season the residues may damage autumn sown crops). The dose rate for most soil-acting herbicides varies with soil type; fine, moist soil conditions are necessary for effective action.

Post-emergence of crop. Metribuzin (see above). "Hoegrass" for wild oats. MCPA for thistles and field bindweed—it may check some varieties, e.g. King Edward. *Dalapon* may be used for couch control after the haulms have died but this might spoil the red colouring of some varieties (see Booklet 2260).

Sugar-beet (mangels, fodder beet)

Couch and other perennial weeds should be controlled in the year prior to planting the crop (pre-harvest "Roundup" in cereals).

Traditionally, weeds were controlled by inter-row cultivations and hand-hoeing combined with singling. Now, the "drilling-to-a-stand" technique has made hand-hoeing unnecessary provided weeds (and pests) are effectively controlled. Most of the troublesome annual weeds can be controlled by one or more herbicides applied pre- and/or post-emergence of the crop provided the correct type and amount of herbicide is applied and the soil conditions are suitable (i.e. fine and moist) for soil-acting types.

The chemicals are usually applied over-all, but may be applied in 15-cm bands over the beet rows where inter-row cultivations are used to prevent soil capping and/or to reduce costs.

The main herbicides which can be applied one or more times are:

Pre-sowing. Tri-allate ("*Avadex BW*")—worked into soil (volatile) for wild oats and some grasses. *Chloridazon* (e.g. "*Pyramin*"), *ethofumesate* ("*Nortron*"), *lenacil* ("*Venzar*") and *metamitron* ("*Goltix*") alone or in mixtures with *tri-allate* for most annual weeds, and worked into seed-bed for better effect in dry conditions.

Pre-emergence. Paraquat ± diquat for fast-growing weed seedlings before beet emerges (risky). *Chloridazon; metamitron; lenacil; ethofumesate* mixtures; *propham/chlorpropham/ fenuron* (PCF).

Post-emergence. Phenmediphan ("*Betanal E*") alone or in mixtures with *barban, chloridazon* or *ethofumesate; metamitron ±* oil; the addition of "*Format*" will improve the control of mayweeds and thistles. One-third dose rates of these chemicals applied with special jets and at higher pressure gives very good results—especially on very small weeds. *Trifluralin* may be worked into soil (volatile) between rows and plants for late weeds, e.g. fathen, knotgrass, redshank. "*Hoegrass*", "*Clout*" or "*Fusilade*" for wild oats and blackgrass. "*Roundup*" for weed beet applied at flowering stage with a rope-wick or roller machine. Nitrate of soda or dendritic salt sometimes used (see Booklet 2254).

Kale, Swede and Turnip

Couch should be controlled in the previous autumn or by early spring treatment with TCA.

Wild oats are suppressed by working "*Avadex*" into the soil at drilling time or TCA about 7 days earlier; "*Hoegrass*" can be sprayed post-emergence.

Many annual broad-leaved weeds can be controlled by *trifluralin* ("*Treflan*") worked into the soil during the fortnight before drilling and/or spraying *propachlor* ("*Ramrod*") on the soil before crop emergence.

Annual weeds are unlikely to be troublesome where these crops are slit-seeded into a chemically destroyed grass sward.

It is possible to control annuals by allowing them to germinate and then destroying them with a contact herbicide such as *diquat* or *paraquat* before drilling the crop.

In kale crops, many troublesome annual weeds, especially fat hen, can be controlled by post-emergence spraying with *desmetryne* ("*Semeron*"). Gas liquor and sulphuric acid have also been used, with reasonable effect, to control weeds such as charlock in the growing crop.

Weed control in other crops is dealt with in Chapter 4, Cropping (see Booklet 2256).

WEED CONTROL IN GRASSLAND

In arable crops, most damage is caused by annual weeds, but in established grassland biennial and perennial weeds cause most trouble. The presence of the weeds causes a *reduction in yield, nutrient quality* and *palatability* of the sward. Stock do not like grazing near buttercups, thistles and wild onions. Some weeds are *poisonous*, e.g. ragwort and horsetails, and some can *taint milk* if eaten, e.g. buttercups and wild onion.

Weeds in grasslands are encouraged by such factors as:

(a) *Bad drainage*, e.g. rushes, sedges, horsetails and creeping buttercup.
(b) *Shortage of lime*, e.g. poor grasses (bents), sorrels.
(c) *Low fertility*: many weeds can live in conditions which are too poor for good types of grasses and clovers.
(d) *Poaching* (trampling in wet weather): the useful species are killed and weeds grow on the bare spaces.
(e) *Over-grazing*: this exhausts the productive species and allows poor, unpalatable plants such as bent, Yorkshire fog, thistles and ragwort to become established.
(f) *Continuous cutting* for hay encourages weeds such as soft brome, yellow rattle, knapweed and meadow barley grass.

TABLE 52

Weed	Control
Bracken	Cut or crush the fronds (leaves) twice a year when they are almost fully opened. If possible, plough deep and crop with potatoes, rape or kale before reseeding. Rotavating about 25 cm deep chops up and destroys the rhizomes. Very good chemical control is possible with asulam applied by air or ULV/CDA sprayer in summer when the fronds are just fully expanded. "Roundup" can be used as an unselective spray before reseeding or selectively with a rope-wick machine.
Buttercups	Spray with MCPA or MCPB. The bulbous buttercup is the most resistant type.
Chickweed	Spray with mecoprop (kills clover) or benazolin mixture.
Docks	Seedlings and curled-leaved type—spray with MCPA or MCPB. Broad-leaved type—plough and take cleaning crop, e.g. kale. Grazing with sheep is helpful. Asulam gives good control when sprayed on the expanded leaves in the spring or autumn.
Horsetails	If possible, improve drainage. Spraying with MCPA or 2,4-D will kill aerial parts only and regrowth occurs—but if it is done 2–3 weeks before cutting, the hay crop should be safe for feeding.
Nettles	Spray with 2,4,5-T alone or in mixtures, or apply "Roundup" with rope-wick.
Ragwort	Cut before buds develop, to prevent seeding. Spray with 2,4-D or MCPA in early May or in autumn. Grazing with sheep in winter is helpful.
Rushes	Improve drainage, if possible. Common or soft rush—spray with MCPA or 2,4-D. Hard and jointed rushes—cut several times per year. Encourage grasses and clovers by good management. Apply "Roundup" with rope-wick when growing well.
Sorrel	Spray with MCPA or 2,4-D. Apply lime.
Thistles	Spray with MCPA or MCPB. The more resistant creeping type should be sprayed in the early flower-bud stage. Avoid over-grazing.
Tussock grass	Improve drainage. Cut off the "tussocks" with flail harvester or topper, or take hay or silage crops. Apply "Roundup" with rope-wick when growing well.

Chemicals are a useful aid to controlling grass-land weeds but should not be regarded as an alternative to good management.

The main chemicals used to control broad-leaved weeds in swards where clover is important are MCPB, 2,4-DB and benazolin; where clover is not important MCPA, 2,4-D, mecoprop, dicamba mixtures, 2,4,5-T and dichlorprop can be used. Asulam is used to control bracken and docks—also kills some grasses. "Roundup", when applied with a rope-wick or roller machine, will selectively kill any tall weeds which are growing actively.

Table 52 is a guide to the control of the more troublesome weeds.

Weed control when reseeding

When reseeding a grass sward, without a cover crop, weeds can be troublesome—especially annuals such as charlock, chickweed and fat hen; annual weeds with upright stems may be killed by mowing, but chickweed (usually the worst) should be sprayed with mecoprop, or a benazolin mixture where clovers are present.

Where grassland has to be reseeded, and ploughing is not possible nor desirable, the old sward can be killed with "Roundup" spray when there is at least 10 cm of actively growing green leaves; this is usually most effective in August; reseeding can follow 2–3 weeks later: the dead surface trash should be burnt—if possible—and if there is a thick mat of decaying surface vegetation it should be lightly rotavated otherwise the new grass seeds may be killed by toxic substances (see also page 148).

Problems can also arise with seedling weed grasses establishing with the sown seeds. This can be overcome by delaying sowing until they establish and then killing them with paraquat before seeding. This may mean leaving the field fallow for some time, e.g. over-winter.

Grass swards which have become weedy can be improved by using selective herbicides, provided there is a reasonable amount of valuable grasses, such as perennial ryegrass, present in the sward; low doses of *dalapon* (3 kg/ha) may be used in June or July to selectively kill weed grasses such as bent, Yorkshire fog, and meadow grasses, but growth of the remainder of the sward is checked for about a month.

Ethofumesate ("Nortron") can be applied in one or two doses between mid-October and February to control many weeds such as chickweed, annual and rough-stalked meadow grasses, blackgrass, sterile and soft brome, wall barley grass and meadow foxtail.

Diquat (2 l/ha) will destroy chickweed which has become established in a grass sward, perhaps after a long dry period. The grass is only checked for a short time and can be grazed about a week after spraying.

(See also page 148, and the following advisory leaflets: ALs 46 (*docks*), 51 (*thistles*), 190 (*bracken*), 280 (*ragwort*), 533 (*rushes*) and Booklet 2056.

SPRAYING WITH HERBICIDES

This is a skilled operation and should be carefully carried out. Some of the more important precautions to take are:

(1) Make a careful survey of the field to determine the weeds to be controlled; choose the most suitable chemical and best time for spraying.

(2) Check carefully the amount of chemical to be applied per hectare and the volume of water to be used (220–330 l/ha is a common range). Make sure the chemical is thoroughly mixed with the water before starting. Soluble and wettable powders should be mixed with some water before adding to the tank. Use the agitator if necessary.

The rate of application is mainly controlled by the forward speed of the tractor (use a speedometer), and the size of nozzle, and to a lesser extent by the pressure (follow the maker's instructions). Always use clean, preferably not hard, water. Always use a filter. An accurate dipstick is necessary when refilling the tank if it is not emptied each time.

(3) When using poisonous chemicals, e.g. dinoseb, it is necessary to wear fully protective clothing (read carefully the instructions issued with such chemicals). Do not blow out blocked nozzles.

(4) Do not spray on a windy day—especially with the hormone type of herbicides and if the spray is likely to blow onto susceptible crops or gardens. Keeping the boom as low as possible and using a plastic spray guard can be helpful; also using higher volume (larger droplets) is better than a very low volume mist for avoiding spray drift.

(5) Spinning disc applicators which produce uniform size droplets only cause drift problems when very small droplet sizes are used.

(6) Make sure that the boom is level and that the spray cones or fans meet just above the level of the weeds to be controlled.

(7) Spray the headlands first; when spraying the rest of the field the drill rows in cereal crops are a useful guide. If using a wide boom it is advisable to use markers; avoid "misses" by slight overlapping.

(8) Wash out sprayer thoroughly on waste ground and leave full of clean water—this avoids scale forming inside the tank which is one of the commonest causes of blockages in the nozzles.

(9) Do not mix pesticides and other chemicals in the sprayer without seeking advice on the matter; only some can be mixed.

See also Appendix 4 (page 221) and MAFF booklet—*Safe and effective spraying*.

As an alternative to *overall* spraying, *band* spraying is often used when applying herbicides to root crops, e.g. sugar-beet. A band about 15 cm wide is sprayed over each row of seeds; the weeds between the rows are controlled by cultivations. This is especially useful when expensive chemicals are being used.

SUGGESTIONS FOR CLASSWORK

1. Examine seedlings of the common agricultural weeds in your area and learn how to identify them. Helpful pictures of weeds are published by the leading firms marketing herbicides.
2. When visiting farms make notes on the sprays used to control weeds, e.g. the main weeds, the chemicals used, time and methods of application. If "tramlines" have been used in cereals, check how these have been laid out and the widths of the drill, sprayer, etc.

FURTHER READING

Weed Control Handbooks, Blackwell.
    Vol. I *Principles of Weed Control.*
    Vol. II *Recommendations.*
Approved Products List, HMSO.
Wild Oats in World Agriculture, ARC.
Weed Research Organisation Reports, ARC.
British Poisonous Plants, Bulletin No. 161, HMSO.
Product literature.
MAFF booklet, *Safe and effective spraying.*
Farmers Weekly supplement, Dec. 1977.

7

PESTS AND DISEASES OF FARM CROPS

PESTS are responsible for millions of pounds of damage to agricultural crops in this country every year.

Before discussing the various methods used to control pests, it is important to understand something of their structure and general habits.

One of the major groups of pests are the *insects*. They are *invertebrates*, i.e. they belong to a group of animals which do not possess an internal skeleton. Their bodies are supported by a hard external covering—the *exoskeleton*. It is composed chiefly of *chitin*, and is segmented so that the insect is able to move.

From the diagram of the external structure of an adult insect (see Fig. 83) it can be seen that the segments are grouped into three main parts:

1. *The head*, on which is found:

(a) The *antennae*—or feelers carrying sense organs—e.g. smelling.

(b) The *eyes*—a number of single and a pair of compound eyes are present in most species.

(c) The *mouthparts*—(see Fig. 84). Two main types are found in insects:
   (i) the biting type;
   (ii) the sucking type—insects in this group suck the sap from the plant and do not eat the foliage.

The type of mouthpart possessed by the insect is of considerable importance in deciding on the method of control.

2. *The thorax*, which bears:

(a) the *legs*—there are always three pairs of jointed legs on adult insects;

(b) the *wings*—found on most, but not on all, species.

3. The *abdomen*, which has no structures attached to it except in certain female species where the egg-laying apparatus may protrude from the end.

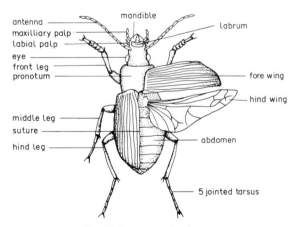

FIG. 83. Structure of an insect.

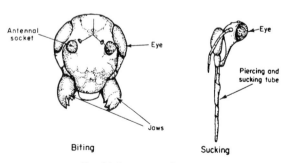

FIG. 84. Insect mouthparts.

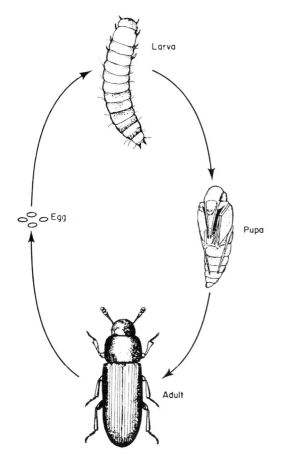

FIG. 85. Four-stage life-cycle.

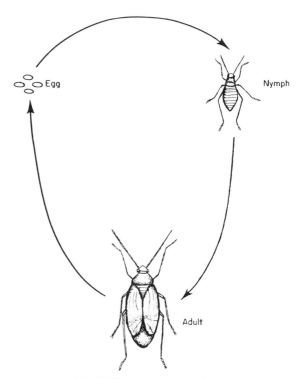

FIG. 86. Three-stage life-cycle.

LIFE-CYCLES

A knowledge of the life-cycles of insects can be of great help in deciding on the best stage at which the insects will be most susceptible to control methods.

Most insects begin life as a result of an egg having been laid by the female. What emerges from the egg, according to the species, may or may not look like the adult insect.

There are two main types of life-cycles:

(1) The "complete" or four-stage life-cycle (see Fig. 85).

(a) The *egg*.

(b) The *larva* (plural larvæ)—entirely different

in appearance from the adult. This is the active eating and growing stage. The larvæ usually possess biting mouthparts, and it is at this stage with many insects that they are most destructive to the crops on which they feed.

(c) The *pupa*—the resting state. The larvæ pupate and undergo a complete change from which emerges—

(d) The *adult* insect—this may feed on the crop, e.g. flea beetle but in many cases it does far less damage than the larvæ, e.g. flies.

(2) The "incomplete" or three-stage life-cycle (see Fig. 86).

(a) The *egg*.

(b) The *nymph*—this is very similar in appearance to the adult, although it is smaller and may not possess wings. It is the active eating and growing stage.

(c) The *adult insect.* Invariably this stage will also feed on and damage the crop, e.g. aphids.

Most insects and/or larvæ and nymphs feeding on crops depend on these crops for part or all of their existence. The crop is the *host* plant, whilst the insect is the *parasite* to the host. Not all insects are harmful to crops; some are beneficial in that they prey on or parasitize crop pests, e.g. the ladybird is particularly useful because, both as the larva and adult, it feeds on aphids which are responsible for transmitting certain virus diseases in plants as well as causing physical damage to plants.

METHODS OF PEST CONTROL

1. Indirect control measures

These aim more at the prevention of the pest attack.

(a) Rotations

As a means of control it is now not so important with more effective chemical control. The principle behind control by rotation is that if the host plant (the crop) is continually grown in the same field for too many successive years, then the parasite will increase in large numbers, e.g. eelworm.

(b) Time of sowing

(i) A crop may sometimes be sown early enough so that it can develop sufficiently to withstand an insect attack, e.g. frit fly on the oat crop.

(ii) A crop can be sown late enough to avoid the peak emergence of a pest, e.g. flea-beetle.

(c) Cultivations

Ploughing exposes pests such as wireworms, leatherjackets and caterpillars, which are then eaten by birds. Well-prepared seed-beds encourage rapid germination and growth. This will often enable a crop to grow away from pest attack.

(d) Encouragement of growth

Good-quality seed should be used which will germinate quickly and evenly. It is also important that the crop is not checked to any extent, say by lack of a plant food. A poor growing crop is far more vulnerable to pest attack than a quick-growing crop. A top-dressing of nitrogen, just as a crop is being attacked, may sometimes save the crop.

(e) Clean farming

Weeds are alternate hosts to a great variety of insects, and, as far as possible, these sources of infestation should be eradicated.

2. Biological control

A parasite or predator is used to control the pest. The method has little application in farming in this country as yet, but research and development to this end is active. An established example of this type of control is in horticulture, where the red spider mite is successfully controlled by predator mites, in cucumber production under glass.

3. Direct control measures

This means chiefly chemical control using a pesticide. These can be used in a number of ways, e.g.

Sprays and dusts.
A granular form for controlling aphids.
Baits for controlling soil pests such as leatherjackets, slugs and snails.
Seed dressing—mainly for the protection of cereals against wireworm, and brassica crops against flea-beetle. Usually the insecticides are combined with a fungicide such as an organo-mercury compound.

Gases, smokes, fumigants are commonly used in greenhouses against aphids chiefly, and in granaries against beetles and weevils.

Basically, there are two ways in which pesticides kill pests:

(a) By contact

The pest is killed when it comes in contact with the chemical, either when:

   (i) it is directly hit by the spray or dust,
   (ii) it picks up the pesticide as it moves over foliage which has been treated,
   (iii) it absorbs vapour,
   (iv) it passes through soil which has also been treated.

(b) By ingestion

As a *stomach* poison the pest eats the foliage treated with the pesticide, or the chemical is used in a bait.

As a *systemic* compound it is applied to the foliage or to the soil around the base of the plant. It gets into the sap stream of the plant and thus the pest is poisoned when it subsequently sucks the sap.

Most pesticides kill by more than one method, which makes them very effective. But many of them are extremely toxic to animals and humans, and by law, certain precautions must be observed by the persons using them.

CLASSIFICATION OF PESTICIDES

Insecticides

(a) The chlorinated hydrocarbons

These insecticides are all stomach and contact poisons and the main ones are:

HCH. As a spray and dust it is used extensively on fruit crops and it controls, amongst other insects, aphids, caterpillars and weevils. It is also a useful soil insecticide for control of wireworm and leatherjackets, but it can taint some crops such as potatoes.

DDT. This is a very persistent insecticide, and its use is now limited by agreement under the Pesticides Safety Precautions Scheme. It should *not* be used for aphid control on any crop, nor should it be used on brassica seed crops, peas or beans (except for weevil control on seedling crops), and grass crops, however utilized.

It should only be used on any other crop where there is no effective and less persistent alternative.

Aldrin. There is agreement for restricted use of this insecticide. It controls cabbage root fly (resistant strains are developing in some areas), wireworms in potatoes only, and leatherjackets (but on spring barley only those varieties which are sensitive to DDT).

Dieldrin. There is agreement for restricted use of this insecticide. It should be used only against cabbage root fly as a dip (resistant strains are developing in some areas) and as a seed treatment on sugar-beet, onion and the French and runner bean.

(b) The organo-phosphorus compounds

As a group the organo-phosphorus compounds and the carbamate compounds are dangerous to use and they should be handled strictly in accordance with the manufacturer's instructions.

The following are examples of some of the organo-phosphorus insecticides in common use.

Non-systemic

Chlorfenvinphos. In granules and liquid form controls cabbage root fly and carrot fly. In granule form only control of frit fly in the maize crop. As a seed dressing helps to reduce wheat bulb fly.

Fenitrothion. For control of aphids and, in a bran bait, leatherjackets and cutworms; also stored grain pests.

Malathion. For control of aphids.

Systemic

Dimethoate. This is used for the control of aphids on many agricultural crops. As an emergency treatment it can be used as a spray against wheat bulb fly.

*Disulfoton. This insecticide controls aphids on brassicas, beans, potatoes and sugar-beet, and also carrot fly. It is used in the granular form.

Formothion. This controls aphids on many agricultural crops, and it can also be used in an emergency against wheat bulb fly.

Menazon. This insecticide controls aphids on brassics, beans, potatoes and sugar-beet. With the latter it is used as a seed dressing against an early attack.

*Phorate. This is used in a granular form chiefly for the control of aphids in beans, potatoes and sugar-beet, and frit fly in maize.

(c) The carbamate compounds

*Carbofuran for control of cabbage root fly, flea beetles and turnip root fly.

*Pirimicarb for control of aphids.

Persistency is a constant cause of concern for food producers, and research is continually directed towards finding safer, less persistent compounds. In this context, as a chemical group, the carbamates can be considered an important development.

(d) Synthetic pyrethroids

Permethrin for control of aphid vectors of BYDV in winter barley and winter wheat. These recently developed chemicals are rapidly being introduced into crop protection. They have a

---

* An asterisk marks those chemicals included in the Health and Safety (Agriculture) (Poisonous Substances) Regulations. Certain precautions (including the use of protective clothing) must by law be taken when using these chemicals and it is advisable to read the official leaflet *The Safe Use of Poisonous Chemicals on the Farm* (ASP/1).

particularly high insecticidinal power whilst generally being safe to man and farm animals. They are efficient contact insecticides with a rapid knockdown action and some are thought to act as anti-feedants. These properties are being seen as valuable in the chemical control of plant viruses by controlling the insect vectors.

Molluscicides

Metaldehyde as a mini-pellet for the control of slugs and snails.

Methiocarb as a mini-pellet for the control of slugs and snails.

Nematicides (including soil sterilants)

Nematicides have been developed for the control of eelworm pests of crops, e.g.

*Aldicarb which also controls Docking disorder in sugar-beet and aphids.

*Oxamyl which also controls Docking disorder in sugar-beet.

Certain chemicals in this group are classified as soil sterilants, and these are now brought into use in field crop production, e.g. *Dazomet*.

It is also important to remember that a certain interval must be observed between the last application of the pesticide and:

(1) harvesting edible crops,
(2) access of animals and poultry to treated areas.

With some pesticides this interval is longer than others. This is another reason for very careful reading of the manufacturer's instructions.

For up-to-date information on pesticides available, and the regulations and advice on the use of these chemicals, reference should be made to the annual publication of the Agricultural Chemical Approved Scheme *Approved Products for Farmers and Growers*.

This is now available from HMSO.

Table 53 indicates the major pests attacking farm crops, and their control.

TABLE 53 MAJOR PESTS AND THEIR CONTROL

Crop attacked	Pest	Description	Life-cycle	Symptoms of attack	Control	Notes
Cereals	*Adult:* Clickbeetle *Larva:* Wireworm	*Adult:* Brown 6–12 mm long *Larva:* Growing to 25 mm long, yellow colour	Larvæ hatch out during summer from eggs laid in the soil. They take 4–5 years to mature, and after pupation in the soil, the adult appears in early autumn	Yellowing of foliage followed by disappearance of successive plants in a row. This is caused by wireworm moving down the row. Larvæ eat into the plants just below soil surface. They are usually found in soil around the plants	Good growing conditions to help the crop grow away from an attack. Wheat and oats more susceptible than barley; they should not be grown where the wireworm count is over 2 million/hectare. All seed should be dressed with gamma HCH	Do not confuse wireworm attack with other pests such as eelworm
	Adult: Cranefly *Larva:* Leatherjacket	*Adult:* Is the "Daddy longlegs" *Larva:* Leaden in colour, 30 mm long	Eggs laid on grassland or weedy stubble in the autumn from which the larvæ soon emerge. They feed on the crop the following spring, pupating in the soil during the summer	Crop dies away in patches, root and stem below ground having been eaten. Larvæ found in soil	If possible, plough the field before August to prevent the eggs being laid. HCH can be applied as a low volume spray, or a bait such as fenitrothion broadcast late in the day	
	Wheat bulb fly	*Larva:* Whitish-grey 12 mm long	Eggs laid on bare soil in the autumn. Larvæ feed on the crop until following May. Pupation then follows either in the soil or plant	Central shoot of plant turns yellow and dies in early spring. Larvæ found in base of tiller	If possible, avoid sowing wheat where the field has lain bare from late summer. Protection includes selection from one of following: (i) Granules or sprays at about sowing time. (ii) Seed treatment, e.g. gamma HCH for late-drilled crops. (iii) Spray at egg hatch, e.g. use chlorfenvinphos. (iv) Spray at first sign of damage, e.g. use formothion	Winter wheat and barley are attacked

(continued overleaf)

TABLE 53—*continued*

Crop attacked	Pest	Description	Life-cycle	Symptoms of attack	Control	Notes
Cereals—*continued*	Frit fly	*Larva:* Whitish, 3 mm long	Three generations in the year; the first when in spring eggs are laid on spring oats and larvae feed on crop in May and June. The second generation damages the oat grain, whilst the third generation over-winters on grass, but when the latter is ploughed for autumn cereals the larvae move on to the cereals	In early summer, the central shoot of the oat plant turns yellow and dies, but the outer leaves remain green; blind spikelets and shrivelled grains are caused by second generation; autumn cereals can have a kink above the coleoptile at single-shoot stage, shoot then turns yellow	Sow spring oats early, and try and get them past the 4-leaf stage as quickly as possible. With *late sown spring oats* can spray at 2-leaf stage at first sign of attack, but advice should be sought. Allow a 6-week interval between ploughing grass and sowing the *winter cereal*. Spray only at first sign of attack if more than 10% of crop is affected; can use chlorpyrifos	Spring oats and maize are particularly susceptible
	Opomyza	*Larva:* Yellowish; slender; about 8 mm long, pointed at both ends	Eggs laid near wheat plant in October and November, hatching early in new year; larvae move down between outer leaves to feed on main tiller. Pupation in early summer, adults appear in June and a month later they migrate to hedgerows before returning to wheat field in late autumn	Circular or short spiral band at base of tiller producing brownish scar and then death of shoots—"deadhearts"	Early sown wheat in the eastern counties at greatest risk. Chemical treatment using triazophos within 3–4 weeks of egg hatch is a possibility, but more trials are necessary.	Winter barley rarely attacked; spring-sown cereals virtually immune
	Gout fly	*Larva:* Legless, yellowish-white, 6 mm long	Two generations in the year, the most important being the first. Larva hatch and feed in plant	Leaf sheath surrounding the ear is swollen and twisted. Poorly developed grain emerges	Sow the crop early. Good growing condi-tions will help to keep it growing	Barley is chiefly affected
	Cereal aphids	Various species of green fly, 1.6–3.3 mm long	Winged females found feeding on cereal crops in May and June. Wingless generations produced which con-tinue feeding during	Depending on the species, the damage caused to the cereal varies from stunted withered growth, occa-sionally with reddish-	The grain aphid causes most concern. Spray in autumn when infestation reaches threshold level. Use organo-	Aphids carry virus diseases from infected to clean plants. See barley yellow dwarf virus disease.

Table 54

Pest	Description	Habits	Symptoms/Damage	Control	Remarks
		summer. Most species move back to winter quarters (woody hosts and some grasses and cereals, depending on aphid species) in autumn, although some may be found on young cereal crops at the end of the year, especially in the milder parts of the country	brown to purple spots on the leaves. The grain aphid causes empty and/or small grain; by puncturing the grain in the milk-ripe stage, the grain contents seep out. This also reduces the weights of the grain	phosphorus or synthetic pyrethroid compound	
Stem and bulb eelworm	Too small to be seen without magnification	Live and breed in the plant. If the plant dies, eelworms become dormant in dead tissue or soil, becoming active again when conditions are suitable	Twisting and swelling, and this normally prevents plants from elongating and producing an ear	Resistant varieties. Rotation to starve out the eelworm. Clean seed	Attacks oats. Eelworms are not insects
Cereal cyst eelworm	Dark-brown lemon-shaped cysts about 1 mm (1/25 in.) long	Live and breed in the roots. White-looking cysts (female containing large numbers of eggs) are found on roots. Later these cysts (now dark brown) become free in the soil to infect the host plant again	Crop shows patches of stunted yellowish-green plants. Root system very bushy. Cysts visible on roots from June onwards	Avoid growing oats too often in the field. Grow resistant varieties when necessary	Oats chiefly infected, but intensive cereal growing will build up eelworm in the soil to affect other cereals
Slugs and snails	Field slug lightish-brown in colour, about 40 mm (1½ in.) long	Wheat grain damaged by being eaten in the ground before it germinates. Young cereals can be completely grazed off by a severe autumn attack. Most active in moist and humid conditions. An attack can be more serious when the seed is direct-drilled if the slit has not been properly covered		Baits containing metaldehyde and methiocarb spread evenly over the field prior to drilling. Extra cultivations in preparing the seed-bed help to check the pests	Winter wheat chiefly attacked
Stored grain — Saw-toothed grain beetle	*Adult:* Dark brown 3 mm long *Larva:* White and flattened	Eggs are laid on the stored grain; larvae feed on the damaged grain. Pupation takes place in the grain or store	The grain heats up rapidly; it becomes caked and mouldy. This is seen with the appearance of the beetles	Pirimiphos-methyl applied as the grain is fed into the store	

(continued overleaf)

TABLE 53—*continued*

Crop attacked	Pest	Description	Life-cycle	Symptoms of attack	Control	Notes
Stored grain— *continued*	Grain weevil	*Adult:* Reddish-brown, about 3 mm long with an elongated snout	During autumn the weevils bore into the stored grain to lay their eggs. The larvæ feed inside the grain where they also pupate	Hollow grains. Sudden heating of the grain. Weevils found a few feet below the surface of stored grain	See the saw-toothed grain beetle	
Maize	Frit fly	*Larva:* Whitish 3 mm long	As for Frit fly on oats	Twisting of leaves surrounding the growing point. In severe attacks this is killed leading to secondary tillers	As a preventive, an organo-phosphorus compound such as phorate in granular form applied at time of sowing	
	Wireworm	See wireworm on cerals			Phorate applied at sowing will give some protection. Gamma HCH applied well before sowing will help to reduce the wireworm population	
	Cyst eelworm	Maize reduces the number of eelworm because it does not reproduce and form cysts		Bronze coloration; severe stunting of plant	Avoid planting maize on fields where eelworm population is relatively high	Maize is susceptible to quite low populations of eelworm; especially on chalk soils following cereals
Beans	Bean aphid (black fly)	Very small oval body, black to green colour	There are many generations in the year. In summer winged females feed on the crop; wingless generations are then produced which continue to feed. Eventually a winged generation flies to the spindle tree on which eggs are laid for over-wintering	On all summer host plants, colonies of black aphids are seen on the stem leaves (especially the underside) and on the flowers. The plant wilts; it can become stunted and with a heavy infestation it may be killed	Apply menazon (spray), or phorate or disulfoton granules in June	It also attacks sugar-beet and mangolds

Crop	Pest	Description	Life cycle	Damage	Control	Remarks
Peas, beans and other legumes	Pea and bean weevil, and striped pea weevil	*Adult:* Yellowish-brown with stripes of lighter, 6 mm long. *Larva:* Legless, white with brown head	During early spring eggs laid in the soil near plants. Larvae feed on roots, whilst adults feed on leaves. Pupation takes place in the soil in mid-summer	Seedling crops checked. U-shaped notches at the leaf margins	Apply DDT or HCH either as a dust or spray when the attack is noticed	
	Pea moth	*Adult:* Dull greyish-brown, about 6 mm long. *Larva:* Yellowish-white with darker head; legless, about 8 mm long	Eggs laid June–mid-August, hatch in a week. Larvae enter pods and feed on peas until fully grown. Larvae leave pod and make way to soil; pupate in spring and adult emerges in early summer	Peas extensively damaged	One or more sprays using an organo-phosphorus insecticide. The timing is important. A pea-spray working service operates in some regions and this can be used as a general guide	(1) Early and very late sown crops suffer less damage. (2) Dried peas for harvesting most vulnerable
Brassicae (cabbage, kale, oil-seed rape, swedes, turnips)	Flea-beetle	A minute black beetle with a yellow stripe down each wing case	Adults emerge from hibernation during late spring to feed on crops. Eggs are laid, but larvae do little damage. Pupation takes place in the soil during the summer	Very small round holes are eaten in the seed leaves of the plants	Sow the crop either early or late, i.e. avoid April and May. Good growing conditions to get the crop quickly past the seed leaf stage. Seed dressing containing gamma HCH should be carried out. A dust or spray containing HCH can be applied as soon as the attack is noticed	Sugar-beet, mangolds and cereal crops can be attacked on occasions
	Cutworm	*Adults:* turnip moth; garden dart moth; heart and dart moth; large yellow under-wing moth	All similar except the garden dart moth; eggs laid and hatched in 10–14 days. After early feeding on leaves, caterpillars go into soil and feed (mostly at night) on stem above and below ground level.	Young plants cut off at base of stem; poor top growth, and plant wilts. Turnip and swede roots can also be damaged by feeding caterpillars	DDT applied as soon as caterpillars are seen; a DDT or fenitrothian bran bait may also be used, broadcast in the evening	Also attacks potatoes—particularly in dry summers. Late sown crops of sugar-beet can also be affected

(continued overleaf)

TABLE 53—*continued*

Crop attacked	Pest	Description	Life-cycle	Symptoms of attack	Control	Notes
Brassica—*continued*	Cutworm—*continued*	*Larva:* Caterpillars —up to 40 mm long; vary in colour according to adult—dull/greyish brown to green; one species has black marks along the back	Most, but not all, are fully fed in the autumn, and over-winter to pupate in the spring in the soil. Earlier fed caterpillars will pupate in the autumn to produce the adult and a second brood of caterpillars in the autumn. The garden dart lays eggs in late summer which hatch out the following spring; caterpillars fully fed in mid-summer then pupate, moths emerging in August			
Brassica seed crops (n.b. oil-seed rape)	Pollen beetle (Blossom beetle)	Metallic-greenish black in colour, 3 mm long	Adults emerge from hibernation during spring to feed on buds and flower parts. Eggs laid and similar damage caused when larvæ emerge	Damaged buds wither and die, and the number of pods set is lessened	When to spray with an organo-phosphorus compound depends on the seed crop concerned and the number of adults present. Two sprays are normally needed, one at the early green bud stage and the other at the early yellow bud stage	Extreme caution should be taken to ensure that no serious damage occurs to pollinating insects
	Pod midge	*Larva:* Whitish-cream	Female can insert eggs in pods through holes in pods. Larvæ feed on developing seed and walls of pod; pupate in soil after 4 weeks. Adults emerge 2 weeks later except for last generation which over-winters in cocoon and pupates in spring	Adults lay eggs in pods; larvæ cause pods to ripen prematurely and seed is shed early	As for seed weevil	Winter oil-seed rape most affected

Crop	Pest	Description	Damage	Control	Remarks
Sugar-beet, mangolds, fodder beet	Seed weevils	Lead-grey in about 2.5 mm long	From hibernation near previous year's seed crops adults lay eggs in young pods. Larvæ feed on seeds in developing pods; they leave the pods and fall to the ground where they pupate in the soil	Seeds destroyed in pods by larvæ. Damage also caused by adult which makes holes for the pod midge to enter	Spray with organo-phosphorus when one or more weevils per plant is seen at end of flowering for autumn-sown crops, and late yellow bud stage for spring crops	As for pollen beetles
	Flea beetle	See flea beetle on brassicae		Seed dressing is not possible	
	Mangold fly	*Larva:* Yellow-white, legless 20 mm long. White oval-shaped eggs are laid in the under-side of leaves in May. Larvæ bore into the leaf tissue and after about 14 days they drop into the soil where they pupate	Blistering of leaf which can become withered. Retarded growth and in extreme cases death of the plant	Good growing conditions to help the crop pass an attack. Spraying carried out using an organo-phosphorus insecticide when more than 25 hatched larvæ, or eggs are counted per plant in the 6–8 leaf stage	
	Aphids (black and green fly)	The green-fly (peach potato aphid) has a very small oval-shaped body of various shades of green to yellow. During spring winged aphids migrate to the summer host crops. They move from one plant to another thus transmitting the virus from an unhealthy to a healthy plant	A severe infestation can cause the death of the plant, but chiefly it will mean a bad attack of Virus yellows, as both aphids are responsible for carrying the virus causing this disease	As for the bean aphid	
	Cyst eelworm	See cyst eelworm on cereals	Crop failing in patches. Plants which do survive are very stunted in growth	See cyst eelworm on cereals	In some areas, by law, sugar-beet may only be grown 1 year in 4 or 5 in fields known to be badly affected. If necessary the soil can be tested for an eelworm count
	Wireworm	See wireworm on cereals	The roots of seedling plants are bitten off	A seed dressing containing organo-mercury plus HCH	

(continued overleaf)

TABLE 53—*continued*

Crop attacked	Pest	Description	Life-cycle	Symptoms of attack	Control	Notes
Sugar-beet—*continued*	Docking disorder	A complex problem. Causes *irregularly* stunted plants with fangy root growth. Often caused by eelworms but soil structure giving poor growing conditions can be a causal factor. The disease is only found in East Anglia on sandy soils, normally alkaline with a low organic matter content. Losses can be minimized by good growing conditions. Success has been achieved by combining measures to improve soil fertility and the incorporation of pesticides such as aldicarb			Soil incorporation of systemic carbemate-type insecticide and nematicide will help	
Carrots	Carrot fly	*Adults:* About 8 mm long; shining black reddish/ brown head and yellowish wings *Larva:* When fully grown creamy-white, 8–10 mm long	Usually two generations a year. Eggs laid in soil surface near carrot. Hatch in 7 days and eventually burrow into root, forming "mines"; after third moult pupate in soil close to tap root. Some of the second, and possibly a third generation (especially in eastern counties) overwinter in the roots, emerging the following spring	Brown and rusty tunnels (mines often with larvae protruding) becoming progressively worse as season proceeds. Foliage of badly infected plants turn red, wilts and dies	(1) Rotation and avoid sowing susceptible crops close together (2) Hygiene round edge of field to cut down shelter for adult (3) Chlorfenvinophos or diazinon incorporated into soil as granules. Gamma HCH treatment not all that effective	Celery and parsnips also attacked
Potatoes	Peach potato aphid (greenfly)	See aphids on mangolds, sugar-beet and fodder beet		A bad infestation will check the growth of the plant, and potato virus diseases are spread	Apply a granular organophosphorus insecticide. This will help to check the spread of disease by killing the aphids. A spray could also be used	
	Wireworm	See wireworm on cereals		Maincrop tubers are riddled with tunnel-like holes	Aldrin dust or spray should be applied to the soil before planting. Lift the crop in early September if possible	

Crop	Pest				
	Cyst eelworm		See cyst eelworm on cereals and L. 284	See cyst eelworm on cereals	Where soil analysis indicates, use either a resistant variety or a nematacide in the seed-bed preparation
	Slugs		See slugs on cereals	Maincrop potatoes damaged by pests eating holes in the tubers	Very difficult. Some varieties resistant
Grass	Leatherjacket		See leatherjacket on cereals	Grass dying off in patches, the roots having been eaten away. Larvæ found in the soil	Spray with gamma HCH or use a bait
	Frit fly		See frit fly on cereals	See frit fly on cereals. The third generation larvæ can reduce the chances of a successful establishment of some autumn-sown grass seeds mixtures	When reseeding allow at least 6 weeks from the destruction of the old sward to the sowing of the new ley. Chemical treatment should shortly be available
Red clover	Stem eelworm	*Adult:* slender and colourless, difficult to see, about 1.2 mm long	Lives and breeds continuously in plant; passes into soil to infect other plants. Can remain dormant in hay made from an infected crop, becoming active again when conditions are suitable	Thickening at base of stem; some distortion of leaves, petioles and stems. Plants are stunted; infested patches increase in size each year	(1) Some varieties of red clover show resistance. (2) Fumigated seed. (3) Rotation—several years' break from red clover	White clover can be affected by a different race, but it is not considered important. Lucerne also affected by different race and the seed is fumigated

OTHER PESTS OF CROPS

Birds

Generally birds are more helpful than harmful, although this will depend on the district and type of farming carried out. To the grassland farmer in the west, birds are not nearly the pest they are to the arable farmer in the Midlands and East Anglia. Many birds in their lifetime will eat some cereal seed, but most of them help the farmer by eating many insect pests and weed seeds, and the diet of some in addition includes mice, young rats and other rodents.

The *wood pigeon* certainly does far more harm than good. Not only does it eat cereal seed and grain of lodged crops, it also causes considerable damage to young and mature crops of peas and brassicas. The only effective ways of keeping this pest down are by properly organized pigeon shoots and nest destruction, and the use of various scarers.

Mammals

Of the wild animals found in the countryside, those which cause most damage to crops are:

(1) *Rabbits* and *hares*—these can be very serious pests. They eat many growing crops— particularly young cereals. Organized shoots can control hares. Clearance of scrub and gassing are helpful in controlling rabbits.
(2) The *brown rat*, the worst pest of all, eats and damages growing and stored crops.
(3) *Mice*, another serious pest, damage many stored crops.

The local rodent officer will give advice on methods of extermination.

The harmless mammals, as far as crops are concerned, are:

The *badger* and *hedgehog*—these eat lots of insects, slugs, mice, etc.
The *fox*—kills rats and rabbits.
The *squirrel*—eats pigeon's eggs.

PLANT DISEASES

Although there are many causes of unhealthy crops, such as poor fertility and adverse weather conditions, the chief cause is disease.

Diseases, like pests, annually cause millions of pounds worth of damage and loss to the agricultural industry.

The four main agencies of disease are:

1. Fungi

Fungi are plants, but they are different from flowering plants in that they do not possess chlorophyll, i.e. the green colouring matter of leaves which is essential for photosynthesis. Therefore, as they cannot manufacture their own carbohydrate, they obtain it from living or dead plants. Thus it is convenient to divide fungi into two main classes:

(a) *Parasitic*. These are dependent on the living host. They are responsible for causing many plant diseases.
(b) *Saprophytic*. These live in dead plants. They play an essential part in helping to break down plant remains into humus.

There are many thousands of different species of fungus, the majority of which are invisible to the naked eye.

A typical fungus is composed of long, thin filaments (made up of single cells) termed *hyphae*. Collectively these are known as *mycelium*. It is through the mycelium that the fungus absorbs nutrients from its host.

With most parasitic fungi, the mycelium is enclosed within the host (only the reproductive parts protruding), although some fungi are only attached to the surface of the host.

Reproduction. Fungi can reproduce simply by fragments of the hyphæ dropping off, but usually reproduction is by *spores*. Spores can be compared to the seeds in ordinary plants, but they are microscopic and occur in immense numbers. The mycelium produces pods which contain the spores, and when the pod is ripe it bursts open, thus scattering the minute spores.

The dispersal of spores. It is important to understand how the spores are dispersed, and so infection spread from one plant to another. And knowing the particular form of dispersal will help in deciding disease prevention and control methods.

Spores can be dispersed by:

(i) *The seed.* The seed is carried from one generation to the next by the spores attaching themselves to the seed, e.g. covered smut of cereals.

(ii) *The soil.* The spores drop off the host plant and remain in the soil until another susceptible host crop is grown in the field. A suitable rotation will go a long way to check diseases caused in this manner, e.g. clubroot of brassicæ.

(iii) *Wind.* Spores carried through the air can spread diseases from an unhealthy to a healthy plant, e.g. cereal smut and rust diseases.

Fungi do show great specialization in that they are only parasitic to one type of host plant or a closely related plant.

The extent of the disease caused by the fungi does depend upon soil and weather conditions and also upon the state of the host crop. A healthy crop which is growing well will withstand an attack far more successfully than a stunted, slow-growing crop.

2. Viruses

The discovery of the virus is fairly recent, and because it is so difficult to isolate little is known about it. It is a very small organism indeed. Something like one million viruses could be contained on an average bacterium. Only by using electron microscopes can it be seen that plant viruses have a sort of crystalline form.

All viruses are parasitic. They are not known to exist as saprophytes. In many virus diseases the disease is not transmitted through the seed.

The virus is present in every part of the infected plant except the seed. Therefore if part of that plant, other than the seed, is propagated, then the new plant is itself infected, e.g. the potato. The tuber is attached to the stem of the infected plant, and infection is carried forward when the tuber is planted as "seed".

With most plant virus diseases, the infection is transmitted from a diseased to a healthy plant by aphids. These are sucking insects which carry the infected sap.

3. Bacteria

Bacteria are very small organisms, only visible under a microscope. They are of a variety of shapes, but those that cause plant diseases are all rod-shaped. Like fungi, bacteria feed on both live and dead material. Although they are responsible for many diseases of humans and livestock, in this country they are of minor importance compared with fungi and viruses as causal agents of crop diseases.

Bacteria reproduce themselves simply by the process of splitting into two. Under favourable conditions this division can take place every 30 minutes or so. Thus bacterial disease can spread very rapidly indeed, once established.

4. Lack of essential plant foods (mineral deficiency)

When essential plant foods become unavailable to particular crops deficiency diseases will appear. In "marginal" situations, where intensive systems of cropping are practised, mineral deficiencies are likely to be more apparent simply because there are insufficient trace elements present to cope with a large crop output. Most of the diseases are associated with a lack of trace elements, but shortage of any essential plant food will certainly reduce the yield, cause stunted growth, and make the crop more vulnerable to pest and disease attack (see also "Chemical Elements Required by Plants", page 13).

TABLE 54   MAIN PLANT DISEASES AND THEIR CONTROL

Crop attacked	Disease	Causal agent	Symptoms of attack	Life-cycle	Methods of control
Cereals	(1) Bunt, covered or stinking smut of wheat (2) Covered smut of barley (3) Covered and loose smut of oats	Fungus	Brown or black spore bodies instead of grain in the ears	Infected grain is planted; seed and fungus germinate together and thus young shoots become infected. The spores are released when the skin breaks, and so combining or threshing contaminates healthy grain	Organo-mercury seed dressing. Or less poisonous but more expensive fungicides can be used
	(1) Leaf stripe of barley (2) Leaf stripe of oats	Fungus	The first leaves have narrow brown streaks. Subsequently brown spots appear on the upper leaves	Infected grain is planted; seed and fungus germinate together and thus young shoots are infected. From the secondary infection, spores are carried to developing grain	As for the covered smuts
	(1) Loose smut of wheat (2) Loose smut of barley	Fungus	Infected ears a mass of black spores. They do not remain enclosed within the grain as with the covered smuts	Similar to the covered smuts, but the fungus develops within the grain. The spores are dispersed by the wind to affect healthy ears	(1) Resistant varieties (2) Clean seed (3) The seed can be dressed with Carboxin
	Net blotch of barley	Fungus	Short brown stripes on older leaves; on younger leaves and adjacent plants, in addition to the striping, irregular-shaped dark brown blotches which can run together. Can spread to ears	Seed-borne disease, but there is evidence that it can spread from previously infected crop	(1) Seed dressing (2) Crop hygiene to clear stubble of previously infect crop (3) Fungicides now available, e.g. Prochloraz, Propiconazole
	Yellow rust	Fungus	Yellow-coloured pustules in parallel lines on the leaves, spreading in some cases to the stems and ears. In a severe attack the foliage withers and shrivelled grain results	The fungus mainly attacks wheat. Infection appears on the plant from May onwards. From the pustules, spores are carried by the wind to infect healthy plants. During winter, spores are dormant on autumn-sown crops	(1) Resistant varieties, although new races appear against which these varieties soon have no resistance (2) Fungicides, e.g. Triadimefon.

Disease	Cause	Symptoms	Spread	Control
Mildew	Fungus	On winter cereals, grey-white and brown mycelium on lower leaves in February. Infection spreads to other leaves and plants. Disease general between May and August. Early infected leaves go yellow and shrivelled. Towards the end of season black spore cases formed among brown fungi	From self-sown cereals in stubble, winter and spring cereals can be infected	(1) Clean-up old stubbles (2) Resistant varieties, although new races of the fungus may appear to nullify previous resistance in a variety (3) Barley seed can be dressed with ethirimol (4) Fungicides available, e.g. tridemorph
Leaf blotch (Rhyncho-sporium)	Fungus	When fully-formed lentil-shaped blotches (light grey with dark brown margins) up to 19 mm long are seen on the leaves. As disease progresses blotches coalesce	Fungus overwinters on self-sown barley plants and on 2-row winter barley crops. From here spores are carried to planted barley crops	(1) Clean stubbles of all self-sown barley plants (2) Systemic fungicides, e.g. Triadimefon *Note*: Attacks only rye and barley, 6-row barley is more resistant
Black stem rust or black rust	Fungus	Reddish-brown lines or spots on the leaves and stems, later succeeded by black streaks on the foliage	Spores are wind-borne—often over long distances—from barberry bushes and then from plant to plant. It does most damage on wheat in south-west counties	Resistant varieties are possible, although this is not now such an important disease
Brown rust of barley and wheat (less important in wheat)	Fungus	Numerous, very small and scattered, orange-brown pustules on the leaves in June. These gradually develop late in the season. A severe attack causes shrivelled grain	The resting spores over-winter on stubbles. From here the pustules are airborne to infect healthy plants	(1) Crop hygiene to clear stubble of volunteer plants (2) Resistant varieties (3) Systemic fungicides, e.g. Triadimefon
Septoria diseases (leaf and glume blotch, i.e. S. Tritici and S. Nodorum)	Fungus	Leaf: bleached or discoloured blotches of varying sizes and shapes (on which appear rows of minute black dots) seen from late autumn to early summer	Leaf spores are liberated in wet weather, and they winter on volunteer crops; they transfer to winter cereals and then move on to spring crops	(1) Crop hygiene to clear stubble of volunteer plants (2) Clean seed (3) Tolerant varieties (4) Systemic fungicides, e.g. Captafol

(continued overleaf)

TABLE 54—*continued*

Crop attacked	Disease	Causal agent	Symptoms of attack	Life-cycle	Methods of control
Cereals— *continued*	Septoria diseases— *continued*	Fungus— *continued*	Glume blotch: becomes prominent in July and August, especially in wet seasons. Irregular, chocolate spots or blotches on glumes, beginning at the tips; later ears become blackened with secondary infection. Leaves also show yellowish areas and fungus can affect stems and leaf sheaths as well. Shrivelled grain results	Glume blotch has a similar life cycle, but the fungus can also be carried on the seed	
	Barley yellow dwarf virus	Virus	Stunted plants in patches or scattered as single plants. Poor root development. Mostly red and yellow colour changes in leaves. Late heading and reduced yield	The crop is infected by cereal aphids	Spray with organophosphorus (see Control of Cereal Aphids, Table 53) *Note*: Wheat also affected
	Barley yellow mosaic virus	Virus	Appears in patches in field. Pale green streaks later turning brown, particularly at leaf tip; leaves tend to roll inwards, remain erect to give plant a spikey appearance. Plants stunted and late to mature	Soil-borne, but probably also carried over from one crop to the next on stubble and other plant debris	Good stubble hygiene. Some varieties, notably *Maris Otter*, are particularly susceptible
	Eyespot	Fungus	Eye-like lesions on stem about 75 mm above ground. Grey "mould" inside stem; straws lodged in all directions. No darkening at base of stem	The fungus can remain in the soil, on old stubble and some species of grasses for several years. It usually attacks susceptible crops in the young stages	(1) Spring-sown barley and wheat are more resistant than autumn-sown crops. Arrange a break of at least 2 years from cereals. Most winter wheat varieties show some resistance (2) Systemic fungicides, e.g. Carbendazim
	Sharp eyespot	Fungus	Lesions more sharply defined than true eyespot. Brown/purple border followed by cream-coloured area and brown-	The fungus is soil-borne but is also found on plant debris	Oats and rye most susceptible, and wheat more than barley. Not a serious disease, but can be accentuated when crop is

		ish centre. Lesions more numerous and occur further up stem than true eyespot. Can cause lodging, "whiteheads" and shrivelled grain	sprayed against true eyespot. Worse in winter crops and mixed farming systems *Note:* Fungus has a large host range, e.g. causes black scurf in potatoes	(1) Stubble hygiene (2) Seed treatment using triadimenol with fuberidazole (baytan) will help (3) Fungicides for control of eyespot and septoria will also help
Fusarium—brown foot rot and ear blight	Fungus	Undefined brown discoloration at base of tillers and lower leaf sheaths. Interior of stem shows pink fungal growth. Premature ripening and "whiteheads" or "blind" ears	A number of fusarium spp. but infection is either seed-borne or from previously infected stubble and crop remains	(1) Stubble hygiene (2) Seed treatment using triadimenol with fuberidazole (baytan) will help (3) Fungicides for control of eyespot and septoria will also help
Crown rust of oats	Fungus	Orange-coloured pustules spread mainly on leaf blade. Later in season black pustules are produced. Severe attack prior to, and including, milk-ripe stage, causes shrivelled grain	Spores are wind-borne from overwintering volunteer plants, and winter oats, to the spring crop	(1) Crop hygiene to clear stubble of volunteer plants (2) Keep winter and spring oat crops as far apart as possible (3) Some varieties show reasonable resistance (4) Some fungicides provisionally recommended
Take-all or whiteheads	Fungus	Black discoloration at base of stem. Grey colour of roots. Ease with which plant can be pulled from the soil. Infected plants ripen prematurely, and produce bleached ears containing little or no grain	Wheat and barley only affected. The fungus survives the soil in root and stubble residues and the host plant is infected when it is grown in the field	Rotation to starve the fungus, but after some years of continuous wheat growing, infection appears to lessen. Extra nitrogen helps the growth of new roots. Bad drainage reduces plant vigour and it is more easily damaged. Direct-drilling appears to lessen the intensity of the disease
Snow mould (mainly of wheat and rye)	Very low temperatures and/or fungus	Patchy crop in autumn. Stunted seedlings occasionally with white-pink mycelium at base. After snow has thawed— withered plants in patches temporarily covered by white-pink mould. Thereafter infected plants stunted, weak root system and shrivelled grain	A seed-borne fungus, but contamination of crop is also possible from the old stubble and other plant debris. Fungus is particularly favoured by low temperatures	(1) Good stubble hygiene (2) Clean seed (3) Seed treatment using triadimenol with fuberidazole

(continued overleaf)

TABLE 54—continued

Crop attacked	Disease	Causal agent	Symptoms of attack	Life-cycle	Methods of control
Cereals—continued	Snow Rot (Typhula Rot) (mainly of barley)	Fungus	Thin, poorly-tillered crop; plants yellowing and withered in patches. Old leaves often covered by white mycelium, young leaves standing erect but eventually yellowed. Weak root system	Soil-borne fungus which can remain dormant for years; when active, infects emerging cereal plants, developing rapidly in dark and humid conditions, i.e. under snow	(1) Sow early in the autumn (2) Consider reducing seed rate if attack is feared (3) Seed treatment using triadimenol with fuberidazole
	Ergot and rye and wheat	Fungus	Hard black curved bodies up to 20 mm long replacing the grain, and protruding from the affected spikelet	Ergots fall to ground and remain until next summer when they germinate and produce short stems with globular heads containing the spores which are then air-borne to the rye and wheat flowers, and certain grass, depending on the species	Although disease has little effect on yield, ergot is poisonous to mammals (but it does contain medicinal properties). Crop rotation and control of grass weeds (especially blackgrass) in the crop will help. Not considered important enough for special control measures
	Manganese deficiency of cereals	Manganese deficiency	Yellowing on leaf veins followed by development of brown lesions. With oats the spots enlarge and can extend cross leaf. Thus the leaf can bend right over in the middle. Older leaves wither and die. Can lead to shrivelled grains		9 kg/ha manganese sulphate in 340 litres/ha water. With a bad attack this may have to be repeated after 3 weeks
Maize	Stalk rot	Fungus	Base of plant attacked in August/September; foliage grey/green colour and wilting. Pith brown/pink at base of stem; leads to premature senescence	Soil-borne fungus	Distinct differences in varietal susceptibility. Grain maize more likely to be affected because disease develops most rapidly in mature crops
	Smut	Fungus	Large black galls on any of the above ground parts of the plant, including the cob	Spores from galls can reinfect other maize plants, or they can remain dormant in the soil, surviving for several years	Would appear to be more serious when maize is cropped more than 1 year in 3. Not a seed-borne disease to the same extent as the other cereal smuts
Beans	Chocolate spot	Fungus	Small circular chocolate-coloured discolorations on leaves and stems; with	Fungus carried over from previous year on debris of old bean haulm and on self-	(1) Clean up stubbles containing remains of old crop of beans

Crop	Disease		Symptoms	Biology	Control
			bad attack symptoms move to flowers and pods. In wet weather spots coalesce	sown plants. But infection can start from almost any dead vegetation with this widespread fungus	(2) Give good growing conditions (3) Fungicide, e.g. Benomyl, can be used Autumn-sown beans are more liable to attack than spring-sown and they suffer more severely
	Ascochyta (leaf spot)	Fungus	Leaves affected by regular brown to black spots, some up to 2 cm in diameter; spots have slightly sunken grey centres (in which can be seen small black spots with brown margins. Pods and seed also affected, the latter covered with brownish-black lesions	Infected seeds when sown may produce seedlings with characteristic disease symptoms on stem at soil level or on lowest leaves. In cold moist conditions the disease will move up the plant and on to other bean plants	(1) Healthy seed—seed can be tested (2) Hygiene—kill any volunteer beans in other fields (3) Seed treatment with benomyl and thiram compound for partial control
Brassicae (Brussels sprouts, cabbage, oil-seed rape, kale, swedes and turnips)	Club root or finger and toe	Fungus	Swelling and distortion of the roots. Stunted growth. Leaves pale green in colour	A soil-borne fungus. The fungus grows in the plant roots and causes the typical swellings. Resting spores can pass into the soil especially if diseased roots are not removed. They can remain alive for several years, becoming active when the host crop is again grown in the field	(1) *Rotation*. With a bad attack advisable not to grow the crop for at least 5 years in the field (2) *Liming and drainage*. The spores are more active in acid and wet conditions (3) *Resistant crops*. Kale is more resistant than swedes or turnips. Some varieties of swedes and turnips are more resistant than others
	Stem canker in oil-seed rape	Fungus	Beige coloured circular spots (0.5–1.0 cm diam.) on leaves; spores spread to produce brownish-black canker at base. Stem splits and rots causing lodging; rapid stem elongation and premature ripening	Air-borne spores produced on the stubble carry infection to young crops in the vicinity. Disease can also be spread from infected seed, but this is less significant	(1) Destroy stubble debris as soon as possible after harvest (2) Rotation; avoid growing rape crops in the same field more often than one year in six (3) Tolerant varieties (4) Use benomyl and thiram fungicide as a seed treatment
	Alternaria (dark leaf and pod spot) on oil-seed rape	Fungus	Circular small brown-to-black leaf spots sometimes coalescing on leaves and later on pods. Premature ripening and loss of seed	Seed-borne disease, although spores can be carried through the air from other infected brassica crops	Seed treatment using iprodione

(continued overleaf)

TABLE 54—*continued*

Crop attacked	Disease	Causal agent	Symptoms of attack	Life-cycle	Methods of control
Brassicae—*continued*	Powdery mildew (particularly Brussels sprouts, swedes and turnips)	Fungus	Upper surface of leaves show blue/black discoloration; sprouts turn black	Spores overwinter on infected plants and are carried by air currents to infect the following year's crop	(1) Varieties differ in their susceptibility to this disease (2) Dinocap may give some control
	Brown heart of swedes and turnips (Raan)	Boron deficiency	No external symptoms, but when the root is cut open a browning or mottling of the flesh is seen. Affected roots are unpalatable		See heart rot of sugar-beet
	Stem rot in kale	Boron deficiency	Cavitation in the pith followed by a brown rot and stem collapse		See heart rot of sugar-beet
Sugar-beet, mangolds fodder-beet	Virus yellows	Virus	First seen in June/early July on single plants scattered throughout the crop—a yellowing of the tips of the plant leaves. This gradually spreads over all but the youngest leaves. Infected leaves thicken and become brittle. The yield is seriously reduced by an early attack	The crop is infected by aphids which have overwintered in mangold clamps and steckling beds. The aphids carry the virus, and they can very quickly infect the whole crop	(1) Good growing conditions to keep the crop growing vigorously (2) All mangold clamps should be cleared by the end of March. If not, they should be sprayed to kill any aphids (3) Seed crop stecklings, should not be raised in the main sugar-beet areas (4) An organo-phosphorus insecticide should be used when the aphids are first seen. This may have to be repeated at least once, possibly using a different insecticide to prevent the development of resistant aphids. A granular organo-phosphorus can be used, or the soil can be dressed with an insecticide to give protection against an early attack (5) Virus yellows tolerant varieties can be grown where the disease is a great hazard

Heart rot	Boron deficiency	In young plants the youngest leaves turn a blackish-brown colour and die off. A dry rot attacks the root and spreads from the crown downwards. The growing point is killed, being replaced by a mass of small deformed leaves		This deficiency is more apparent on dry and light soils and can be made worse by heavy liming. Apply borax at 22 kg/ha as soon as the disease is seen. Use a boronated compound fertilizer on suspected soils	
Potatoes	Blight	Fungus	Brown areas on leaves. Whitish mould on the underside of leaves. Leaves and stems become brown and die off	Infected tubers (either planted, ground keepers, or throw outs from clamps) produce blighted shoots. From these shoots the fungus spores are carried by the wind to infect the haulms. From the haulms the spores are washed into the soil to infect the tubers. Infection can also take place at harvest. The fungus cannot live on dead haulm	Blight spreads rapidly in warm, high humidity, weather and warnings of such conditions are given by ADAS. Early preventative spraying, followed by repeated sprays every 10–14 days is advisable—especially in areas where blight is a problem; a wide range of fungicide may be used such as maneb, mancozeb, copper and "Patafol". The haulms should be destroyed chemically before harvest to prevent the tubers becoming infected whilst being lifted
	Leaf roll	Virus	Lower leaves are rolled upwards and inwards; they feel brittle and crackle when handled. The other leaves are lighter green and more erect than normal. Yield is lowered	The virus is transmitted by aphids from plant to plant. Infected tubers (which show no signs of the disease) are planted and thus the disease is carried forward from year to year	(1) Use certified "seed" which is grown in areas such as Scotland and Northern Ireland where aphids are not prevalent due to the colder climate. Thus the "seed" is usually free from virus infection (2) Systemic sprays or granules will control the aphids, and thus reduce the spread of the virus
	Mosaics	Virus	May range from a faint yellow mottling on leaves to a severe distortion of the leaves and distinct yellow mottling. Yield can be seriously reduced by the severe forms	Some of the viruses responsible are spread by aphids but some, e.g. Virus X, are spread by contact between leaves and roots and on machinery and clothing	As for leaf roll. Some varieties resistant

(continued overleaf)

TABLE 54—continued

Crop attacked	Disease	Causal agent	Symptoms of attack	Life-cycle	Methods of control
Potatoes—*continued*	Common scab	Actinomycete	Skin-deep irregular-shaped scabs on tuber: these can occur singly or in masses. With a severe attack cracking and pitting takes place with secondary infection by insect larvæ and millipedes	The soil-borne organism attempts to invade the growing tuber which responds by development of corky tissue to restrict the parasite to the surface layers. Organism re-enters soil when infected seed is planted	(1) Avoid liming just prior to planting potatoes. Disease is particularly prevalent on light sandy, alkaline soils (2) Irrigation of dry, light soils an advantage. Dry conditions favour the spread of the disease (3) Some varieties are more resistant than others
	Powdery scab	Fungus	Appearance can be similar to common scab but the spots are rounder and formed as raised pimples under the skin which burst; sometimes cankers and tumours develop	Spore balls can remain in the soil for many years or may be planted on infected tubers and the zoospores attack the new tubers by way of the lenticles, eyes or wounds—resulting in scab development	Usually more troublesome in wet seasons and lime-rich soils. Potatoes should not be planted after a severe attack for at least 5 years. Avoid using infected seed or contaminated FYM; some varieties are very susceptible
	Dry rot	Fungus	Infected tubers are usually first noticed in January and February. The tuber shrinks and the skin wrinkles in concentric circles. Blue-pink or white pustules appear on the surface	The soil-borne fungus enters the tuber from adhering soil. Infection can only enter through wounds and bruises caused by rough handling at harvest. The disease can be easily spread during storage	(1) If the potatoes are handled carefully, infection is considerably reduced. Some varieties are more resistant than others to the disease (2) Use "Fusarex" or "Storite" when storing
	Spraing	Tobacco rattle virus (TRV) or potato mop-top virus (PMTV)	Foliage—very variable: TRV—stem mottling; PMTV—yellow blotches and bunching of leaves on short stems—like a mop. Tubers—primary (after soil infection): wavy or arc-like brown, corky streaks in flesh of cut tuber. Secondary—from infected tubers: PMTV—badly formed and cracked tubers. TRV—brown spots in flesh	TRV spread by nematodes in soil, especially in light sandy soils. PMTV—spread by powdery scab fungus and can remain in the soil for years in fungal resting bodies	Plant only resistant varieties, e.g. *King Edward*, *M. Piper*, on infected soils. Do not plant tubers showing spraing symptoms or carrying much powdery scab. Seek expert advice
	Blackleg	Bacteria	Plants stunted and pale green or yellow foliage: easily pulled out of the ground	The bacteria move to tubers via the stolons and in wet soil to healthy tubers to enter via	Rogue or reject seed crops where the disease shows on foliage. Do not plant

Crop	Disease	Cause	Symptoms	Spread	Control
			and stem base is black and rotted. Infected and neighbouring tubers develop a wet rot in the field or in store, especially in damp and badly ventilated (warm) conditions	lenticles or damaged areas. Carried on seed tubers	infected tubers. Tuber dips may spread the disease to healthy tubers
	Gangrene	Group of Phoma fungi	A serious tuber rot which develops in storage, usually late; it shows as grey "thumb-mark" depressions on the tubers and the flesh beneath is rotted; also, pinhead black spore cases	The fungus remains alive in the soil and on trash, and can infect tubers in the soil and from tuber to tuber when handling	Do not plant diseased seed. Assist tuber wounds to heal by keeping them warm (about 17°C) and humid up to 14 days after any handling operation. Treat seed with butafume or "Storite"
	Skin spot	Fungus	Tuber symptoms develop during late storage and appear as pimple-like, dark brown, shrunken spots with raised centres. The worst damage is the destruction of the buds in the eyes of seed tubers	Mainly spread by infected tubers. Tuber infection occurs at lifting and is worse in cold, wet seasons	As for gangrene
Grass	Barley yellow dwarf virus	Virus	Leaves turn yellow, red or brown, discoloration starting at tips and going down leaves. Plants generally stunted, but can produce more tillers. Disease more conspicuous in single plants than in whole sward	Spread by aphid	Very difficult; not economic to spray. Some ryegrass varieties more resistant than others. Note: Affects other grasses, wheat and barley, see page 198
	Ryegrass mosaic	Virus	Yellowish/green mottling or streaking of leaves. Severe infection can show more general browning of leaf	Spread by mites which are favoured by hot, dry weather	Some ryegrass varieties are more resistant than others
Red and white clover	Clover rot	Fungus	Foliage turns olive green and then black and eventually dies. The root can also die	Resting bodies of fungus produced on affected plants in winter and spring. They are small (size of clover seed), white at first and then turning black. Bodies remain dormant in summer but in autumn produce seed-borne spores which affect other plants	(1) Clean seed (2) Use resistant varieties where possible (at present fewer resistant varieties with white clover) (3) Rotation; may have to be an interval of at least 6 years. Note: Also affects lucerne and sainfoin

(continued overleaf)

ITCH – J

TABLE 54—continued

Crop attacked	Disease	Causal agent	Symptoms of attack	Life-cycle	Methods of control
Lucerne	Verticillium wilt	Fungus	Usually seen in fairly isolated patches in first harvest year; in next 2 years spreads to many parts of the field. Normally after first cut, lower leaves turn pale-yellow colour and then white, and eventually shrivel from base upwards. Whole plant finally dies	Disease can be introduced by contaminated seed; spores can also be transported by air, as well as being spread by contaminated fragments of the crop moving from plant to plant and then from field to field by machine	(1) Where suspected use tolerant varieties (2) Use clean seed which, as an insurance, can be treated with thiram (3) Harvest healthy crops first before moving onto older infected crops

Physiological disorders (sometimes referred to as stress)

This can be due to external or environmental conditions which will normally only temporarily upset the plant's metabolism. There may be occasions when the effect is more permanent.

Temporary, e.g. a high water table in the early spring. This will cause yellowing of the cereal plant as its root activity is restricted considerably reducing, therefore, its oxygen and plant food requirements. When the water table goes down the plant is able to grow normally once more, assuming a healthy green colour.

Permanent, e.g. where the soil has become compacted the root activity of the plant can become restricted. This will result in poor stunted growth with the plant far more vulnerable to pest and disease attack. The yield will certainly be reduced.

THE CONTROL OF PLANT DISEASES

Before deciding on control measures it is important to know what is causing the disease. Having ascertained, as far as possible, the cause, the appropriate preventative or control measure can then be applied.

1. Crop rotations

A good crop rotation can help to avoid an accumulation of the parasite. In many cases the organism cannot exist except when living on the host. If the host plant is not present in the field, in a sense the parasite will be starved to death, but it should be remembered that:

(a) Some parasites take years to die, and they may have resting spores in the soil waiting for the susceptible crop to come along, e.g. club root of the brassicæ family.

(b) Some parasites have alternative hosts, e.g. fungus causing take-all of wheat is a parasite on some grasses.

2. Removal of weed host

Some parasites use weeds as alternative hosts. By controlling the weeds the parasite can be reduced, e.g. cruciferous weeds such as charlock are hosts to the fungus responsible for club root.

Both these preventative measures (1 and 2) illustrate the importance of having a sound knowledge of the parasites attacking crops.

3. Clean seed

The seed must be free from disease. This applies particularly to wheat and barley which can carry the fungus causing loose smut deeply embedded in the grain. Seed should only be used from a disease-free crop.

With potatoes it is essential to obtain clean "seed", free from virus. In some districts where the aphid is very prevalent, potato seed may have to be bought every year.

4. Resistant varieties

In plant breeding, although the breeding of resistant varieties is better understood, it is not by any means simple.

For some years plant breeders concentrated on what is called single or major gene resistance. However, with few exceptions, this resistance is overcome by the development of new races of the fungus to which the gene is no longer resistant.

Breeding programmes are now concentrating on multigene or "field resistance" which means that a variety has the characteristics to *tolerate* infection from a wide range of races with little lowering of yield. Emphasis is now on tolerance rather than resistance.

5. Varietal diversification

By choosing cereal varieties with different disease resistance ratings for growing in adjacent fields or in the same field in successive years, the risk of a serious infection in any year of yellow rust and mildew in winter wheat and mildew in spring barley can be considerably reduced. See NIAB Leaflet No. 8—*Cereals.*

6. Cereal mixtures (blends)

This is a natural extension of diversification in that varieties from different diversification groups are grown together in the same field. In this way a disease-carrying spore from a susceptible variety, but within a blend of varieties (two or three varieties) making up the crop, has less chance of successfully infecting a neighbouring plant, than in a pure crop. Yields of blended crops appear to be more reliable than pure crops and this is achieved with the use of fewer fungicides. More work still needs to be done on the growing of cereal mixtures.

7. The control of insects

Some insects are carriers of parasites causing serious plant diseases, e.g. control of the greenfly (aphid) in sugar-beet will reduce the incidence of virus yellows. Furthermore, fungi can very often enter through plant wounds made by insects.

8. Remedying plant food deficiencies

A deficiency disease can often be overcome if the deficient plant food is remedied at an early stage. It is, however, sometimes difficult to diagnose the disease in time. Tissue analysis will help in this situation.

9. The use of chemicals

Broadly speaking, chemical control of plant disease means the use of a fungus killer—a fungicide. A fungicide may be applied to the seed, the growing plant, or to the soil. It can be used in the form of a spray, dust or gas. To be effective, it must in no way be harmful to the crop, nor after suitable precautions have been taken, to the operator or others, and it must certainly repay its cost.

The dressing of seed with a fungicide—seed disinfection. This is carried out to prevent certain soil and seed-borne diseases. In many cases an insecticide is added to help prevent attacks by soil-borne pests. Various fungicides can be used, depending upon the disease to be controlled:

(a) *Organomercury*—used as a preventive measure against many seed-borne diseases of cereals, fodder beet and mangels (dry treatment only).

This compound has been used for many years, but concern is now being felt about mercury levels in food. Reliable alternatives, not containing mercury, are now becoming available commercially, although they are much more expensive.

Organomercury can be combined with other chemicals to give a wider range of control of plant diseases, and control of some pests.

(b) *Carboxin with organomercury*—for controlling loose smut in wheat and barley in addition to disease controlled by organomercury on its own.

(c) *Guazatine*—for controlling covered smut in wheat and partial control of fusarium and septoria. In combination with *imazalil* it also covered smut, net blotch and leaf stripe of barley.

(d) *Triadimenol with fuberidazole*—used to control the important seed-borne diseases in spring barley in addition to the foliar diseases mildew and rhynchosporium, as well as giving some protection against yellow and brown rust. Winter barley can also be treated with this compound. This will protect the plant from the same diseases over the autumn and winter and, as at the same time it will reduce considerably many sources of innoculum, there should be less of a disease problem in the crop the following spring and summer. A broad-spectrum fungicide, but with a different mode of action, should certainly be able to deal with it. Seed treated with these compounds should not be directdrilled.

(e) *Ethirimol*—for helping to prevent mildew

on winter and spring barley. It should not be used on seed which has already been treated with an organomercury compound.

(f) *Thiram*—this is used as a dry seed dressing to prevent seed decay and pre-emergence damping-off in beans, peas, maize and linseed. It also controls leaf spot in oats. See also Table 55, Cereal Fungicide Groups.

The composition of these fungicides (particularly the organomercury) will vary slightly according to the commercial brand, but they all consist of a very small proportion of the active ingredient, plus a large percentage of the carrier. A dye is added so that dressed seed can easily be recognized and it must always be labelled as such. The dressing of the seed is preferably carried out by the seed merchant.

Organomercury and/or carboxin dressed seed should be used as soon as possible a week after treatment. Seed treated with other compounds listed can be used immediately following treatment. If storage is necessary it should be kept in dry airy conditions. It should not be used for human or animal consumption. It can generally be kept until the next year (doubtful possibly with guazatine-treated seed), but a germination re-check is always advisable.

There is evidence that seed dressing is not always being carried out (this particularly applies to the farmer's own seed), and diseases such as the covered smuts of all the cereals, loose smut of oats, leaf stripe of barley and oats, and net blotch of barley, are increasing again, although there are additional reasons for the last-mentioned disease.

Application to the plant. A good example is the control of potato blight. The fungicide can be used as a dust or spray, and it may have to be repeated at 10–14-day intervals, according to the season.

Copper, and *organic* compounds, are the chemicals used for controlling blight. *Organic tin* compounds are now being used as well, and because of more persistency they give better protection for the tubers against the blight spores.

Systemic fungicides are now being developed to control potato blight.

Many systemic fungicides can now be used for the control of cereal diseases—see below.

Soil use of fungicide. Has little application in agriculture, but *dazomet* as a soil sterilant is used for controlling club root on brassicæ either by dusting the seedling roots or dipping them in a paste.

Fungicides and cereal diseases

Because of the extreme difficulties, if not impossibility, of breeding varieties resistant to cereal diseases, fungicides are now playing a very important part in the growing of cereals, although tolerance against certain diseases should not be ignored (page 201).

It should be understood that, apart from drought, disease is the biggest single factor influencing yield, certainly in the south and west of the country, and on the coastal fringes, with the possible exception of yellow rust in wheat although, in this respect, the distribution of the wheat crop in the country must influence the incidence of yellow rust (see Fig. 87). The major effect of these diseases is to lower the specific weight of the grain. Grain size plays a very important part in determining the yield of the crop.

Different treatment systems involving the use of fungicides can now be considered as an essential part of a cereal-growing programme. There are three main systems.

1. *Assessment of the disease risk*

Fungicides are used only when it is considered that a specific disease has developed to a point (the threshold) which will actually cause a loss of yield, e.g. mildew on spring barley; it is recommended that the crop should be sprayed when 3% of the second or third leaf is affected by mildew.

2. *Prophylactic control*

This system involves the use of a fungicide or fungicides before the disease or diseases are

TABLE 55   CEREAL FUNGICIDE GROUPS
(based on work of Dr. M. J. Griffin)

Group	Chemical group/ mode of action	Common names	Trade names	Diseases controlled (V. good—in bold)
1	Benzimidazoles (MBCs) inhibit or disrupt mitosis	benomyl	Benlate,	**E**, **Rh**, S, M (not N)
		carbendazim	e.g. Bavistin, Derosal, Focal and in mixtures	**E**, **Rh**, S, M (not N)
		thiophanate-methyl	Cercobin	**E**, **Rh**, S, M (not N)
		fuberidazole + triadimenol	Bayton (st)	cs, **ls**, **lf**, **N**, M, f, Rh, R
2	Sterol biosynthesis inhibitors, mainly for leaf-disease control	triadimenol		
		propiconazole	Tilt, Radar	**M**, **N**, **R**, **Rh**, E
		prochloraz	Sportak	**M**, **N**, **R**, **Rh**, E, **S**
		triadimefon	Bayleton	**M**, **R**, **Rh**, S (not N)
		triforine	Saprol	**M**
		nuarimol	Triminol	**M**
		imazalil	in Muridal and Murbenine Plus (st)	
3	Morpholines	tridemorph	Calixin, Bardew, Ringer	**M**, R
		fenpropimorph	Corbal, Mistral	**M**, **R**, Rh
4	Hydroxypyrimidines	ethirimol	Milstem, Milgo	**M**
5	Organic phosphates	ditalimfos	Farmil	N
6	Carboxamides interferes with respiration	carboxin	in Murganic RPB and Safeguard (st)	
7	Guanidines	guazatine	Murbenine (st)	**bu**, f, s
8	Captafol	captafol	Captafol, Sanspor	**S**, **Rh**
9	Chlorthanil	chlorothanil	Bravo	**S**, Rh
10	Organic-mercurials	organomercury	various seed treatments	
11	Sulphur		Thiovit, Elosal	M
12	Dithiocarbamates	maneb, mancozeb (Dithane), managanese-zinc-dithocarbamate (Merolan), propineb (Antracol), thiram and zineb are mainly used in mixtures		

Disease abbreviations: E—eyespot, M—mildew, N—net blotch, R—rusts, Rh—rhynchosporium, S—septoria, st—seed treatment, bu—bunt, cs—covered smut, ls—loose smut, f—fusarium, lf—leaf stripe.

(a) Groups 1–7 are *site specific*, i.e. they act at one or a few sites only.

(b) Groups 8–12 are *multi-site*, i.e. they can act at many sites and so are less likely to allow fungicide-resistant strains to develop.

For further information on the use of fungicides for the control of cereal diseases see Ministry Booklet 2257.

present. It can be described as a programmed approach whereby factors such as minimum varietal resistance and other predisposing circumstances could bring about a disease situation, e.g. carbendazim to control eyespot, and partially control mildew and septoria in winter wheat. Obviously in some years this treatment will not be cost effective.

3. *Managed disease control*

This treatment involves both prophylactic and specific control (according to the assessment of the disease risk), e.g. seed treatment for spring barley using ethirimol to control mildew followed by a spray such as triadimefon when 3% of the area of the oldest green leaves are affected.

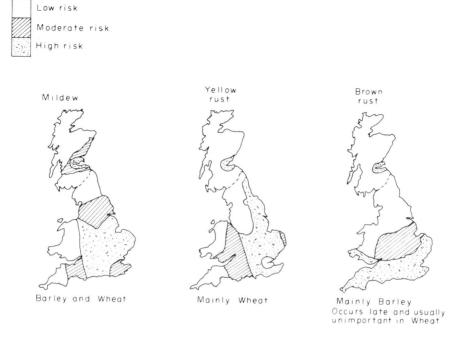

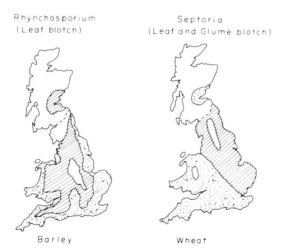

FIG. 87. Generalized maps of regions where leaf diseases of cereals are most likely to occur.

Fungicide resistance

When a fungal disease is controlled effectively by a fungicide, the fungus is "sensitive" to the chemical. However, other strains of the fungus can and do occur over a period of time, and some of these may be resistant ("insensitive" or "tolerant") to the fungicide which means that the disease is then not controlled adequately. The more often the same chemical is used, the greater are the chances of resistant strains developing. There is also an increased risk of this happening with fungicides which act at one specific site in the fungus compared with multi-site fungicides, i.e. those fungicides which act at many sites in the fungus.

There are other ways of avoiding a build up of resistance by a fungus:

1. A reduction of disease levels by good husbandry.
2. Use of resistant varieties; diversifying varieties in adjacent fields and in successive years; cereal mixtures.
3. Where possible, fungicides with different modes of action should be used when more than one has to be used on the same crop.
4. A late application (after flowering) of a fungicide should, as far as possible, be avoided.
5. Where feasible, the use of appropriate fungicide mixtures. These mixtures must have clearance under the Pesticide Safety Precautions Scheme.

Unsatisfactory disease control following the use of fungicides is, at present, not often due to fungicide resistance. There are several other reasons, the main ones being, wrong timing, the use of too low a dose and poor application.

SUGGESTIONS FOR CLASSWORK

1. Make a collection of the most important pests attacking crops.
2. Look out for any possible pest and disease attack on crops growing in the district. Examine closely the symptoms of attack, and find out the most suitable remedy.
3. When handling insecticides and fungicides, pay every attention to the manufacturers' intructions.

FURTHER READING

Sugar Beet Pests, Bulletin No. 162, HMSO.
Insecticide and Fungicide Handbook, Blackwell.
The Use of Fungicides for the Control of Cereal Diseases, MAFF Booklet 2257.
Dr. D. G. Hessayon, *The Cereal Disease Expert*, PBI Publication.
Potato Diseases, NIAB/PMB.
Potato Diseases, HMSO.

APPENDIX 1

WORLD CROP PRODUCTION

THE DEMAND for food supplies is continually increasing to supply the increasing world population and a higher standard of living in developing countries. Agricultural scientists are likely to play an increasingly important role in this extra production, but the situation is bedevilled by political, social, economic and religious considerations.

The earth's land surface (about 30% of total surface) can be divided approximately as follows: one-fifth is under polar ice; one-fifth desert; one-fifth mountain; one-fifth pasture and forest; one-tenth bare rock; and one-tenth is *arable land*, i.e. about 3% of the total surface area.

The most intensive agriculture in the world is in Imperial Valley, California—an area of about 200,000 hectares.

LAND USE IN THE UNITED KINGDOM

Total area: 24 million ha consisting of approximately:

4.8 million ha (20%)—mountains, forests, urban areas, motorways, etc.
7.2 million ha (30%)—rough grazing, including deer forest.
4.8 million ha (20%)—permanent grassland (7 years +).
2.2 million ha (10%)—temporary grass (leys).
5.0 million ha (20%)—arable crops.

The areas of the various arable crops are approximately:

3.95 million ha (78%)—cereals (1.5 m wheat, 2.3 m barley, 0.15 m others).
0.19 million ha (3.8%)—potatoes.
0.21 million ha (4.2%)—sugar-beet.
0.16 million ha (3.2%)—forage crops (kale, swedes, turnips, etc.).
0.20 million ha (4.0%)—field beans and oil-seed rape.
0.18 million ha (3.6%)—vegetables.
0.06 million ha (1.2%)—fruit.
0.10 million ha (2.0%)—other crops and fallow.

APPENDIX 2

METRICATION

IT HAS been assumed that, when this edition is published, metrication will be well established in agriculture and so only metric units are used throughout the text. There is still considerable doubt about the use of certain units, for example, when metrication was introduced it was clearly understood that centimetres should only be used in very exceptional circumstances, but they are now being commonly used in many cases, e.g. spacing of crop plants. In the text, centimetres are used where it would seem they are more sensible than millimetres (note, 1 centimetre = 10 millimetres).

Many farmers and other agriculturists are likely to use Imperial units for many years to come and so the following simplified conversion factors may be helpful. They are correct to within 2% error, which is acceptable in agriculture!

The following factors are approximate conversions for easy mental calculations; for accuracy, reference should be made to conversion tables. To convert metric to Imperial the multiplying factors should be inverted.

imperial	metric	multiply by	example
inches	to millimetres	$100/4$	$12\,\text{in} = 300\,\text{mm}$
feet	to metres	$3/10$	$30\,\text{ft} = 9\,\text{m}$
yards	to metres	$9/10$	$100\,\text{yd} = 90\,\text{m}$
square feet	to square metres (m^2)	$1/11$	$55\,\text{sq ft} = 5\,\text{m}^2$
square yards	to square metres (m^2)	$10/12$	$60\,\text{sq yd} = 50\,\text{m}^2$
cubic feet	to cubic metres (m^3)	$11/400$	$800\,\text{cu ft} = 22\,\text{m}^3$
cubic yards	to cubic metres (m^3)	$3/4$	$12\,\text{cu yd} = 9\,\text{m}^3$
nos. per foot	to nos. per metre	$10/3$	$30\,\text{per ft} = 100\,\text{per m}$
nos. per sq ft	to nos. per sq metre	11	$20/\text{sq ft} = 220/\text{m}^2$
lb per cu ft	to kg per cu metre	$16\,(4 \times 4)$	$50\,\text{lb/ft}^3 = 800\,\text{kg/m}^3$
lb per bushel	to kg per hectolitre	$10/8$	$60\,\text{lb/bush} = 75\,\text{kg/hl}$
pounds	to kilogrammes	$9/20$	$100\,\text{lb} = 45\,\text{kg}$
chain (22 yards)	= chain (20 m)		
acres	to hectares	$4/10$	$16\,\text{acres} = 6.4\,\text{ha}$
nos. per acre	to nos. per hectare	$10/4$	$40/\text{acre} = 100/\text{ha}$
fert. units/ac	to kg per hectare	$10/8$	$40\,\text{units/ac} = 50\,\text{kg/ha}$
cost per unit	to cost per kilogramme	2	$9\text{p/unit} = 18\text{p/kg}$

imperial	metric	multiply by	example
tons	to tonnes	1	for accuracy, $\times$ by 1.016
cwt	to kilogrammes	50	for accuracy, $\times$ by 50.8
tons per acre	to tonnes per hectare	$10/4$	2 tons/ac = 5 tonnes/ha
cwt per acre	to tonnes per hectare	$1/8$	32 cwt/ac = 4 tonnes/ha
cwt per acre	to kg per hectare	$1000/8$	2 cwt/ac = 250 kg/ha
pounds per acre	to kg per hectare	$11/10$	10 lb/ac = 11 kg/ha
pints per acre	to litres per hectare	$7/5$ or $11/8$	5 pints/ac = 7 litres/ha
gallons per acre	to litres per hectare	11	20 gal/ac = 220 l/ha
tons per cu yard	to tonnes per cu metre	$4/3$	1 ton/yd^3 = 1.3 t/m^3
pints	to litres	$4/7$	7 pints = 4 litres
gallons	to litres	$9/2$	1100 gal = 5000 litres
lb per gallon	to kg per litre	$1/10$	4 lb/gal = 0.4 kg/litre
pence per lb	to pence per kg	$20/9$	18p/lb = 40p/kg
horsepower	to kilowatts	$3/4$	100 h.p. = 75 kW
acres per hour	to hectares per hour	$4/10$	5 ac/hour = 2 ha/hour
lb ft in^2 (psi)	to kilopascals (kPa)	7	28 psi = 196 kPa (2 bars)
lb ft in^2 (psi)	to bars	$7/100$	200 psi = 14 bars

Irrigation inch/acre = cm/ha = 100 m^3

APPENDIX 3

AGRICULTURAL LAND CLASSIFICATION IN ENGLAND AND WALES

THE 11 million hectares of agricultural land in England and Wales has been classified into five grades by the Land Service of the Ministry of Agriculture—for use by planners and agriculturists.

This survey work is published on coloured O.S. maps on two scales:

(a) 1:63,360, i.e. 1 inch to 1 mile.
(b) 1:250,000, i.e. 1 inch to 4 miles. There are seven of these, each covering approximately one MAFF region.

The details shown in (b) are more generalized than in (a).

The classification is based mainly on:

climate—rainfall, transpiration, temperature and exposure.
relief—altitude, slope, surface irregularities.
soil—wetness, depth, texture, structure, stoniness and available water capacity.

These characteristics can affect:

the range of crops that can be grown,
the level of yield,
the consistency of the yield and
the cost of obtaining it.

The following is a brief description of the five grades:

Grade 1 land has very little or no physical limitations for producing high yields of most agricultural and horticultural crops. It occupies less than 3% of all agricultural land in England and Wales and is coloured *dark blue* on the maps.

Grade 2 (14% and coloured *light blue*) has some minor limitations and may have some restrictions in the range of horticultural and arable root crops which may be grown successfully.

Grade 3 (50% and coloured *green*) has moderate limitations due to soil, relief and/or climate. Grass and cereals are the principal crops, the range of horticultural and root crop is restricted. This grade is being reclassified into three sub-grades.

Grade 4 (20% and coloured *yellow*) has severe limitations due to adverse soil, relief and/or climate and is only suitable for low-output enterprises.

Grade 5 (14% and coloured *light brown*) has very severe limitations due to adverse soil, relief and/or climate and is generally under grass or rough grazing.

Urban areas are coloured *red*, and

Non-agricultural areas are coloured *orange* on the maps.

Additionally, the MAFF is undertaking a physical classification of the hill and upland areas of grades 4 and 5, to be divided into two main categories: (see Booklet 2358)

The Uplands—the enclosed and wholly or partially improved land; there will be five sub-grades.

The Hills—unimproved areas of natural vegetation; there will be six sub-grades.

The physical factors to be taken into account will be: vegetation, gradient, irregularity, and wetness.

The object of this classification is to help the farmer to plan possible improvement schemes, and also, to be a guide to those concerned with conservation and amenity who are trying to preserve the scenery and ecology.

Booklets giving full details of all the classifications and grades and maps can be obtained from the MAFF.

SOIL SERIES AND SOIL SURVEY MAPS

In the grading of land for the Land Classification Maps and Reports, use was made of the maps which show the Soil Series classification of England and Wales. A "soil series" is a group of soils formed from the same, or similar, parent materials, and having similar horizons (layers) in their profiles. Each soil series is given a name, normally that of a place near where it is commonly found and usually where it was first recognized and described. The name is used for the soil group, however widespread, throughout the country, and is also used on the maps which are produced on 1:63,000 and/or 1:250,000 scales. A few examples of the named soil series are:

Romney series. These soils are deep, very fine, sandy loams found in the silt areas of Romney Marsh, and also in parts of Cambridgeshire, Lincolnshire and Norfolk. They are potentially very fertile soils.

Bromyard series. Red-coloured, silt loam soils (red marls) found in Herefordshire and other parts of the West Midlands and in the South-west. They should not be worked when wet. Suitable crops are cereals, grass, fruit and hops.

Evesham series. Lime-rich soils formed from Lias (or similar) clays, found in parts of Warwickshire, Gloucestershire, Somerset, and East and West Midlands. They are normally very heavy soils and are best suited to grass and cereals.

Sherborne series. Shallow (less than 25 cm deep), reddish-brown, loam-textured, soils of variable depth and stoniness (and correspondingly subject to drought). They are mainly found on the soft Oolitic limestones of the Cotswolds, and in part of Northamptonshire, and the Cliff region of Lincolnshire. The soils of this series are moderately fertile, easy to manage, and mainly grow cereals and grass, but also some root crops and potatoes, e.g. the good quality *King Edwards* in Lincolnshire.

Worcester series. Red silt loam (or silty clay loam) formed from Keuper Marl and found in the East and West Midlands and in the South-west. They are slow-draining, require subsoiling regularly, and best suited to grass and cereals.

Newport series. Free-draining, deep, easy-working, sandy loams over loamy sands and with varying amounts of stones. They are formed from sands or gravels of sandstone or glacial origin and found in many areas of the Midlands and North as well as Shropshire. They are well suited to arable cropping.

Several different soil series may be found on the same farm and sometimes in the same field.

The *Land Use Capability Classification* in the U.K. has been modelled on the United States Department of Agriculture Classification Scheme and uses the soil series groups in conjunction with limitations imposed by wetness (w), soils (s), gradient and soil pattern (g), erosion (e) and climate (c).

The Classification divides all land into seven classes (classes 1–4 are very similar to grades 1–4 of the Land Classification Maps prepared by the Land Service of ADAS (see page 216); classes 6 and 7 are virtually useless for agriculture). Each class has sub-divisions and limitations as indicated by the letters w, s, g, e or c, after the class number, e.g. class 2w. This classification is being carried out by staff of the Soil Survey which is based at Rothamstead Experimental Station, Harpenden, Herts., with the assistance of ADAS, and is considered to be of greater value to the farm adviser (and planners) than the Land Classification Maps! Maps and booklets are published as the survey work is completed.

SOIL TEXTURE ASSESSMENT IN THE FIELD

The texture of a soil (i.e. the amount of sand, silt, clay and organic matter present) can be measured by a mechanical analysis of a representative sample of the soil in a laboratory, but this may take a long time to organize. It is therefore very important for the farm adviser (and possibly the farmer himself) to be able to assess the texture of the soil in the field, not only as a guide for cultivations and general management, but also because soil-acting pesticides (mainly herbicides) are becoming increasingly important. Many of these chemicals are adsorbed by the clay and/or organic matter in the soil and so higher dose rates may be required on soils which are rich in these materials, and, also, on sandy soils, surface-applied residual herbicides may be washed into the root zone of the crop too easily and so cause damage.

With practice, it is possible for persons with sensitive fingers to become reasonably skilled at assessing soil texture by feeling the soil in the field. The following is a general guide only to how this is done:

Carefully moisten a handful of stone-free soil until the particles cling together (avoid excess water) and work it well in the hand until the structure breaks down; now rub a small amount between the thumb and finger(s) to assess the texture according to how gritty, silky or sticky the sample feels. The handful of moist soil can also be assessed by the amount of polish it will take, and the ease or difficulty it takes to get into a ball and other shapes.

Sands feel gritty, but are not sticky when wet (loose when dry) and do not stain the fingers. Four grades may be distinguished from the coarseness or fineness of the gritty material in the sample.

Clays (at the other extreme of particle sizes) take a high polish when rubbed, are very sticky, bind together very firmly, and are very difficult to mould into shapes. Three grades may be distinguished, depending on the amounts of sand (grittiness) and/or silt present.

Silty soils have a smooth silky feel and the more obvious this is, the greater is the amount of silt present. The amount of polish the sample takes, and its grittiness, are guides to the amounts of clay and sand present.

Loams have a fairly even mixture of sand, clay and silt, and, because these tend to balance one another, these soils are not obviously gritty, silky or sticky, and take only a slight polish. A ball of moist loam soil is easily formed, and the particles bind together well.

These are main texture grades, but a wide range of intermediate grades exist, each having different amounts of sand of various sizes, silt and clay particles. All this can be complicated by the amount of organic matter present (it has a soft silky feel and is usually dark brown or black in colour).

Loamy sands and *sandy loams* are intermediate groups between sands and loams; they feel gritty because of the sand present but vary in the amount of stain left on the fingers, and ease of moulding and stickiness; there are four grades of each, based on the size of the sand particles.

Clay loams are intermediate between clays and loams and are sticky, bind strongly together and are difficult to mould.

Skill in recognizing all the grades of soil, by feeling and rubbing, usually only comes from long experience and by comparing unknown soils with standard samples of known texture. Further details on assessing soil texture are given in soil textbooks (MAFF leaflet 796).

ADAS classify soils for their herbicide advisory work into the following textural groups (clay % in brackets):

Sands (0–5%) including coarse sand, sand, fine sand, very fine sand and loamy coarse sand.
Very light soils (5–10%), loam sand, loamy fine sand and coarse sandy loam.
Light soils (10–20%), loamy very fine sand, sandy loam and fine sandy loam.
Medium soils (20–25%), very fine sandy loam, silty loam, loam and sandy clay loam.
Heavy soils (30–40%), clay loam, silt loam and silty clay loam.
Very heavy soils (40–55%), sandy clay, clay and silty clay.

With the addition of *organic soils*, and the exclusion of the sands and very heavy soils group (usually only found in subsoils), the above groups are used by advisers for dose rate recommendations (see *Approved Products Handbook* or booklets 2254, 2256).

These groups are also widely used in advisory work for assessing available-water capacity, suitability for mole drainage, workability and stability of soils.

APPENDIX 4

FACTORS AFFECTING THE APPLICATION (AND MIXING) OF CROP PROTECTION CHEMICALS

(a) *Weather conditions.* Normally, dry settled conditions are best for most spray chemicals. The effects of rain after spraying vary according to the chemical involved—some, such as paraquat ("Gramoxone"), are not affected by rain shortly after application, whereas others such as glyphosate ("Roundup") and contact herbicides require at least 6–8 hours of dry weather if they are to act effectively. Some wild oat herbicides may benefit from some very light rain after application to concentrate the herbicide on the lower part of the leaf blade where it acts more effectively. Frost on leaves at time of spraying could affect the intake of the chemical and cold, and poor-growing conditions reduce the effectiveness of growth-regulating herbicides such as MCPA, 2,4-D, etc.

Windy conditions prevent uniform application and may cause problems with drift onto neighbouring crops. Following an application of some ester-formulated herbicides, a day or two of warm weather may cause these to become volatile and drift onto and damage susceptible neighbouring crops, e.g. oil-seed-rape.

(b) *Soil conditions.* Soil-acting residual herbicides act best when the soil is in a damp, finely-divided condition.

Volatile chemicals, e.g. "Avadex BW", "Eptam", trifluralin, must be worked into the soil which should be in a free-working condition and as free of stones and clods as possible.

Large flat stones on the surface of the soil may cause soil-applied herbicides to accumulate in crop-damaging concentrations around the edges of the stones.

Wet conditions may prevent the use of ordinary ground sprayers and so aircraft (for approved chemicals) or special low ground pressure vehicles (LGPV) may have to be used.

(c) *Formulation.* Most spray chemicals are formulated in water solutions to be diluted with water, but the less soluble ones are often formulated as wettable powders which have to be carefully mixed with some water before adding to the tank (these materials are now being marketed ready-mixed with water to make it easier for the operator to use them). Some chemicals which are only soluble in oil are formulated as emulsions.

A few chemicals, e.g. "Avadex", may also be formulated as very small granules to be applied on the soil surface by a special applicator.

Some are formulated as very fine powders which are made to stick to the leaves by electrostatic charges.

Some herbicides, e.g. "Envoy", are formulated as very soluble free-flowing "dry liquid" granules.

Many factors influence the formulation of a product, e.g. the properties of the active ingredient, transport and storage stability, ease of application, activity against weeds, pests or diseases, crop tolerance, all aspects of safety, ease of manufacture and cost. Pesticide formulations are usually carefully guarded secrets.

(d) *Mixing two or more chemicals.* The application of a single product for the control of weeds, pests or diseases is a fairly simple operation provided the weather and soil conditions are suitable and the sprayer is in good working order.

Problems arise where several chemicals are required about the same time—such as one or more herbicides, fungicides, insecticides and, possibly, chlormequat ("Cycocel"), some trace elements, and even liquid fertilizer. Some of these can be mixed without reducing the desired effects (in some cases an improved effect may be obtained) but in many cases it is very risky to mix chemicals either because of reduced effect or because of damaging effects on the crop such as scorch, yield reduction, toxicity, etc.

Information is usually supplied with each product about its compatibility with some other products but this is of limited value. Some advisers have a lot of experience with mixing various chemicals but this knowledge is not readily available to everyone.

Where products from two or more firms are mixed and something goes wrong, it is unlikely that any of the firms will accept any responsiblity for any damage which may have been caused unless the mixture was approved on the leaflets.

Independent advisers and salesmen making recommendations on mixing chemicals are usually insured against possible damage occurring. There is very little independent research being done on this and usually where there is no sign of damage and the weeds, pests and/or diseases are controlled, then it is often assumed that all is well. It may be that yields are reduced without it being realized. It is usually safe to mix the chelated forms of trace elements with other chemicals.

One of the main causes of scorch damage resulting from mixing chemicals is the excessive quantity of wetting agents present in the mixture.

(e) *Method of application.* The ordinary agricultural sprayers are fitted with hydraulic nozzles to control the flow and disperse the liquid into drops. There are two types of nozzle; the fan type (commonest) and the hollow cone (or swirl) type. In both types the liquid is forced through the orifice as a sheet which then breaks into drops of various sizes—some so small that they drift away, and others so large that they bounce off the plants and are useless with chemicals which should remain on the foliage. The farm sprayer is used for applying many different chemicals for many different purposes and the results are reasonably acceptable because, for most purposes, there will be some drops of the correct size for the application of the chemical being used. This is wasteful of chemicals in cases where only a part of the spray is retained on the foliage but the dose rates recommended allow for this loss. The need for very accurate, uniform coverage is not so important with systemic chemicals which can move through the plant or with soil-acting residual chemicals, but it is important with contact chemicals such as some herbicides, insecticides and fungicides which only act at the point of contact.

The research and development work with tooth-edged spinning discs, starting with the Micron Herbi and Ulva hand-held battery-operated applicators, and the NIAE types, showed that this method of application could produce droplets of very uniform size and that the size of droplet could be varied, as required, by changing the speed of the discs. This is known as CDA (controlled droplet application) and much development work is now concentrated on producing sophisticated tractor-mounted or light self-propelled machines. When more of these machines are in use it is likely that chemicals will be specially formulated for them (see Leaflet 792).

Applicators are being developed which produce charged droplets which readily stick to leaves—these are of very limited use for herbicides but have great possibilities for fungicides and insecticides.

Chemicals such as "Roundup"—sometimes specially formulated—may be selectively applied by special roller machines, e.g. "Adville", or wick rope, e.g. "Weed-Wiper", to tall weeds above the level of the crop leaves, e.g. weed beet in sugar-beet, tussock grass and thistles in grassland; there are no drift problems with these machines and chemical is only taken up by the patches of weeds.

The volume rates normally used for spraying farm chemicals are:

high volume	650–1100 litres/hectare
medium volume	220–650 litres/hectare
low volume	55–220 litres/hectare
very low volume	up to 55 litres/hectare
ultra low volume (ULV) and CDA	10–50 litres/hectare

Farm sprayers are available in many sizes and forms. Those with very wide spray booms are difficult to use on undulating ground. Many ingenious devices have been developed to prevent the booms swinging about and bouncing up and down excessively. The "tramline" system (see STL 189) for spraying cereal crops has solved many of the problems of covering a field quickly and accurately (avoiding overlaps and misses). Reference has also been made to the practice of spraying on page 178.

(f) *Safety and approval* of spray chemicals. It is very important that the instructions on the leaflets which accompany spray chemicals should be followed as carefully as possible, especially when dealing with the more poisonous types. Anyone who has been working with pesticides and feels ill should see a doctor or go to hospital as soon as possible and bring the spray leaflets so that proper treatment can be given at once; many of the leaflets include instructions for doctors.

The *Pesticides Safety Precautions Scheme (PSPS)* is a voluntary arrangement between the chemical industry and the government whereby no pesticide product is put on the market until conditions have been agreed by the government which ensure that there should be no risk to users or consumers of treated crops or foodstuffs and that the risks to wild life are minimal. Chemicals included in the scheme should only be used on named crops and in the manner specified. Mixtures should also have PSPS clearance.

The *Agricultural Chemicals Approval Scheme (ACAS)* for herbicides, fungicides and insecticides is a voluntary scheme operated on behalf of the Departments of Agriculture in the U.K. by the Agricultural Chemicals Approval Organization at Harpenden. Approval is granted by the Organization for specific uses under U.K. conditions when it is satisfied that the product fulfils the claims made on the label (this can be reviewed from time to time). Approval is not given to a product which is not cleared under PSPS.

Further details of these schemes and other regulations covering the sale and use of poisonous substances are given in the Approved Products booklet which is revised annually and obtainable from HMSO. It is sometimes known as the "orange bible" in the trade!

(See also MAFF leaflet APS/1, *The safe use of poisonous chemicals on the farm.*)

The increasing research and development work on PGRs (plant growth regulators—auxins, gibberellins, cytokinins, abscisins, ethylene) to control or alter the normal growth of crop plants in many different ways is likely to increase the complexities of crop spraying in the next decade; chemicals such as chlormequat (an anti-gibberellin) which has straw-strengthening and other effects on cereals, "Cerone", and "Terpal", are just the beginning!

Computers might help to solve some of the problems, and create others!

APPENDIX 5

CROPS

Crop names	Botanical (Latin) names
Cereals: wheat	*Triticum aestivum (T. vulgare)*
barley	*Hordeum sativum*
oats	*Avena sativa*
rye	*Secale cereale*
maize	*Zea mays*
Potato	*Solanum tuberosum*
Sugar-beet, mangel, fodder beet	*Beta vulgaris*
Cabbage group	*Brassica oleracea*
Savoy	*Brassica oleracea* var. *bullata*
drumhead	*Brassica oleracea* var. *capitata*
red (white)	*Brassica oleracea* var. *rubra (alba)*
Brussels sprouts	*Brassica oleracea* var. *bullata* s.var. *gemmifera*
Cauliflower	*Brassica oleracea* var. *botrytis*
Sprouting Broccoli, Calabrese	*Brassica oleracea* var. *botrytis* var. *italica*
Kohlrabi	*Brassica oleracea* var. *gongylodes*
Kale—marrow stem	*Brassica oleracea* var. *acephala* s.var *medullosa*
thousand head	*Brassica oleracea* var. *acephala* s.var *millicapitata*
curly	*Brassica oleracea* var. *acephala* s.var *laciniata*
Turnip group	*Brassica rapa* var. *rapa*
turnip oil-seed rapes	*Brassica campestris* var. *oleifera*
Swede group	*Brassica napus*
Swedes	*Brassica napus* var. *napobrassica*
Swede forage rape	*Brassica napus* var. *oleifera*
Swede oil-seed rapes	*Brassica napus* var. *oleifera*
hungry gap and rape kale	*Brassica napus*
Mustard—brown	*Brassica juncea*
white	*Sinapsis alba (Brassica alba)*
Fodder radish	*Raphanus sativus* var. *campestris*
Buckwheat	*Fagopyrum esculentum*
Carrot	*Daucus carota*
Parsnip	*Pastinaca sativa*
Celery	*Apium graveolens*
Onion	*Allium cepa*
Pea	*Pisum sativum*
Beans—field and broad	*Vicia faba*
green, dwarf, French	*Phaseolus vulgaris*
runner	*Phaseolus coccineeus*
soya	*Glycine max*
Vetch (tares)	*Vicia sativa*
Lupins—yellow (white)	*Lupinus luteus (L. albus)*
pearl (blue)	*Lupinus mutabilis (L. augustifolius)*
Lucerne (alfalfa)	*Medicago sativa*
Sainfoin	*Onobrychis viciifolia*
Linseed and Flax	*Linum usitatissimum*
Sunflower	*Helianthus* spp.
Grasses: ryegrass—Italian	*Lolium multiflorum*
hybrid	*Lolium (multiflorum* X *perenne)*
perennial	*Lolium perenne*

Grasses:	cocksfoot	*Dactylis glomerata*
	Timothy	*Phleum pratense*
	meadow fescue	*Festuca pratense*
	tall	*Festuca arundinacea*
	red	*Festuca rubra*
Clovers:	red	*Trifolium pratense*
	white	*Trifolium repens*
	alsike	*Trifolium hybridum*
	crimson	*Trifolium incarnatum*

APPENDIX 6

WEEDS

(A) annual, (B) biennial, (P) perennial

Common names	Botanical (Latin) names
Allseed (A)	*Chenopodium polysperum*
Amphibious bistort (P)	*Polygonum amphibium*
Autumn crocus (P)	*Crocus nudiflorus*
Barley grass—meadow (P)	*Hordeum secalinum*
wall (A)	*Hordeum murinum*
Bent—black (P)	*Agrostis gigantea*
common (P)	*Agrostis tenuis*
creeping (P) (Watergrass)	*Agrostis stolonifera*
Bindweed—black (A)	*Polygonum convolvulus*
field (P)	*Convolvulus arvensis*
Birdsfoot trefoil—common (P)	*Lotus corniculatus*
Blackgrass (A)	*Alopecurus myosuroides*
Bracken (P)	*Pteridium aquilinum*
Bristly oxtongue (A) or (B)	*Picris echioides*
Brome—soft (A) or (B)	*Bromus mollis*
sterile (A) or (B)	*Bromus sterilis*
field (A) or (B)	*Bromus arvensis*
erect (upright) (P)	*Bromus erectus*
Broomrape—common	*Orobanche minor*
Burdock—greater (B)	*Arctium lappa*
Burnet—salad (P)	*Poterium sanguisorba*
fodder (P)	*Poterium polygamum*
Buttercup—bulbous (P)	*Ranunculus bulbosus*
corn (A)	*Ranunculus arvensis*
creeping (P)	*Ranunculus repens*
meadow (crowfoot) (P)	*Ranunculus acris*
Campion—white (B)	*Silene alba* (also red (B) and bladder (P))
Campion (A)	*Silene noctiflora*
Carrot—wild (A)	*Daucus carota*
Cat's ear (P)	*Hypochaeris radicata*
Chamomile—corn (A) or (B)	*Anthemis arvensis*
Charlock (yellow) (A)	*Sinapsis arvensis*
Chervil—rough (B)	*Chaerophyllum temulentum*
Chickweed—common (A)	*Stellaria media*
mouse-ear (A)	*Cerastium holosteoides*
Cleavers (A)	*Galium aparine*
Coltsfoot (P)	*Tussilago farfara*
Cornflower (A)	*Centaurea cyanus*
Corn mint (P)	*Mentha arvensis*
Cornsalad—common (A)	*Valerianella locusta*
Couch—common (P)	*Agropyron repens*
onion (false oat-grass) (P)	*Arrhenatherum elatius* (var. *bulbosum*)
Cowbane (water hemlock) (P)	*Cicuta virosa*

Common names	Botanical (Latin) names
Cow parsley (wild chervil) (P)	*Anthriscus sylvestris*
Cranesbills (several) (A)	*Geranium* species
Cuckooflower (Ladies' Smock) (P)	*Cardamine pratensis*
Daisy (P)	*Bellis perennis*
—oxeye dog (P)	*Chrysanthemem leucanthemum*
Dandelion (P)	*Taraxacum officinale*
Darnel (A)	*Lolium temulentum*
Dead-nettle—red (A)	*Lamium purpureum*
Docks—broad-leaved (P)	*Rumex obtusifolius*
—curled (P)	*Rumex crispus*
Duckweed	*Lemna minor*
Fat-hen (A)	*Chenopodium album*
Fescue—red (P)	*Festuca rubra*
rat's tail (A)	*Festuca* or *Vulpia myuros*
Field speedwell (Buxbaums) (A)	*Veronica persica*
Fleabane—common (P)	*Pulicaria dysenterica*
Flixweed (A)	*Descurainia sophia*
Fool's parsley	*Aethusa cynapium*
Forget-me-not—field (A)	*Myosotis arvensis*
Foxglove (B) or (P)	*Digitalis purpurea*
Foxtail—meadow (P)	*Alopecurus pratensis*
Fritillary (P)	*Fritillaria meleagris*
Fumitory—common (A)	*Fumaria officinalis*
Gallant soldier (A)	*Galinsoga parviflora*
Garlic—field (P)	*Allium oleraceum*
Gromwell—corn or field (A)	*Lithospermum arvense*
Ground elder (Bishop's weed) (P)	*Aegopodium podagraria*
Ground ivy (P)	*Glechoma hederacea*
Groundsel (A)	*Senecio vulgaris*
Hawkbit—autumn (P)	*Leontodon autumnalis*
rough (P)	*Leontodon hispidus*
Hawksbeard—rough (A) or (B)	*Crepis biennis*
smooth (A) or (B)	*Crepis capillaris*
Heart's ease (A)	*Viola tricolor*
Heather (P)	*Calluna vulgaris*
—bell (P)	*Erica cinerea*
Hedge mustard (A)	*Sisymbrium offininale*
Hedge parsley (A)	*Torilis japonica*
Hemlock (A) or (B)	*Conium maculatum*
Hemp nettle (A)	*Galeopsis tetrahit*
Henbane (A)	*Hyoscyamus niger*
Henbit (A)	*Lamium amplexicaule*
Hoary cress or pepperwort (P)	*Lepidrum draba*
Hogweed (B)	*Heracleum sphondylium*
Horsetail—field (P)	*Equisetum arvense*
marsh (P)	*Equisetum palustre*
Knapweed (P)	*Centaurea nigra*
Knawel—annual (A)	*Scleranthus annuus* (also *perennial*)
Knotgrass (A)	*Polygonum aviculare*
Marigold—corn (A)	*Chrysanthemum segetum*
Mayweed—scented (A)	*Matricarla recutita*
scentless (A)	*Tripleurospermium maritimum inodorum*
Meadow grass—annual (A)	*Poa annua*
rough-stalked (P)	*Poa trivialis*
smooth stalked (P)	*Poa pratensis*
Medick—black (A)	*Medicago lupulina*
Mercury—annual (A)	*Mercurialis annua*
dog's (P)	*Mercurialis perennis*
Mignonette—wild, cut-leaved (A) or (P)	*Reseda lutea*

Common names	Botanical (Latin) names
Mugwort (P)	*Artemisia vulgaris*
Mustard—black (A)	*Brassica nigra*
white (A)	*Sinapsis alba*
Nettle—common stinging (P)	*Urtica dioica*
small, annual (A)	*Urtica urens*
Nightshade—black (A) or (B)	*Solanum nigrum*
deadly (P)	*Atropa belladonna*
Nipplewort (A)	*Lapsana communis*
Oat—bristle (greys) (A)	*Avena strigosa*
spring wild (A)	*Avena fatua*
winter wild (A)	*Avena ludoviciana*
Oat grass—downy (P)	*Helictotrichon pubescens*
false (onion couch) (P)	*Arrhenatherum elatius* (var. *bulbosum*)
Onion—wild (crow garlic) (P)	*Allium vineale*
Orache—common (A)	*Atriplex patula*
Pansy—field (A)	*Viola arvensis*
wild (A)	*Viola tricolor*
Parsley—cow (B) or (P)	*Anthriscus sulvestris*
fool's (A)	*Aethusa cynapium*
Parsley-piert (A)	*Aphanes arvensis*
Pearlwort (A) or (P)	*Sagina procumbens*
Pennycress—field (A)	*Thlaspi arvense*
Persicaria—pale (A)	*Polygonum lapathifolium*
Pineappleweed (A)	*Matricaria matricariodes*
Plantain—greater (P)	*Plantago major*
ribwort (narrow-leaved) (P)	*Plantago lanceolata*
Poppy—corn (A)	*Papaver rhoeas*
opium (A)	*Papaver somniferum*
californian (A)	*Eschscholzia californica*
Primrose (P)	*Primula vulgaris*
Radish—wild (white charlock) (A) or (B)	*Raphanus raphanistrum*
Ragwort—common (B) or (P)	*Senecio jacobia*
marsh (B)	*Senecio aquaticus*
Ramsons (P)	*Allium ursinum*
Redshank (A)	*Polygonum persicaria*
Reed—common (P)	*Phragmites australis*
Restharrow—common (P)	*Ononis repens*
spiny (P)	*Onions spinosa*
Rush—jointed (P)	*Juncus articulatus*
soft (common) (P)	*Juncus effusus and conglomeratus*
hard (P)	*Juncus inflexus*
heath (P)	*Juncus squarrous*
Saffron—meadow (P)	*Colchicum autumnale*
St. John's Wort (P)	*Hypericum perforatum*
Scabious—field (P)	*Knautia arvensis*
Scarlet pimpernel (A)	*Anagallis arvensis*
Sedges (P)	*Carex* spp.
Selfheal (P)	*Prunella vulgaris*
Shepherd's-needle (A)	*Scandix pecten-veneris*
Shepherd's-purse (A)	*Capsella bursa—pastoris*
Silverweed (P)	*Potentilla anserina*
Soft-brome (A) or (B)	*Bromus mollis*
Soft-grass—creeping (P)	*Holcus mollis*
Sorrel—common (P)	*Rumex acetosa*
sheep's (P)	*Rumex acetosella*
Sow-thistle—corn (P)	*Sonchus arvensis*
common (A)	*Sonchus oleraceus*

Common names	Botanical (Latin) names
Speedwell (common, field, Buxbaum's) (A)	*Veronica persica*
germander (P)	*Veronica chamaedrys*
procumbent (A)	*Veronica agrestis*
ivy-leaved (A)	*Veronica hederifolia*
Spurge—sun (A)	*Euphorbia helioscopia*
dwarf (A)	*Euphorbia exigua*
Spurrey—corn (A)	*Spergula arvensis*
Stork's-bill—common (A) or (B)	*Erodium cicutarium*
Thistle—creeping (field) (P)	*Cirsium arvense*
spear (Scotch) (B)	*Cirsium vulgare*
Toadflax—common (P)	*Linaria vulgaris*
Travellers'-joy (old man's beard) (P)	*Clematis vitalba*
Treacle mustard (A)	*Erysimum cheiranchoides*
Trefoil hop (A)	*Trifolium campestre*
Tussock grass (P)	*Deschampsia cespitosa*
Venus's-looking-glass (A)	*Legousia hybrida*
Vetch—common (tares) (A) or (B)	*Vicia sativa*
kidney (P)	*Anthyllis vulneraria*
Viper's bugloss (B)	*Echium vulgare*
Water-dropwort—hemlock (P)	*Oenanthe crocata*
Watergrass (P)	*Agrostis stolonifera*
Willow-herb—rosebay (P)	*Epilobium angustifolium*
Woodrush—field (P)	*Luzula campestris*
Yarrow (P)	*Archillea millefolium*
Yellow borage	*Amsinckia intermedia*
Yellow Iris or Flag (P)	*Iris pseudacorus*
Yellow rattle (A)	*Rhinanthus minor*
Yorkshire fog (P)	*Holcus lanatus*

APPENDIX 7

CROP DISEASES

Crop	Common names	Latin names
Cereals		
All	mildew	*Erysiphe graminis*
Wheat (barley)	yellow rust	*Puccinia striiformis*
Barley	brown rust	*Puccinia hordei*
Wheat	brown rust	*Puccinia recondita*
Oats	crown rust	*Puccinia coronata*
Barley	leaf blotch (Rhyncosporium)	*Rhynchosporium secalis*
Wheat (barley)	glume blotch	*Septoria nodorum*
Wheat (barley)	leaf spot	*Septoria tritici*
Oats	dark leaf spot, speckle blotch	*Leptosphaeria avenaria*
Oats	stripe, leaf spot, seedling blight	*Pyrenophora avenae*
Barley	leaf stripe	*Pyrenophora graminae*
Barley	net blotch	*Pyrenophora teres*
Oats	halo blight	*Pseudomonas coronafaciens*
Rye, wheat, barley	ergot	*Claviceps purpurea*
Wheat	bunt, covered, stinking smut	*Tilletia caries*
Barley (oats)	bunt, covered, smut	*Ustilago hordei*
Oats	bunt, covered, smut	*Ustilago avenae*
Maize	bunt, covered, smut	*Ustilago maydis*
Wheat, barley	loose smut (different races on wheat and barley)	*Ustilago nuda*
Wheat, barley, oats	eyespot	*Pseudocercosporella herpotrichoides*
Wheat, barley, oats	sharp eyespot	*Corticium (Rhizoctonia) solani*
Wheat, barley	take-all	*Gaeumannomyces graminis* v. *graminis*
Oats	take-all	*Gaeumannomyces graminis* v. *avenae*
Wheat, barley, oats	brown foot rots (and ear blight)	*Fusarium species*
Wheat, barley	black (sooty) mould	*Cladosporium herbarum*
Wheat (barley), oats, maize	scab	*Gibberella zeae*
Maize	stem rot	*Fusarium species*
Potatoes		
	blight	*Phytophthora infestans*
	pink rot	*Phytophthora erythroseptica*
	common scab	*Streptomyces scabies*
	powdery scab	*Spongospora subterranea*
	gangrene	*Phoma exigua* var. *foveata*
	watery wound rot	*Pythium ultimum*
	wart disease	*Synchytrium endobioticum*
	sclerotinia	*Sclerotinia sclerotiorum*
	blackleg	*Erwinia (carotovora)* v. *atroseptica*
	skin spot	*Oospora pustulens*
	black scurf and stem canker	*Corticium (Rhizoctonia) solani*
	dry rot	*Fusarium caeruleum*
	silver scurf	*Helminthosporium solani*
	ring rot	*Corynebacterium sepedonicum*

Crop	Common names	Latin names
Sugar-beet	black leg	*Pleospora bjoerlingii*
	downy mildew	*Peronospora farinosa*
	powdery mildew	*Erysiohe* species
	leaf spot	*Ramularia beticola*
	rust	*Uromyces betae*
	violet root rot	*Helicobasidium purpureum*
Legumes		
Peas, beans	downy mildew	*Peronospora vicia*
Peas	leaf and pod spot	*Ascochyta pisi*
Beans	leaf and pod spot	*Ascochyta fabae*
Beans	chocolate spot	*Botrytis cinerea. B. fabae* (severe)
Beans,	stem rot (bean sickness)	*Sclerotinia trifolium*
red clover	clover rot (sickness)	*Sclerotinia trifolium*
All	damping-off of seedlings	*Pythium* species
Peas	pea wilt	*Fusarium oxysporum*
Lucerne	wilt	*Verticillium albo-atrum*
Peas, beans	grey mould	*Botrytis cinerea*
Peas	powdery mildew	*Erysiphe pisi*
Peas	halo blight	*Pseudomonas phaseolicola*
Peas	anthracnose	*Colletotrichum lindemuthianum*
Brassicas	club root (finger and toe)	*Plasmodiophora brassicae*
	powdery mildew	*Erysiphe cruciferarum*
	downy mildew	*Peronospora parasitica*
	white blister	*Cystopus candidus*
	light leaf spot	*Gloeosporium concentricum*
	dark leaf spot	*Alternaria* species
	stem canker	*Phoma lingam*
	ring spot	*Mycosphaerella brassicicola*
	soft rot	*Pectobacterium cartovorum*
	wilt	*Fusarium oxysporum*
Onion	white rot	*Sclerotium cepivorum*
	downy mildew	*Peronospora destructor*
	neck rot	*Botrytis allii*
	leaf spot	*Botrytis* species
Grasses	choke	*Epichloe typhina*
	blind seed disease	*Gloeitinia temulenta*
	ergot	*Claviceps purpurea*
	mildew	*Erysiphe graminis*
	rusts	*Puccinia* species
	leaf fleck	*Mastigosporium rubricosum*

APPENDIX 8

INSECT PESTS

Crop	Common names	Latin names
Cereals		
Wheat, barley, oats	cyst nematode (root eelworm)	*Heterodera avenae*
Oats	stem and bulb eelworm	*Ditylenchis dipsaci*
Wheat	bird cherry aphid	*Rhopalosiphum padi*
	grass aphid	*Metopolophium festucae*
	rose grain aphid	*Metopolophium dirhodum*
	grain aphid	*Macrosiphum avenae*
All	leatherjackets	*Tipula* spp. and *Nephrotoma maculata*
All	wireworms	*Agriotes* species
All	slugs	*Agriolimax reticulatus* (mainly)
Wheat	wheat bulb fly	*Leptohylemyia coarctata*
Wheat	opomyza flies	*Opomyza florum*
Barley	gout fly	*Chlorops pumilionis*
Oats, barley, wheat	frit fly	*Oscinella frit*
	saddle-gall midge	*Haplodiplosis equestris*
	grain weevils	*Sitophilus* species
	saw-toothed grain beetle	*Oryzephilus surinamensis*
	mites	*Acarus siro*
Potatoes	root eelworm (pathotype A)	*Heterodera rostochiensis*
	yellow cyst nematode (pathotype Ro 1)	*Globodera rostochiensis*
	white cyst nematode	*Globodera pallida*
	aphid—peach potato	*Myzus persicae*
	aphid "false top roll"	*Macrosiphum euphorbiae*
	Colorado beetle	*Leptinotarsa decemlineata*
	cutworms (Noctuid moths)	
	e.g. turnip moth	*Agrotis segetum*
	heart and dart moth	*Agrotis exclamationis*
	large yellow underwing	*Noctua pronuba*
	garden dart moth	*Euxoa nigricans*
	slugs, leatherjackets, wireworms—cereals	
Sugar-beet	cyst nematode (root eelworm)	*Heterodera schachtii*
	aphid—peach potato	*Myzus persicae*
	aphid black bean (blackfly)	*Aphis fabae*
	mangold fly	*Pegomyia hyoscyami* var. *betae*
	millipedes	*Blaniulus guttulatus*
	millipedes	*Brachydesmus superus*
	Docking disorder	*Longidorus* and *Trihodorus* species
	cutworms, slugs, wireworms, leatherjackets, see Potato and Cereals	
Peas and Beans	pea and weevil	*Sitona* species
	pea cyst nematode (pea root eelworm)	*Heterodera gottingiana*
	stem and bulb eelworm	*Ditylenchus dipsaci*
	black bean aphid (blackfly)	*Aphis fabae*

Crop	Common names	Latin names
Brassicae	flea beetle	*Phyllotreta* species
	cabbage root fly	*Erioischia brassicae*
	blossom (pollen) beetle	*Meligethes aeneus*
	seed weevil	*Ceuthorhynchus assimilis*
	mealy cabbage aphid	*Brevicoryne brassicae*
	cabbage stem flea beetle	*Psylliodes chrysocephala*
	bladder pod midge	*Dasyneura brassica*
	cabbage white butterflies	*Pieris* species
Onions	onion fly	*Hylemyia antiqua*
Carrots	carrot fly	*Psila rosae*
	carrot willow aphid	*Cavariella aegopodii*

APPENDIX 9

CROP SEEDS

THE FOLLOWING are average figures and are only intended as a general guide and for comparisons. Precision drilling of crops requires seed counts per kilogramme to be known for the stock of seed being sown and merchants will usually supply these figures.

Crop	1000 seeds weight (g)	Seeds per kilogramme (000's)	Seeds per m² for every 10 kg/ha sown	Seeds per m² for every kg/hl sown
Cereals: wheat	48	21	21	75
barley	37	27	27	68
oats	32	31	31	52
maize	285	3.5	3.5	75
Grasses: ryegrass—Italian	2.2	455	455	29
ryegrass—Italian tetra.	4	250	250	
ryegrass hybrid	2.15	465	465	
ryegrass perennial S 23	1.65	606	606	
ryegrass perennial S 24	2	500	500	35
ryegrass perennial tetra.	3.3	303	303	
cocksfoot	1	1000	1000	33
Timothy	0.3	3333	3333	62
meadow fescue	2	500	500	37
tall fescue	2.5	400	400	30
Clovers: red	1.75	571	571	80
white—cultivated	0.62	1613	1613	82
wild	0.58	1724	1724	82
Lucerne (alfalfa)	2.35	425	425	77
Sainfoin—milled	21	48	48	
Peas—marrowfats	330	3	3	
large blues	250	4	4	78
small blues	200	5	5	
Beans—broad	980	1	1	
winter and horse types	670	1.5	1.5	
tick	410	2.4	2.4	80
dwarf, green, French	500	2	2	
Vetches (tares)	53	19	19	77
Linseed and flax	10	100	100	68
Carrots	1.5	660	660	
Onions—natural seed	4	246	246	
mini-pellets	18	55	55	
Sugar-beet—rubbed and graded	14	71	71	
pelleted	62	16	16	
Cabbages	4.1	240	240	
Kale	4.5	220	220	

Crop	1000 seeds weight (g)	Seeds per kilogramme (000's)	Seeds per m² for every 10 kg/ha sown	Seeds per m² for every kg/hl sown
Swedes	3.6	280	280	
Turnips	3	330	330	
Oil-seed rape	4.5	220	220	
Brussels sprouts	4.7	210	210	
Cauliflower	4.1	240	240	

APPENDIX 10

THE DECIMAL CODE FOR GROWTH STAGES OF ALL SMALL GRAIN CEREALS (ZADOK'S AND OTHERS)

THIS precise international system for identifying growth stages of cereals is likely to replace other systems for use in all pesticide experimental work, and, to a lesser extent in advisory work.

In the code, ten principal growth stages are distinguished and each is sub-divided into secondary stages.

Code   0   *Germination.* Sub-divided—00 dry seed to 09 when first leaf reaches tip of coleoptile.

Code   1   *Seedling growth*
  10   1st leaf through coleoptile
  11   1st leaf unfolded
  12   2 leaves unfolded
  13   3 leaves unfolded
  14   4 leaves unfolded
  15   5 leaves unfolded
  16   6 leaves unfolded
  17   7 leaves unfolded
  18   8 leaves unfolded
  19   9 or more leaves unfolded

Code   2   *Tillering*
  20   main shoot only
  21   main shoot and 1 tiller
  22   main shoot and 2 tillers
  23   main shoot and 3 tillers
  24   main shoot and 4 tillers
  25   main shoot and 5 tillers
  26   main shoot and 6 tillers
  27   main shoot and 7 tillers
  28   main shoot and 8 tillers
  29   main shoot and 9 or more tillers

Code   3   *Stem elongation*
  30   leaf sheath erect and exeeds 5 cm in length
  31   1st node detectable (seen or felt after removing outer leaf sheaths)
  32   2nd node detectable (seen or felt after removing outer leaf sheaths)

33   3rd node detectable (seen or felt after removing outer leaf sheaths)
34   4th node detectable (seen or felt after removing outer leaf sheaths)
35   5th node detectable (seen or felt after removing outer leaf sheaths)
36   6th node detectable (seen or felt after removing outer leaf sheaths)
37   flag leaf just visible
38   —
39   flag leaf ligule/collar just visible

For example, a plant having leaves unfolded (15), a main shoot and three tillers (23), and two nodes detectable (32), would be coded as: 15, 23, 32.

Code 4.   40–49 *Booting* stages in development of ear in leaf sheath to when awns are visible (49).
Code 5.   50–59 stages in development of the inflorence.
Code 6.   60–69 stages in development of anthesis (pollination).
Code 7.   70–79 milk development stages in grain.
Code 8.   80–89 dough development stages in grain.
Code 9.   90–99 ripening stages in grain (including dormancy stages).

See also, drawings of some vegetative stages to illustrate the coding system (page 171) and MAFF publications.

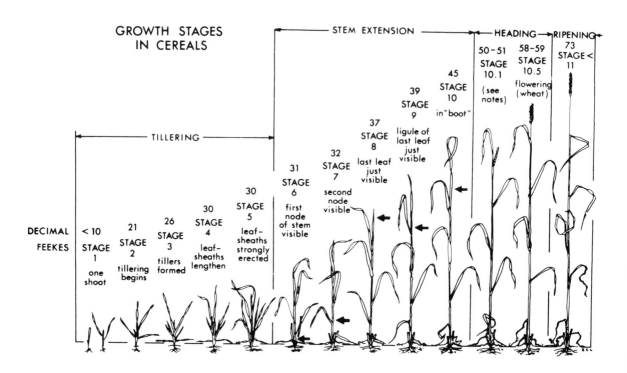

FIG. 88
(Reproduced by permission of HMSO)

THE GROSS MARGIN SYSTEM OF ANALYSIS

TO ENABLE farmers to compare the financial performances of their farms with those of similar farms, and, as an aid to farm planning, it is now commonplace to use the Gross Margin (GM) system. It is a reasonably straightforward and easily understood system which has stimulated the study of farm management, but there can be many pitfalls in the interpretation of the figures. Great care and experience are required in making full and correct use of the system.

The Gross Margin for a crop or livestock enterprise is usually given as a "per hectare" figure and it is the difference between the Gross Output of the enterprise and its Variable (direct) costs.

Gross Margin = Gross Output minus Variable Costs.

For example, for a crop, the Gross Output/hectare is the total sales for all parts of the crop (e.g. cereal grain and straw) and including deficiency payments, grants, etc., together with the market value of produce retained on the farm—divided by the number of hectares grown.

The Variable Costs are those directly applicable to the crop, such as seed, fertilizers, sprays, casual labour, contract work and miscellaneous costs relating solely to the crop, e.g. baler twine, marketing expenses, etc.

The Gross Margin is not a profit figure; from it must be deducted the other cost items (common or fixed costs) on a total farm (or per hectare) basis, to give the Net Farm Income, i.e. the return to the

GROSS MARGINS FOR GOOD CROPS (PER HECTARE)

	Crop						
	Winter wheat	Winter barley	Spring barley	Maincrop potatoes	Sugar-beet	Oil-seed rape	Dried peas
Yield (tonnes)	8	7.5	6.5	50	55	4	4.5
Price per tonne (£)	120	110	110	80	26	270	166
Gross Output (£)	960	825	715	4000	1430	1080	747
Variable costs (£)							
Seed	43	40	40	300	45	25	90
Fertilizers	110	100	70	200	140	125	30
Sprays	70	70	45	120	100	50	60
Others, casual labour				250			
sundries				175			
transport					130		
windrowing						25	
Total	223	210	155	1045	415	225	180
Gross Margin (£)	737	615	560	2955	1015	855	567

farmer (and his wife) for their manual labour, their management and interest on all capital other than land and buildings. The Fixed Cost items are those which cannot be easily allocated to one particular enterprise and include:

rent (actual, or estimated for owner-occupiers),
regular labour (including paid management),
machinery (depreciation, repairs, and fuel),
general overheads such as general maintenance, repairs, office expenses, insurance, fees, subscriptions, etc.

The total Fixed Costs divided by the area of farmland gives the Fixed Costs per hectare.

The tables show some examples of current Gross Margins, Variable and Fixed Costs. These are only intended as a guide and can vary considerably from farm to farm and from year to year.

SOME EXAMPLES OF FIXED COSTS ON MEDIUM-SIZED FARMS
(PER HECTARE)

	Arable—mainly cereals	Arable (intensive)	Arable and dairying	Arable with pigs and poultry	Arable with sheep and beef cattle
	£	£	£	£	£
Regular labour	90	180	150	250	95
Machinery	140	250	150	260	85
Rent and rates	70	100	80	90	70
General overheads	30	80	40	90	30
Total	330	610	420	690	280

(See also Nix, *Farm Management Pocketbook*, Wye College and Norman and Coote, *The Farm Business*, Pub. Longmans.)

APPENDIX 12

ESTIMATING CROP YIELDS
FROM RANDOM SAMPLE WEIGHINGS

EVERY kg per 10 m^2 is equivalent to 1 tonne per hectare (10,000 m^2/ha). For example, for roots, potatoes, kale, etc., if rows are 0.5 m apart and four lengths of 5 m each are taken as random samples, i.e. 20 × 0.5 = 10 m^2, and if the total weight of the samples is 40 kg, then the estimated yield would be 40 tonnes per hectare. An allowance for headland and other wastage must be made when estimating field yields.

Row width (mm)	Length of row to dig, or cut, so that 1 kg is equivalent to 1 tonne per hectare (m)
1000	10 (e.g. 4 × 2.5 m)
900	11
800	12.5
750	13.3
700	14.3
600	16.7
500	20.0
400	25.0 with these longer lengths, an alternative would be to dig half these lengths and then every kilogramme would be equivalent to 2 tonnes per hectare
300	33.3
200	50.0

An indication of the likely accuracy of these estimates will be obtained from the similarity of the yields in each of, say, four samples taken at random.

APPENDIX 13

FARMING AND WILDLIFE CONSERVATION

ALTHOUGH modern farming and conservation can be compatible to a very great extent, there will always be a degree of contention between the farmer and/or landowner and the conservationist.

However, the majority of farmers are conscious of the responsibility they have as custodians of the land, and they appreciate the aspirations, as well as the fears, of those concerned with amenity and conservation. The responsible conservationist, on his part, realizes the needs of the farmer who is helping to feed the nation.

At present our farms are producing some £7000 million of food each year but this, in fact, only represents some 70% of the food consumed in the United Kingdom. If standards of living in this country are to be maintained and eventually improved again, then we should, as far as possible, be insulated from the pressure on world food resources which will inevitably increase as other countries (particularly from the third world) demand a higher standard of living. There is a very delicate balance between supply and demand of food in the world, and it is the shortages (even if only temporary) which have such a devastating effect on the spiral of inflation.

All this means that our farming industry must strive for even greater efficiency. It is here that the clash between food production, amenity and conservation, must arise. There are some amenity societies which believe in multiple land use, but they do not, in fact, accept that food production has priority over all claims upon the land. However, it must be argued that the case for agriculture is an economic one. It is not possible to have efficient farming and a beautiful countryside on the cheap; it has to be paid for—out of rates, taxes and higher food prices.

Undoubtedly, too many hedges (many of great ecological value) were removed, particularly in the eastern counties, 20–30 years ago, although the rate of hedgerow removal has now slowed down. Large fields are not inevitable in modern arable farming. The size should obviously blend with the particular part of the countryside, but generally it is difficult to justify more than 20 hectares on the one hand, and less than 8 hectares at the other end of the scale. In the grassland areas it is reasonable to assume that, as dairy farms become larger, so too will field size with the consequent loss of hedgerow. However, no stock farmer can ignore the value of the hedge for shelter, and it is difficult to see very large fields dominating the rural landscape.

The Countryside Commission, whilst concerned, accepts that a changing agricultural landscape is inevitable, but it contends that, by education, a compromise can be reached between the need for optimum food production and rural conservation. It also requested our Farmers' Unions to formulate a Code of Practice for farmers and landowners so that further changes in landscape should make the best possible compromise for both food production and conservation.

The Nature Conservancy Council makes out a very good case for conservation being important, not only for farming, but for the nation as a whole. All living organisms are interrelated, and either directly or indirectly all plants and animals, having descended from wildlife, are still dependent on wild species. This is the only way that evolutionary processes can continue. The plant breeder, for example, often turns to the wild species in his endeavours to improve crop potential.

The Council recommends that the best land should always be used for food production, but that uncropped land (and this does not necessarily mean grass) should be left for conservation. Certainly, even under lowland conditions, there are areas of land which are uneconomic to farm. High performance per hectare is obviously necessary but, because of increasing production costs, the emphasis has to be on optimum, rather than on maximum production. This should help the conservationist because the incentive to cultivate every corner of every field on the farm is no longer there for the farmer.

The Wildlife and Countryside Act 1981 is a serious attempt by the Government to help bridge the gap between the farming industry and the conservationist. In general the industry should have few complaints about the Act. Voluntary co-operation (with a limited number of exceptions) is the key to the many provisions which more vitally concern the farmer. Provided farmers are seen to be making an effort to achieve some form of compatibility between conservation (to include both wildlife and landscape) and food production on the farm they should have little to fear. However, if the Act is not seen to be working, if voluntary control is not achieving the desired results (and this will need an objective appraisal), then there will be pressure on the Government of the day to bring in statutory control.

The Act encourages Management Agreements to be drawn up between the farmer and the Local Authority. It is intended that any such Agreement will cover the whole farm with the idea of managing it on the basis of a conservation plan. Compensation will be paid if the plan affects the profitability of the farm.

There is a statutory obligation where Sites of Special Scientific Interest (SSSIs) are involved. If there is an SSSI on the farm, the farmer has to give 3 months' notice to the Nature Conservancy Council (NCC) if he wishes to carry out any operation on the site which may affect its special characteristics. If the NCC objects it will then offer a Management Agreement to the farmer for the whole farm which will include a restriction of the proposal objected by the Council. It can always have the final say because, following consultations, an Order by the Secretary of State for the Environment can prevent any projected work being carried out on the SSSI, although compensation will be paid by the NCC. A fine will be imposed if the Order is not complied with and the NCC can, if necessary, purchase the site by Compulsory Order.

The Farming and Wildlife Advisory Group is doing much to foster the essential compromise between farming and conservation. It is important that its activities are encouraged, particularly in view of the new Act.

FURTHER READING

New Agricultural Landscapes—Issues, Objectives and Action, published by the Countryside Commission.
Landscape—The Need for a Public Voice, published by the Council for the Protection of Rural England.
Nature Conservation and Agriculture, published by the Nature Conservancy Council.
Caring for the Countryside (N.F.U. and C.L.A.).
Wildlife and Countryside Act, HMSO.

APPENDIX 14

THE EUROPEAN ECONOMIC COMMUNITY (EEC) OR EUROPEAN COMMUNITY (EC) OR COMMON MARKET

THE EEC and EURATOM (European Atomic Energy Community) came into being on 1 January 1958 (under the Treaty of Rome 1957) with the same six founder members—France, West Germany, Italy, The Netherlands, Belgium and Luxembourg—as set up the European Coal and Steel Community (ECSC) in 1952 (under the Treaty of Paris 1951) and known as "The Six".

The United Kingdom, Denmark and Ireland joined the Community on 1 January 1973 (under Treaty of Accession 1972) and the group was known as "The Nine".

Greece joined the Community in January 1981; Spain and Portugal may join later.

The ten states (over 250 million pop.) make a larger unit than the U.S.A. or U.S.S.R. These states have united, or are uniting, to form one trading, farming and industrial system. The aims are to develop an economic union, a common currency, and a common foreign policy, and ultimately, political unity.

Brussels is the "Capital" of the Common Market. The European Parliament meets in Strasbourg, but also has offices in Luxembourg.

The three Communities (EEC, EURATOM and ECSC) are managed through two main institutions, namely:

(a) The Council of Ministers—the supreme decision-making body. The Foreign Ministers usually represent member states but other ministers may join them sometimes. They meet in Brussels or Luxembourg: normally, decisions have to be unanimous, so a veto is possible.

(b) The Commission—this initiates and puts proposals to the Council, and is the executive body for implementing Council decisions. It also ensures that the Treaties are not violated by states, institutions, or individuals (its decisions can be challenged in the European Court of Justice in Luxembourg). At present there are thirteen members who swear to act as "Europeans"; two are British. There is a staff of over 9000.

   The Commission controls the common budget and the Common Agricultural Policy (CAP), and collects funds from tariffs and taxes.

The Committee of Permanent Representatives (COREPER) can act to protect vital national interests.

COPA (the EEC's NFU) and COGECA (the COPA for co-operatives) also try to influence the Council and the Commission. Representatives, with their own national interest in mind, scrutinize and can challenge Commission proposals before the Council considers them.

Enforcement of decisions lies with authorities within member states and this gives some latitude in the process!

Important decisions made by the Council of Ministers are embodied in: "regulations"—these

242

automatically become law in all member states; "directives"—which state aims which are binding on member states but the means of implementing them is left to National Governments, and they usually require changes in national legislation within a certain time.

A "decision" is addressed to a government, organization or individual and is binding in every respect on those named.

"Recommendations" and "opinions" are not binding but express the views of the Commission on policy.

THE COMMON AGRICULTURAL POLICY (CAP)

The CAP is a very important part of the Common Market; its aims were laid down in the Treaty of Rome, viz.:

1. Increased productivity by:
   (a) technical progress,
   (b) rational development of agricultural production,
   (c) optimum use of land, labour and capital.
2. Improved incomes for those engaged in agriculture.
3. Stability of markets.
4. Certainty of supply of products.
5. Guarantee of reasonable prices for consumers.

To achieve these aims there are common methods of trading, pricing and financial support, and there are common organizations in the market for: cereals, milk (powdered), butter, cheese, beef, sheep meat, veal, pork, bacon, sugar, olive oil, other edible oils, oilseeds, poultry, eggs, hops, tobacco, rice, wine, fruit, vegetables, fish, flax and hemp. Potatoes are being considered; early potatoes may come under vegetables.

Farming income comes from produce sold in the Community market, and also from exports which may be subsidized. To ensure a reasonable income to farmers, surplus farm produce can be offered to National Intervention Agencies when prices fall below a certain level. This level (the intervention price) is normally agreed annually by the National Ministers of Agriculture who form the Agriculture Market Council of the Council of Ministers.

The cost of agricultural support comes from import levies, tariffs, the equivalent of up to 1% VAT, and some additional funds from the Common Budget.

The central financing arrangements, and the powers which the Council of Ministers and the Commission have, form what has been described as a Western European Ministry of Agriculture.

Member states could change the whole of the CAP by a unanimous vote in Council. There are often differences of opinion (and crises) when deciding on aid to different farming areas, and support prices.

Financial matters (including prices) are now dealt with in European currency units (e.c.u.) which are slightly different from the previously used units of account (1 u.a. = 1.21 e.c.u.); for sterling there were originally 2.4 u.a. to the £, now there is a "reference parity" (the Green Pound) of approx. 1.61641 e.c.u. to the £. This is used for calculating intervention prices, levies, etc. There is also a system of "monetary compensatory amounts" (MCAs) to allow for fluctuating values of European currencies. Green currency values in u.a. can be changed by unanimous agreement by the Council of Ministers.

The European Agricultural Guarantee and Guidance Fund (FEOGA) is the part of the Common Budget which deals with the CAP.

Cereals are a very important part of EEC farm production (28% of area) and their price also affects livestock costs and prices.

METHODS OF PRICE CONTROL

Cereals

A "target price" (not a guaranteed price) for each type of cereal, except oats, is the price it is hoped that producers will get on the open market in the area of shortest supply (Duisburg, Ruhr Valley, West Germany). It is a delivered (not a farm-gate) price: increments are added monthly up to May for barley, maize and rye, and to June for wheat.

A "threshold price" (minimum import price) is set for all cereals and applies to all ports around the Community. It is approximately the target price less the cost of transport from Rotterdam to Duisburg, and is adjusted seasonally to allow for storage costs of the home crop.

The variable levies, calculated daily in Brussels, are the differences between the lowest c.i.f. (cost, insurance, freight) offers on world markets and the threshold prices for each cereal. They are payable by importers on each shipment into the Community from non-member countries. If world prices are higher than Community prices, subsidies may be paid on imports.

When home production exceeds demand, intervention (support) buying is required. The basic "intervention price" for each type of cereal except oats is about 12–20% below the target price and changes seasonably with the target price. Intervention prices are agreed annually. There are principal market centres (the main buying-in centres) in member countries (twelve in the U.K.) and the intervention price is the delivered price for grain of specified quality and quantity standards (see page 69); a premium is paid for bread-quality wheats.

Oilseeds (oil-seed rape, sunflowers)

These crops are encouraged in order to save imports. They have guide prices which act as target prices and intervention prices are related to them. Producers' returns may be made up to the guide prices by deficiency payments and the intervention buying system.

Sugar-beet

The support system for sugar is complicated by the fact that agreements have been reached with ten developing countries (the Lomé Convention) for the supply of 1.3 million tonnes of sugar per year (about the same as the U.K. production). Some French overseas territories are included in the common sugar market. An intervention price is set for white sugar and sugar-beet production is based on quotas (A and B) which allow for a two-tier pricing system.

Import subsidies protected the consumers from very high world prices in 1974–5.

Beans and peas

To encourage the production of protein foods for livestock feeding there is a special scheme for beans and peas which involves the payment of a subsidy (related to the world price of soya bean meal) to feed compounders—provided the compounders pay farmers the minimum price fixed by the Community.

Fruit and vegetables

Producer organizations are given more responsibility for these products than for other products because of storage problems and the demand for fresh produce.

There are customs duties of 7–25% for fruit and 10–21% for vegetables, and a countervailing duty on imports if the price falls for two successive days below a minimum import price based on cost of production and marketing.

Basic prices are fixed annually by the Council of Ministers for cauliflowers, tomatoes, table grapes, peaches, apples and pears.

Member states may fix intervention prices between 40% and 70% of the basic price and when market price falls below this intervention level for 3 days the member states must buy in.

Producer organizations can also fix commodity reserve prices and, with some assistance, need not market produce at lower prices.

There are strict grading standards for fruit and vegetables.

Crop seeds

New legislation to implement EEC Directives has been introduced to deal with the marketing of crop seeds (and propagating material) and cataloguing of varieties, e.g.

(a) minimum standards of purity, germination and weed content;
(b) official certification of trueness of varietal description;
(c) restriction of the generations of seed crops outside the breeders' control;
(d) restrictions of the varieties for which seed could be marketed throughout the EEC to those in a common catalogue.

Responsibility for the National Lists of varieties and seed certification is shared between the National Institute of Agricultural Botany (NIAB) and the Ministry of Agriculture, Fisheries and Food (MAFF).

Other directives from the EEC concern disease and pest control, e.g. potato cyst (root) eelworm (PRE) and wart disease of potatoes. New regulations deal with the declaration of nutrients in fertilizers.

Many more directives and regulations are likely to follow as the EEC becomes more unified. It would appear that many of these restrictions are more tolerable to some European farmers in countries where the enforcement of the law is more lax than in the U.K.

INDEX

ACAS (Agricultural Chemicals Approval Scheme) 184, 222
acetic acid 42, 58
Acidity 31
Actinomycetes 23, 204
Adult (insects) 181–182
Adventitious roots 6
Aeration (soil) 22
Aggregation (soil particles) 25
Aldicarb 184
Aldrin 183, 192
Algae 23
Alkalinity 31
Alluvium 18
Alpha-amylase 75
Alternaria 201
Alternate husbandry 30, 59, 63, 64
Altitude 62, 63
Aminotriazole (Weedazol T-L) 138, 175
Ammonium compounds 51, 52
Ammonium nitrate 51
Analysis
  fertilizer 50–54
  silage 162
  soil 31, 49
Anhydrous ammonia 51, 54
Animals (soil) 24
Annual meadow grass 173
Annual nettle 170
Annual weeds 45, 88, 91
Annuals 4
Aphids 96, 98, 186, 188, Table 53
Apical dominance 96
Approved products see ACAS
Aqueous ammonia 51
Ascochyta 201
Aspect 62
Asulam ('Asulox') 177
Atrazine 82, 169
Auricles 71, 122, 123, 125
"Avadex" 170, 220
Available water 22, 37
Awns 70, 124

Baars irrigator 39
Bacteria 15, 23, 195
Badger 194
Baking quality 74, 75
Bales, baling 150–154, 161

Barban 170, 174
Barberry 199
Barley 63–73, 77–79, Tables 53 and 54
Barley grass 148
Barley yellow dwarf virus 78, 184, 198, 205
Barley yellow mosaic virus 198
Barling, D. M. 73
Barn hay-drying 152–154
Basalt 18
Basic slag 14, 52, 60
Bast 3
BCM see Carbendazim
Bean (broad) 5–7, 83, 85
Bean stem rot 84
Beans (field) 27, 64–65, 83–84, 188, 200, 201
Bed system 104
Beef 147
Bees 84, 118, 119
Beetles (saw-toothed, grain) 67, 187
Benazolin 91, 169, 177
Benomyl ("Benlate") 148, 248, 250, 253
Bents 31, 121, 131
Benweed see Ragwort
Berkankamp Scale 91
BHC see HCH
Biennials 4, 176
Big bales 151, 153
Big Bale silage 161
Bindweed
  black 170
  field 41, 84, 167, 170
Biological control 182
Birds 82, 194
Biscuits 75
Black fen 18, 30
Black fly (beans) 84, 188
Blackgrass 41, 174
Blackleg 117, 204
Black rust 197
Blight (potato) 98, 203, 209
Blossom beetle 190
Blowing (soil) 28, 30, 40
Blueprints 73, 95, 100
Bolting (sugar beet) 13, 101
Boron (borax) 14, 31, 202, 203
Botanical names see Latin names
Bracken 44, 56, 166, 168, 177
Bran 5, 74, 75

Brassicae   189, 201
"Break" crops   66
Broad beans   5–7, 83, 85
Broad red clover   131, 137, 138
Brome grasses   131, 172, 173, 175
Brown-heart (swedes, turnips)   14, 107, 202
Brown rust   194, 197, 211
Brussels sprouts   9, 51, 112–114, 201
Buckrake   157, 158
Bud   9
Bulk fertilizer   54
Bunt   196
Burnet   133, 134
Burnt lime   32
Bushel weights   66
Buttercups   166, 168, 170, 177
Butyric acid   158, 160

Cabbage   112
Cabbage root fly   113, 183
Calcareous soils   29
Calcium   14, 32
Calcium carbonate   32, 33
Calcium oxide   32
Calomel   116
Calyx   11
Canned potatoes   100
CAP (Common Agriculture Policy)   243–245
Capillary water   22
Capping   45
Captan   84, 88
Carbamates   184
Carbendazim ("Bavistan", "Derosal")   210
Carbofuran   184
Carbohydrates   1, 14, 66, 158
Carbon   14
Carbon dioxide   1–3, 13, 32, 68
Carotene   1
Carpel   1
Carrot fly   192
Carrots   8, 114, 115
Catch cropping   111
Caterpillars   182, 189
CCC (cycolel)   72, 73, 77
CDA   221
Celery   31
Cellulose   1
Cellulose xanthate   45
Cereals   40, 66–82, Tables 53 and 54
   continuous   64–66
   diseases   Table 53
   growth stages   171, 235
   in EEC   244
   mixtures (blends)   208
   pests   Table 54
   recognition   69–71
   straws   58, 68
Certification (seed)   95, 117–120
Centre-pivot irrigation   39
Chaff   67
Chalk   18, 29, 32, 78
Charlock   91, 166, 169, 170, 177
Chlorfenvinphos   183
Chelates   49

Chemical pans   44
Chickweed   169, 170, 175, 177
Chicory   133, 134
Chilled grain storage   68
Chisel ploughing   42
Chitting (sprouting)   96
Chlorinated hydrocarbons   183
Chlormequat (CCC)   72, 73, 77
Chlorophyll   1, 2, 14, 23, 221, 242
Chocolate spot (beans)   84, 200
CIPC (chlorpropham)   100, 176
Citric solubility   52
Clay   21, 24, 26, 30
Clay soil   21, 26, 27, 218, 219
Clay-with-flints   29
Claying   30, 40
Cleavers   91, 167, 170
Click beetle   185
Climate   18, 62
"Clout"   88, 176
Clover rot   88, 205
Clovers   44, 131, 132, 170, 177
Club-root (finger and toe)   107, 201, 207
CMPP (mecoprop)   169, 170, 175
Cobalt   13
Cocksfoot   128, 137, 138
Coleoptile   7
Coleorhiza   5
Colloid   20, 26
Coltsfoot   41, 46, 213
Combine-drill   29, 55, 76
Common scab (potatoes)   23, 98, 204
Common Market (EEC)   242–245
Compound fertilizers   53, 54
Concentration (fertilizer)   54
Conservation
   herbage   149–164
   wild life   240, 241
Contact herbicide   168
Contact insecticide   183
Copper   14, 31, 203, 209
Corms   10
Corn   66
Corn marigold   31, 170
Corn mint   41, 166
Corolla   11
Cotswold Cereal Centre   73
Cotswolds   29, 72, 73, 119, 217
Cotton-grass   30, 121
Cotyledons   4–7, 11
Couch   9, 41, 43, 45, 46, 166, 175, 176
Covered smut   196
Crane-fly   185
Creeping buttercup   166, 168, 170, 177
Creeping thistle   166, 170, 177
Crested dogstail   130
Crispbread   80
Crisps   100
Cropping   26–31
Cross-pollination, cross-fertilization   12, 80, 95, 118–120
Crown rust (oats)   199
Crumbs, crumbs structure   25, 26, 40
Cultivars   see Varieties

Cultivations   40–46
Cultivators   42, 43
Culm   122
Cutworms   189
Cycocel   72, 73
Cyst nematodes (eelworms)   *see* Table 53, 64, 96, 97

2,4-D   83, 169, 170
2,4-DB   82, 169, 170, 177
Daddy-long-legs   185
Dalapon   91, 147, 148, 175, 178
Damp grain storage   67, 68
Dazomet ("Basamid")   184, 209
DDT   93, 189
Deep-ploughing   19, 30, 42
Deficiency symptoms
    boron   14, 102
    copper   14
    magnesium   48, 102
    manganese   14, 31, 88, 102
    nitrogen   48
    phosphate   48
    potash   48
Deflocculate   26
Denitrifying bacteria   15
Depth of drains   35
Desmetryne ("Semeron")   109, 176
Dewatering green crops   164
Dicamba   169, 170
Dichlorprop (2,4-DP)   169, 170
Dicotyledons   4–7
Dieldrin   116, 183
Diffusion   1, 4
Digestibility (herbage)   133, 135, 155
Dimethoate ("Rogor")   184
Dinoseb (DNBP)   88, 99, 119, 168, 170, 178
Diploid   102, 128, 131
Diquat   91, 99, 178
Direct drilling   41, 42, 138
Direct re-seeding   126, 136
Disc harrows   43
Disulfoton   184
Ditches   34, 36
Docking disorder   184, 192
Docks   45, 170, 177
Dolomitic limestone   32
Dorset (Waltham) wedge silo   160, 161
Downland   29, 121
Drag harrows   43
Drainage
    silos   159
    soils   27, 33–36, 168
Dredge corn   80
Driers
    grain   67, 82
    grass   150–154, 162
Drills and drilling   40, 41, 84, 138, 139
Drilling-to-a-stand   102, 175
"Drummy" soil   25
Dry liquid   220
Dry rot (potatoes)   204
Dry valleys   29
Drying
    cereals   67, 82

grass   150–154, 162
peas   89
Dutch barn   152, 153, 156
Dutch systems
    hay dring   152, 153
    potato cults   97
D-value   133, 135

Earthworms   19, 20, 24, 31, 41
EDTA   49
EEC   *see* Common Market
Eelworms   Table 53
Effluent   159
Eggs (insect)   Table 53, 181
Electric fence   140
Embryo   5, 7
Endosperm   5, 7
Ensilage   83, 158–162
EPTC ("Eptam")   175
"Eradicane" (EPTC + antidote)   82
Ergot   200
Erosion   19, 25, 30, 217
Ethirimol ("Milstem, Milgo")   208, 209
Ethofumesate ("Nortron")   148, 176, 178
Evaporation   22, 33, 37
Exoskeleton   180
Eyespot   64, 65, 166, 198

"Fairy rings"   23
Fallowing   46
False seedbed   45
Farming and Wildlife Advisory Group   241
Farmyard manure (FYM)   55, 56, 84, 97, 139
Fathen   166, 169, 170, 176
Fattening cattle   146
Feel (soil)   24, 218
Fermentation (silage)   158–162
Ferric hydroxide   25
Fertility (soil)   25, 64, 176
Fertilization   11, 12, 84
Fertilizers   47–55,   *see also* Separate crops
Field bindweed   41, 84, 167, 170
Field capacity   21, 22
Field drainage   33–36
Field resistance (tolerance)   207
Fields   27–31
Flax   92, 93
Flea-beetle   106, 189
Flints   29
Flocculate   26
Flour   74, 75
Flowering   11, 12
Flowers   11, 12
Fodder beet   106
Fodder radish   110
Foggage   135, 141
"Follow-N" system   147
Foot-rot   82, 200
Forage crops   110, 111, 155
Forage harvester   158
Forage peas   111
Formalin (with sulphuric acid)   159, 160
"Format"   176
Formic acid   159

Formulation   220
Forward creep grazing   147
Fox   194
Foxgloves   166
Free-draining   33
French beans   6, 15, 83, 85
Frit fly   79, Table 53, 130, 186, 188
Frits   49
Fronds   177
Frost   13, 18, 24, 27, 62, 63, 96, 136
"Frost-pocket"   62
Fruit   5, 12
Fumitory   170
Fungi
   diseases   table 54, 194, 195
   soil   23, 207
Fungicides   Table 54, 208–210
Fungicide resistance   212
Fungicide groups   210
Funicle   5
Furrow press   44
"Fusarex" (TCNB)   96, 100
Fusarium   199
"Fusilade"   88, 176
FYM   see Farmyard manure

Gangrene   205
Gas liquor   52, 176
Germination   6, 7, 13
Gibbs slot-seeder   148
Glacial drift   18
Glashouses   96
Gleying (gleys)   19
Glucose   1
Glume blotch   197
Glumes   124
Glyphosate ("Roundup")   41, 100, 138, 175, 176, 177, 220
"Goltix" (metamitron)   176
Gorse   121
Gout fly   186
Grading (potatoes)   100
Grain   5, 6, 12, 69
Grain drying   67, 68
Grain quality standards   69
Grain storage   67, 68
Grain weevils   188
Granite   18, 19
Grass diseases   48, 205
Grass drying   162, 164
Grass identification   126–130
Grassland   chapter 5 (121), 176–178, 241, 257, 258
Grassland improvement   190, 191, 223–225
Grasses   125–130
Grazing
   barley   78
   oats   79
   rye   80, 143
   systems   141–147
   wheat   76
Green beans   83, 85
Green-crop drying   162, 164
Green-fly   see Table 53
Green manuring   59
"Greening" (potatoes)   94, 97

Grid irrigation   39
Gross margins   237, 238
Ground-keepers   100
Ground rock phosphate   72
Growth inhibitors   169
Growth regulators   77, 169, 222
Growth stages   91, 171, 235, 236

Hares   194
Harrows   43
Harvesting   see Separate crops
Hay, haymaking   149–155
Hay additives   154
HCH   183, Table 53
Heart-rot (sugar beet and mangolds)   203
Heather   18, 30, 121
Heating (grain)   67, 68
Heat units   87
Heavy soils   see Clays
Hedges   27–29
Hedgehog   194
Hemlock   166
Hempnettle   170
Herbicides   168, 169, 170, 174–178
Herbs   122, 133
High dry matter silage   162
Hill drainage   34
Hill grazing   121, 129
Hilum   5
H.I. short rotation ryegrass   127
"Hoegrass"   174, 175, 176
Hoes   43
Horizons (soil)   17, 18, 217
Hormones   169
Horsetails   176, 177
Hose-reel irrigator   39
Host plant   194
Humus   19, 20, 30
Hydrated lime   32
Hydrogen   14
Hydrogen sulphide   56, 159
Hydroponics   49
Hyperphosphate   52
Hyphae   194
Hypomagnesaemia   32, 48, 139

Identification
   grasses   122–125
   legumes   125–126
Igneous rock   18
Inhibitors (silage)   159
Inflorescence   12, 123, 124
Insect structure   18
Insecticides   183, 184
Internode   9
Intervention standards (cereals)   69
Iron   13, 41
Iron pans   44
Irrigation   36–39, 88, 98, 99, 104
Isoproturon   174
Italian ryegrass   127, 143

Kainit   53, 102
Kale   42, 108, 109, 176
Keiserite   48, 102

Kernals 69
Knapweed 176
Knot-grass 169, 170, 176

Lactic acid 158–160
Ladybird 182
Laloux system 72
Larvae Table 53, 181
Late-flowering red clover 131, 137, 138
Latin names
   crops 223, 224
   diseases 229, 230
   pests 231–232
   weeds 225–228
Leader-follower system 145, 146
Leaf diseases 65, 211
Leaf (leaves) 1, 7, 10, 11
Leaf blotch (rhynchosporium) 197
Leaf protein concentrate (LPC) 164
Leaf-roll 117, 203
Leaf sheath 122–125
Leaf-stripe 196
Leaflets 11, 94
Leatherjackets 96, 185, 193
Legumes 13, 82, 125, 126
Ley 64, 122, 135–138
Ley farming 63
Lichens 19
Life-cycles (insects) 181
Light
   sunlight 13, 63, 96
   soil see Sandy soil
Ligules 71, 123
Lime requirement 31, 33
Limestone 18, 30–33, 63
Liming 28, 31–33, 139
Linen 92
Linseed 92, 93
Linuron 93, 169, 175
Lippiatt, Peter 73
Liquid fertilizer 54, 55
Liquid manure 56–58
Litter 56
Liver fluke 34
Loam (soil) 28, 29, 218
Lodging 77, 118
Lodicules 124
Loose smut 196
Lucerne (alfalfa) 29, 132, 155, 206
Lupins 28, 32, 89

Magnesium 14, 32, 48, 102
Magnesium limestone 32
Maize 81, 82, 155
Major nutrients 14
Malathion (insecticide) 67, 183
Malting (barley) 67, 77
Management
   leys 140–144
   soils 25
Manganese deficiency 88, 102, 200
Manganese 20, 42, 44, 102, 143, 188
Manfold fly 191
Mangolds (mangels) 105, 106, 175

Manures 55–60
   for grassland 139, 140
Market gardening 28, 31, 116,, 117
Marl 40
Marsh-spot (peas) 14
Mashlum 80
Mat 31
Mayweeds 91, 169, 170, 176
MCPA 169, 170, 175, 177
MCPA/dicamba 169
MCPA/TBA ("18:15") 169
MCPB 88, 118, 169, 170, 177
Meadow barley grass 176
Meadow fescue 129, 138
Mechanical analysis (soil) 24
Mecoprop (CMPP) 169, 170, 175
Menazon ("Saphicol") 184
Metaldehyde 184
Metamitron ("Goltix") 176
Metamorphic rocks 18
Metribuzin ("Sencorex") 175
Metrication 214, 215
Mice 194
Micro-organisms 15, 20
Micropyle 5
Mildew 65, 197, 211
Miling quality 74, 75
Mineral deficiency 14, 195
Mineral matter 18, 20, 25
Mineral oils 115
Minimal cultivations 42
Minor nutrients 13, 14
Mixing chemicals 220, 221
Moisture content
   grain 67–69
   hay 149–154
Moisture holding capacity 21
Molasses 160
Mole 24, 41
Mole-drainage 27, 35, 36
Mole plough 36
Molinia 30, 21
Molybdenum 13
Monocotyledons 5–7
Monoculture 65, 66
Monogerm seed 102
Morrey system 144
Mosaic viruses 117, 203
Mosss 19, 30
Mouldboards 42
Mould inhibitor 159
Mouth parts (insect) 180
Muck soils 30
Mucronate tip 124, 126
Multigene resistance 207
Multigerm seed 102
Muriate of potash 53
Mushrooms 23
Mustard (white) 59
Mustards 170
Mycelium 194

Nardus 121
Navy beans 83, 86

Neck-rot (onions)   116
Nematicides   184
Nematodes   Table 53
Net blotch (barley)   196
Nettles   170, 177
Neutralizing value (N.V.)   32
Nightshade   88, 166
"Nitram"   50
Nitrates   15, 49
Nitrate pollution   15, 16
"Nitro-chalk"   50
Nitrogen   14–16, 47–52, 139–143
Nitrogen cycle   15
Nitrogen fertilizers   51, 52
Nitrogen fixation   15, 49, 82
Node   9, 235
Nodules   15
Norfolk 4-Course rotation   64, 65
"Nortron" (ethofumesate)   148, 176, 178
Nutrients   4, 12–14, 47–53
Nymph (insect)   181

Oatmeal   67, 79
Oats   32, 67, 71, 79, 80, Tables 53 and 54
Oil   1, 90, 92, 93
Oil-seed rape   90–92, 244
"Onion" couch   78, 166, 173
Onion (bulb)   115, 116
Onion fly   116
Opomyza   186
Orache   170
Organic farming   60, 61
Organic manures   20, 52, 55–58
Organic matter   18, 20, 27
Organo-mercurial seed dressing   74, 208
Organo-phosphorous insecticides   183, Table 53
Osmosis   3, 4
Ovary   11, 124
Over-grazing   176
Ovules   11
Oxamyl ("Vydate")   184
Oxygen   1–3, 14, 22, 23, 68

Paddock grazing   144, 145
Palatibility   127–131, 176
Pales   124
Panicle   123. 124
Pans   44
Paraplow   45
Paraquat   168, 175, 176, 220
Parasite   182
Parent rock   17
Pasture improve   147–148, 177, 178
Peach potato aphid   192
Peas   38, 87–89
Pea and bean weevil   189
Pea-moth   189
Peat bog   30
Peats   20, 30
Pelleted sed   102
Pennycress   170
Pentanochlor   168
Perennial ryegrass   121, 127, 128, 139, 141, Fig. 56
Perennial weeds   45, 166, 170, 175

Perennials   4
Permanent grass   27, 29, 121
Pest (insect) control   Table 53
Petals   11
pH   31, 32, 75
Phloem   3, 10
Phorate   184
Phosphate fertilizers   52
Phosphorus (phosphate)   14, 47, 48, 49, 52, 60
Phosphorus pentoxide (P_2O_5)   52
Photosynthesis   1, 2, 13
Physiological age   96
Pick-up reel   77
Pigeons   82, 83
Placement (fertilizer)   55
Plantains   133, 134, 170
Plant diseases   Table 54
Plant food   see Nutrients
Plant food ratios   54
Plant groups   4
Plant populations   72, 76, 78, 82, 84, 86, 88
Plant structure   8–12
Plants   1
Plasmolysis   4
Plastic pipes (drainage)   35
Plough pans   44
Ploughing   27, 30, 42, 40, 97, 128
Ploughs   42
Plumule   5, 6
Poaching   28, 30, 176
Pod midge   190
Poisonous weeds   166, 167
Polygonum weeds   169
Polyploid seed   102
Polythene bags (fert.)   54
Pollen beetle   190
Pollen grains   11, 12
Pollination   11, 12, 84, 119
Poppy   169, 170
Pore space (soil)   21
Porous back-fill (drains)   35
Potash fertilizer 48, 53
Potassium oxide (K_2O)   53
Potassium (potash) (K)   14, 27, 48, 53, 60, 139
Potato tuber   9, 94, 99
Potato weed control   41, 98, 175
Potatoes   41, 94–100, 117, 118, 175, Tables 53 and 54
Potential transpiration   37
Poultry manure   59, 60
Powdery mildew   113, 202
Powdery scab   204
Precision drills   40, 102
Productivity (soil)   25
Prometryne   88, 169, 175
Propachlor ("Ramrod")   113, 169, 176
Propham   169, 176
Prophylactic control   209, 210
Propionic acid   68, 160
Protein   1, 41, 77, 83, 162
Protozoa   23
PSPS (Pesticides Safety Precautions Scheme)   222
Pulling peas   87, 89
Pulse crops   82, 83
Pupa (insect)   181

"Pyramin" 176

Quality
   cereals 68, 74–77
   potatoes 94
   silage 162, 163

Raan (swedes, turnips) 14, 107, 202
Rabbits 19, 28, 41, 58, 89, 120, 194
Radicle 5, 6
Ragwort 166, 176, 177
Rainfall 25, 37, 62
Rain guns 38, 45
Rape 90–92, 110
Raphanobrassica 110, 111
Rat 194
Red clover 64, 131, 155
Red fescue 129
Redshank 170, 176
Reeds 18, 30
Re-seeding 136–138
Residual herbicides 168, 169
Residual values (fert.) 60
Resistant varieties 207
Respiration 3, 68, 158
Rhizome 9, 45, 166, 167
Rhynchosporium 197, 211
Ribgrass 133, 134
Riddling 100
"Ridge and furrow" 27, 36
Ridging 41, 42, 97
Ripening *see* Separate crops
Rock 17–19
Rolling 22, 31, 44, 136, 179
Rolls 44
Rooks 82
Root eelworms Table 53
Root hairs 3, 8
Root systems 8, 37
Rooting depth (crops) 37, 95
Root tip 3, 17, 22, 24, 37
Roots 1, 3, 6, 8
Rope-wick 104, 176, 177, 221
Rotary cultivators (rotavator) 43–46, 97
Rotary sprinklers 38
Rotational grazing 143, 144
Rotations 63–65, 182
Rough-stalked meadow grass 129
"Roundup" *see* Glyphosate
Rubbed and graded seed 102
Runch 170
Rushes 18, 33, 168, 177
Rusts *see* Table 54
Rye 32, 80, 111, 142
Ryegrass mosaic 205

Sacks (bags) 54, 67, 100
Salt 53, 59, 102
Sainfoin 133
Sand 21, 28
Sandstones 18
Sandy soil 24, 28, 218, 219
Saphrophytic 194
Saw-toothed beetles 67, 187

Scab (common potato) 23, 98, 204
Scales 11
Schleswig-Holstein "system" 72, 73
Scutellum 5
Sealed silage 160
Seaweed 59, 97
Sedge 18, 30, 168
Sedimentary rock 18
Seedbeds 40, 41; *see also* under crops
Seed disinfection 208
Seed dressings 74, 182
   cereals Tables 53 and 54, 74
   kale 108
   swedes 107
Seed mixtures 135–138
Seeds
   in EEC 245
   1000 grain weight 233, 234
Seed production 117–120
   cereals 117
   clovers 118, 119
   grasses 118
   kales and swedes 120
   potatoes 95, 96, 117, 118
   sugar beet 119, 120
Seed rates
   barley 78
   beans 86
   cabbage 112
   carrots 115
   cereals 73
   grass and clover 137–138
   kale 108
   linseed 93
   lupins 89
   maize 82
   mangolds 106
   oats 79
   oil-seed rape 90
   onions 116
   peas 88
   potatoes 95
   rye 80
   sugar beet 101
   turnips, swedes 107, 110
   wheat 76
Seed structure 4, 5
Seed weevils 191
Seedling 6, 7
Seeds 4, 5, 11, 12, 126, 166, 212, 233
Selective herbicides 168
"Semeron" 176
Seminal roots 6
Semi-intensive system 147
"Sencorex" 175
Sepals 11
Septoria 197, 211
Set-stocking 145, 146, 147
Shakaerator 45
Shales 18
Sharp eyespot 198
Sheep 76, 147, 177
Sheep's fescue 121
Sheep's sorrel 31, 177

Shepherd's purse   170
Shoddy   52, 60
Shoot   5, 6
Short-duration ryegras   125, 127, Fig. 55, 130
Sieve tubes   3
Silage   155–163
   additives   159, 160
   crops   155
Silage effluent   157, 159
Silos
   grain   67, 68
   grass   156, 157
Silt   21, 29, 30, 218
Silt soils   29, 218
Simazine   84, 169
Singling   103, 107, 109
Skin spot   205
Slit-seeding   41, 58
Slugs   41, 187, 193
Slurry   56–58, 109, 140
Smooth-stalked meadow grass   130
Smut   198, 200
Snow mould   199
Snow rot   200
Sodium   13, 27, 53
Soft brome   176
Soil
   aeration   22, 23
   air   20, 22, 23
   analysis   31, 49
   animals   24
   capping   45, 176
   classification   216–219
   condition   25
   fertility   25
   formation   18, 19
   loosening   44
   micro-organisms   23
   pans   44
   particles   18–21
   productivity   25
   profile   17
   reaction   31
   series and surveys   217
   sterilants   184
   structure   24
   texture   24, 218
   types   26–31, 216–219
   water   21, 22, 33–39
Sorrel   31, 177
Sourness   31
Soya bean   90
Space-drilled   113
Spacing (drains)   34, 35
Speckled yellows   14
Speedwell   170
Spike   12, 123, 124
Spores   194, 195
Spraing   204
Spray chemicals
   safety   222
   application   178, 220–222
Spraying
   diseases   Table 54

pests   Table 53
   weeds   178, 220–222
Spraylines   38
Sprouting
   cereals   69, 80
   potatoes   96
Spurrey   31, 170
Squirrel   194
Stalk-rot (maize)   82, 200
Stamens   11, 124
Starch   1, 10, 94
Stecklings   119
Stem   9, 10, 122, 123
Stem and bulb eelworm   187
Stem canker   201
Stem rot (kale)   202
Sterile brome   78, 167, 175, 220, 225
Stigma   11, 124
Stipules   10, 125, 126
Stocking rate   143–147
Stolon   9, 166
Stoma, stomata   1, 2
Stomach poison (insect)   183
Stopping (cocking)   114
"Storite" (thiabendazole)   95, 100
Straight fertilizers   51–53
Strains   7, 122
Straw   58, 59, 68, 72
Straw burning   58, 69
Strip-grazing   143, 144
Striped pea weevil   189
Stubble cleaning   45, 46
Stubble turnips   109, 110
Style   11
Sub-soil   17, 44, 219
Sub-soiler   44
Sugar   1, 3, 100–105, 158
   in EEC   244
Sugar beet   27, 38, 65, 100–105, 175, 176, 191, 192, 244
Sulphate ofammonia   32, 51
Sulphate of ammonia   32, 51
Sulphur   14
Sulphuric acid   168, 176
Summer fattening   146
Sunflowers   93, 94
Superphosphate   14, 52
Surface soil   17
Swath   80, 150, 151
Swedes   107, 108, Tables 53 and 54, 176
Sweetcorn   81, 82
Systemic insecticide   84, 184, Table 53

T-sum   142
Taint
   crops   183
   milk   166, 176
Take-all   41, 65, 166, 199
Tall fescue   129
Tap root   8
TCA   169, 176
Temperature   12, 13, 63, 67, 68, 96, 100, 101, 156
Tendril   10, 11
Tetrachlornitrobenzene (TCNB, "Fusarex")   96, 100
Tetraploid clovers   131

Tetraploid ryegrasses 128
Thiram (TMTD) 84, 88, 209, Table 54
Thistles 166, 169, 170, 177
Threshing 67
Thunderstorms 15
Tile drains 34–36
Tillage 40–46
Tillering
  cereals 73, 77, 191
  grasses 140
Tilth 25
Timothy 125, 128, 129
Topping
  pasture 140, 145
  sugar beet 105
Topography 19, 25
Topsoil 17
Tower silo 157
Trace elements 13, 49
"Tramlines" 72, 175, 222
Translocated herbicides 168
Translocation 3
Transplanting 112, 113, 119
Transpiration 3, 4, 13, 37
Trees 26–30, 47
Tri-allate ("Avadex BW") 84, 88, 170
Tridemorph 107, 197
Trifluralin ("Treflan") 91, 176
Triple superphosphate 52
Tripoding
  hay 155
  peas 89
Triticale 81
TRV (PMTV) 204
Tuber 9, 94
Tunnel (floor) drying 67, 153, 154
Turnips 4, 64, 107–109, 111, 176, Tables 53 and 54
Tussock grass 33, 167, 177
Two-sward system 145
Twitch see Couch
Types of grassland 121, 122
Types of soil 26–31, 216–219
Types of silage 163
Typula rot (barley) 200

Underground drains 33–36
Undersowing 126, 139, 169, 170
Unit cost (fert.) 50
Units (fert.) 33, 49, 50
Urea 51, 52
Urine 55, 57

Vaporizing oils 115
Varieties see Separate crops
Vegetables
  in EEC 245
  on farms 116, 117
Ventilation 96, 99, 109, 152, 154
Verticillum wilt 206
Vetches 89
Vining peas 87, 88
Virus diseases see Table 54
Virus yellow 119, 202
Viruses 195

Volume rate (spraying) 222
VTSC 96, 117, 118

Wall barley grass 178
Waltham wedge silage 160
Ware (potatoes) 94
Warmth 12
Warping 39, 40
Waste lime 32
Water
  in the plant 1–4, 13
  in the soil 21, 22, 33–39
Watergrass 166
Water-holding capacity 21
Water-logged 21, 33
Water meadows 29
Water requirements (crops) 37
Water-table 33
Weather 62, 149, 168
Weathering (rocks) 18, 19
Weed beet 103, 104
Weed control Chapter 6, see also under various crops and 45
Weed grasses 130, 131
Weeds
  control in
    beans 84, 85
    Brussels sprouts 113
    carrots 115
    cereals 169–175
    grassland 148, 176–178
    kale 109, 176
    linseed 93
    maize 82
    oil-seed rape 91
    onions 116
    peas 88
    potatoes 41, 98, 175
    sugar beet 175, 176
    swedes, turnips 176
  harmful effects of 166
  seeds 166
  spread of 166
Weed wiper 104, 177, 221
Weevil (grain) 188
Westerwolds 127, 137
Wetting and drying 19, 24, 27, 40
Wheat 5, 6, 32, 40, 64–77, 169–175, Tables 53 and 54
Wheat bulb fly 74, 185
White clover 119, 131, 132, 137, 138
White mustard 59
Whiteheads see Take-all
White-rot (onions) 116
Whole crop 155
Wick rope 104, 221
Wild life conservation 240, 241
Wild oats 41, 167, 170, 172, 174
Wild onion 46, 166, 167, 176
Wilting point 22
Wilting (silage) 158, 159
Windrowing (swathing) 80, 91
Wireworms 185, 188, 191
Wolds 29
World crop production 213
WRO pasture improvement 148

Wye College grazing system   144

Xylem   3

Yarrow   133, 134
Yellow dwarf virus   198
Yellow rattle   176
Yellow rust   196, 211
Yields (crops)   47
  barley   78
  beans   85, 86
  Brussels sprouts   112, 114
  carrots   114
  kale   108
  maize   82
  mangolds   106

  oats   79
  oil-seed rape   92
  onions   115
  peas   87
  potatoes   95
  rye   80
  sugar beet   101
  swedes and turnips   107, 110
  wheat   73, 75
Yield, estimation of   239
Yorkshire fog   131, 176

Zadok's growth stages   235, 236
Zero grazing   146
Zero tillage   41
Zinc   13, 31